Principles of Seed Science

Principles of Seed Science

Editor: Lisa Swinton

R CALLISTO REFERENCE

www.callistoreference.com

Callisto Reference,
118-35 Queens Blvd., Suite 400,
Forest Hills, NY 11375, USA

Visit us on the World Wide Web at:
www.callistoreference.com

ISBN: 978-1-64116-149-7 (Hardback)

Cataloging-in-Publication Data

Principles of seed science / edited by Lisa Swinton.
    p. cm.
Includes bibliographical references and index.
ISBN 978-1-64116-149-7
1. Seeds. 2. Seed technology. 3. Seed pathology. I. Swinton, Lisa.
SB113.7 .P75 2019
631.521--dc23

Table of Contents

Preface..IX

Chapter 1 **Mobilization of reserves and vigor of soybean seeds under desiccation
with glufosinate ammonium**.. 1
Carolina Maria Luzia Delgado, Cileide Maria Medeiros de Coelho and
Gesieli Priscila Buba

Chapter 2 **Morphoanatomy of fruit, seed and seedling of *Ormosia paraensis* Ducke**................. 9
Breno Marques da Silva e Silva, Camila de Oliveira e Silva, Fabíola Vitti Môro and
Roberval Daiton Vieira

Chapter 3 **Biometric description of fruits and seeds, germination and imbibition pattern
of desert rose [*Adenium obesum* (Forssk.), Roem. & Schult.]**................................ 16
Ronan Carlos Colombo, Vanessa Favetta, Lilian Yukari Yamamoto,
Guilherme Augusto Cito Alves, Julia Abati, Lúcia Sadayo Assari Takahashi and
Ricardo Tadeu de Faria

Chapter 4 **Seed germination of Brazilian guava (*Psidium guineense* Swartz.)**....................... 24
Márcia Adriana Carvalho dos Santos, Manoel Abílio de Queiróz,
Jaciara de Souza Bispo and Bárbara França Dantas

Chapter 5 **Effect of sodium nitroprusside (SNP) on the germination of *Senna macranthera*
seeds (DC. ex Collad.) H. S. Irwin & Baneby under salt stress**............................. 32
Aparecida Leonir da Silva, Denise Cunha Fernandes dos Santos Dias,
Eduardo Euclydes de Lima e Borges, Dimas Mendes Ribeiro and
Laércio Junio da Silva

Chapter 6 **Viability of Brazilwood seeds (*Caesalpinia echinata* Lam.) stored at room
temperature in controlled atmospheres**... 40
Nestor Martini Neto and Claudio José Barbedo

Chapter 7 **Physiological potential of stylosanthes cv. Campo Grande seeds coated
with different materials**.. 49
Priscilla Brites Xavier, Henrique Duarte Vieira and Cynthia Pires Guimarães

Chapter 8 **The effect of drying temperatures and storage of seeds on the growth
of soybean seedlings**... 57
Cesar Pedro Hartmann Filho, André Luís Duarte Goneli, Tathiana Elisa Masetto,
Elton Aparecido Siqueira Martins and Guilherme Cardoso Oba

Chapter 9 **Germination and initial development of *Brachiaria brizantha* and *Brachiaria
decumbens* on exposure to cadmium, lead and copper**... 66
Karine Sousa Carsten Borges, Raquel Custódio D'Avila, Mari Lúcia Campos,
Cileide Maria Medeiros Coelho, David José Miquelluti and
Natiele da Silva Galvan

Chapter 10   **Protective action of nitric oxide in sesame seeds submitted to water stress**................................ 75
Raquel Maria de Oliveira Pires, Genaina Aparecida de Souza,
Denise Cunha Fernandes dos Santos Dias, Leonardo Araujo Oliveira and
Eduardo Euclydes de Lima e Borges

Chapter 11   **Physiological changes in osmo and hydroprimed cucumber seeds germinated
in biosaline water**... 83
Janete Rodrigues Matias, Renata Conduru Ribeiro, Carlos Alberto Aragão,
Gherman Garcia Leal Araújo and Bárbara França Dantas

Chapter 12   **Action of nitric oxide in sesame seeds (*Sesamum indicum* L.) submitted
to stress by cadmium**... 92
Raquel Maria de Oliveira Pires, Genaina Aparecida de Souza,
Amanda Ávila Cardoso, Denise Cunha Fernandes dos Santos Dias and
Eduardo Euclydes de Lima e Borges

Chapter 13   **Tetrazolium test for viability estimation of *Eugenia involucrata* DC. and *Eugenia
pyriformis* Cambess. Seeds**.. 100
Fernanda Bernardo Cripa, Laura Cristiane Nascimento de Freitas,
Andrieli Cristine Grings and Michele Fernanda Bortolini

Chapter 14   **Behavior of coffee seeds to desiccation tolerance and storage**................................ 107
Luciana Aparecida de Souza Abreu, Adriano Delly Veiga,
Édila Vilela de Resende Von Pinho, Fiorita Faria Monteiro and
Sttela Dellyzette Veiga Franco da Rosa

Chapter 15   **Methodology of the tetrazolium test for assessing the viability of seeds
of *Eugenia brasiliensis* Lam., *Eugenia uniflora* L. and Eugenia pyriformis Cambess**......... 115
Edmir Vicente Lamarca and Claudio José Barbedo

Chapter 16   **Physiological maturation of cowpea seeds**... 123
Narjara Walessa Nogueira, Rômulo Magno Oliveira de Freitas,
Salvador Barros Torres and Caio César Pereira Leal

Chapter 17   **Methods for overcoming seed dormancy in *Ormosia arborea* seeds,
characterization and harvest time**... 129
Aparecida Leonir da Silva, Denise Cunha Fernandes dos Santos Dias,
Liana Baptista de Lima and Glaucia Almeida de Morais

Chapter 18   **Morphological aspects of fruits, seeds, seedlings and *in vivo* and *in vitro*
germination of species of the genus *Cleome***... 137
Tatiana Carvalho de Castro, Claudia Simões-Gurgel, Ivan Gonçalves Ribeiro,
Marsen Garcia Pinto Coelho and Norma Albarello

Chapter 19   **Standard germination test in physic nut (*Jatropha curcas* L.) seeds**....................... 147
Glauter Lima Oliveira, Denise Cunha Fernandes dos Santos Dias,
Paulo Cesar Hilst, Laércio Junio da Silva and Luiz Antônio dos Santos Dias

Chapter 20   **Fruit processing and the physiological quality of *Euterpe edulis* Martius seeds**............. 155
Patrícia Ribeiro Cursi and Silvio Moure Cicero

Chapter 21   **Physiological analysis and heat-resistant protein (LEA) activity in squash
hybrid seeds during development**.. 166
Patricia Pereira da Silva, Antônio Carlos Souza Albuquerque Barros,
Edila Vilela de Resende Von Pinho and Warley Marcos Nascimento

Chapter 22   **Treating sunflower seeds subjected to ozonization**..173
Vitor Oliveira Rodrigues, Fabiano Ramos Costa, Marcela Carlota Nery,
Sara Michelly Cruz, Soryana Gonçalves Ferreira de Melo and
Maria Laene Moreira de Carvalho

Chapter 23   **Seed germination of Brazilian *Aldama* species (Asteraceae)**....................................182
Aline Bertolosi Bombo, Tuane Santos de Oliveira, Beatriz Appezzato-da-Glória and
Ana Dionísia da Luz Coelho Novembre

Permissions

List of Contributors

Index

Preface

I am honored to present to you this unique book which encompasses the most up-to-date data in the field. I was extremely pleased to get this opportunity of editing the work of experts from across the globe. I have also written papers in this field and researched the various aspects revolving around the progress of the discipline. I have tried to unify my knowledge along with that of stalwarts from every corner of the world, to produce a text which not only benefits the readers but also facilitates the growth of the field.

The study of seed is undertaken by the branch of seed science. The seed is an embryonic plant that is enclosed within a protective covering. It is formed during reproduction in seed plants, including gymnosperms and angiosperms. Seeds are of immense economic significance to humans. They can be used as edibles such as cereals, legumes and nuts. Seeds are also used for extracting cooking oils, food additives, spices and beverages. They have also been used for medicinal purposes such as castor oil and tea tree oil. Hybrid seed production is crucial for modern agriculture. Hybrids are developed to improve traits in crops such as high yield, improved disease resistance, better color, etc. Open pollination and clonal propagation are alternative techniques to hybridization that are used in agriculture. This book discusses the fundamentals as well as modern approaches of seed science. It unravels the recent studies in this field. It will serve as a valuable source of reference for graduate and postgraduate students, as well as for experts.

Finally, I would like to thank all the contributing authors for their valuable time and contributions. This book would not have been possible without their efforts. I would also like to thank my friends and family for their constant support.

Editor

Mobilization of reserves and vigor of soybean seeds under desiccation with glufosinate ammonium

Carolina Maria Luzia Delgado[1], Cileide Maria Medeiros de Coelho[1*]
Gesieli Priscila Buba[1]

ABSTRACT – The physiological quality of seeds depends on the cellular organization and their capacity to mobilize reserves. The goal of this study was to assess the germination and vigor of soybeans seeds of Benso1RR and NA 5909 RG cultivars from desiccated plants or not, by mobilizing reserves. Cultivars from desiccated plants or not, by mobilizing reserves. The plants were desiccated with glufosinate ammonium (GLA) in reproductive stage R7.1 or not (TST). The physiological seed quality was assessed by germination tests, accelerated aging and seedling length. For the mobilization of reserves were assessed within 48 hours of soaking: phytate, protein and soluble sugar contents. In 48 hours of soaking, seeds from desiccated plants in cultivar NA 5909 RG showed lower mobilization of soluble protein and soluble sugar, reflecting low vigor seedlings compared to their controls. For cultivar Benso1RR, in 48 hours of soaking, the application of GLA did not affect the mobilization of soluble protein when compared to control and did not differ as to vigor. It is concluded that the use of the desiccant has negatively influenced the mobilization of soluble protein and soluble sugar for cultivar NA 5909 RG, reflecting a low percentage of germination and low vigor.

Index terms: vigor, herbicide, protein, soluble sugar.

Mobilização de reservas e vigor de sementes de soja sob dessecação com glufosinato de amônio

RESUMO – A qualidade fisiológica em sementes depende da organização celular e da sua capacidade de mobilização de reservas. O objetivo do trabalho foi avaliar a germinação e vigor das sementes de soja das cultivares Benso1RR e NA 5909 RG proveniente de plantas dessecadas ou não, através da mobilização de reservas. As plantas foram dessecadas com glufosinato de amônio (GLA) no estádio reprodutivo R7.1 ou não (TST). A qualidade fisiológica das sementes foi avaliada através dos testes de germinação, envelhecimento acelerado e comprimento de plântula. Para a mobilização de reservas foram avaliados em até 48 h de embebição, teores de fitato, proteína e açúcar solúvel. Em 48 h de embebição sementes provenientes de plantas dessecadas da cultivar NA 5909 RG apresentaram menor mobilização de proteína solúvel e açúcar solúvel refletindo em plântulas de baixo vigor em relação a sua testemunha. Para a cultivar Benso1RR em 48 h de embebição, a aplicação do GLA não afetou a mobilização de proteína solúvel quando comparada à testemunha, não diferindo quanto ao vigor. Conclui-se que a utilização do dessecante influenciou negativamente na mobilização de proteína solúvel e açúcar solúvel para a cultivar NA 5909 RG refletindo em baixo percentual de germinação e baixo vigor.

Termos para indexação: vigor, herbicida, proteína, açúcar solúvel.

Introduction

The soybean seeds period of stay in the field after physiological maturity is crucial in the deterioration, influencing the fall of vigor. An alternative to prevent the seeds from remaining in the field after reaching physiological maturity is the application of herbicides aiming to quickly desiccate the plant, bringing forward its seeds harvest. Thus, application of the desiccant becomes an important factor to maintain the seed physiological quality.

The seeds physiological quality depends on the cellular organization and reserves mobilization capacity for the formation of more vigorous seedlings. According to Henning et al. (2010) and Pereira et al. (2015) more vigorous soybean seeds have higher soluble protein, starch and soluble sugars contents as well as higher reserves mobilization capacity

[1]Centro de Ciências Agroveterinárias, Universidade do Estado de Santa Catarina Luiz de Camões, 88520-000 – Lages, SC, Brasil.
*Corresponding author <cileide.souza@udesc.br>

during germination, resulting in soybean seedlings with better initial performance.

Currently several studies have been relating desiccant use with seeds final quality, where the herbicide mode of action, weather conditions and phenological stage in which the plant is found may maintain or negatively affect the seeds quality, and even not bring forward the harvest. According to Guimarães et al. (2012) the glufosinate ammonium herbicide reduces soybean seed germination when applied in the R6 stage and the herbicide paraquat promotes the best seeds germination rates and vigor when used in stages R6 and R7.2. Pinto et al. (2014) have found that desiccation with glufosinate ammonium in bean seeds had a lower percentage of normal seedlings in the germination test, accelerated aging and first count.

Glufosinate ammonium is a contact herbicide, with the ability to move at a short distance. Its mode of action inhibits the enzyme glutamine synthetase (GS) blocking glutamine synthesis via glutamate and thus the assimilation of NH_4^+, which is accumulated in the plants leaves (Carvalho, 2013). According to the same author, glufosinate ammonium might still affect the synthesis of some amino acids, production of free radicals and blocking photosynthesis, leading to plant death. However, few studies report what changes occur in the reserve components during germination in seeds from plants desiccated with the herbicide.

The seeds germination process corresponds to three stages; initially, the first stage of water absorption, which occurs quickly and passively; afterwards, the second stage, which is characterized by delayed water absorption; and the third phase, where again the seeds begin to absorb water quickly (Teulat-Merah et al., 2011; Weitbrecht et al., 2011)

From the precise moment when each of the three stages starts and ends in the germination process of soybean seeds, it becomes easier to infer the possible effects of desiccant application in soybean plants in reproductive stage R7.1, on the seeds physiological quality and their subsequent effect on the germination process and vigor.

In this sense, the objective of this study was to assess the germination and vigor of soybean cultivars Benso1RR and NA 5909 RG by mobilizing reserves during the germination process from desiccated plants or not, with glufosinate ammonium (GLA) in reproductive stage R7.1 (preharvest).

Material and Methods

The seeds used in the experiment are from the 2012/2013 harvest, derived from plants desiccated or not with glufosinate ammonium in stage R7.1 (Ritchie et al., 1982). The work sample was performed according to Regras para Análise de Sementes (Rules for Seed Analysis) (Brasil, 2009).

The seeds physiological quality was assessed by the germination percentage, seedling length and accelerated aging test. The reserves mobilization during germination (phytate, sugar and soluble protein) was determined in soaking times 0 h, 3 h, 6 h, 16 h, 24 h and 48 h, characterizing the soaking three-phase pattern.

The germination test consisted of four subsamples of 50 seeds, germinated between three sheets of germitest paper, moistened with distilled water 2.5 times the weight of the paper. The rolls with the seeds remained in germination at 25 °C temperature. Assessments were performed on the fifth day, considering only normal seedlings, according to Brasil (2009).

To assess the seedlings length, the methodology proposed by Nakagawa (1999) was used. Assessments were performed five days after sowing, by measuring the total length of normal seedlings with the help of a caliper. The results were expressed in centimeters.

For assessment of seed vigor, accelerated aging test was conducted, in which four subsamples of 50 seeds for each treatment were used. Seeds were distributed on aluminum screens fixed on the inside of plastic boxes and 40 mL of water were added. The boxes were closed and kept in an aging chamber for 48 h at 42 °C, according to Marcos-Filho (1999). After this period, the seeds were germinated at 25 °C and on the fifth day after the test installation normal seedlings counting was carried out, as in the germination test.

The soaking curve was obtained according to Ma et al. (2004) and Han et al. (2013) with some adjustments. For each soaking time, two replicates with 50 seeds on roll paper were used, which was formed using three sheets of germitest paper by repetition. The paper was moistened with distilled water (2.5 times the weight of the paper) and the rolls were taken to the germinator in a vertical position at 25 °C, C in the presence of natural light, remaining for 3 h, 6 h, 16 h, 24 h and 48 h. At each time, a sample of 5 g of seeds was taken and dried at 105 °C for 24 h to determine moisture. By means of the moisture difference between the times, the soaking curve was obtained. The moment the radicle protrusion reached 50% + 1 was considered the end of phase II and the beginning of phase III.

Phytate content was determined by the method of Latta and Eskin (1980), which is based on the formation of an iron-sulfosalicylic acid compound of dark blue staining (Wade reagent). In the presence of phytate, iron is sequestered and becomes unavailable to react with sulfosalicylic acid, which results in decreased color intensity. Readings were taken in a spectrophotometer at each time of the soaking curve, and three repetitions were performed at each time.

For soluble protein analysis, three replicates were

performed with 1 g of soybean seed from all selected points of the soaking curve. The samples were ground in liquid nitrogen to form a flour. The material was homogenized in 10 mM potassium phosphate buffer (pH 7.5) containing 1 mM of EDTA (Ethylenediamine tetra acetic acid, widely abbreviated as EDTA (ethylenediamine tetra acetic acid), 3 mM of DTT (dithiothreitol) and about 4% of PVPP (polyvinyl polypyrrolidone), according to Azevedo et al. (1998) and Garcia et al. (2006). The homogenate was centrifuged at 3500 rpm for 30 minutes. The total soluble proteins were determined in a spectrophotometer according to the method described by Bradford (1976) and results were expressed in mg.g^{-1} of seeds dry weight.

For total soluble sugar analysis, samples of 0.250 g of dried and ground seeds of each point of the soaking curve were weighed. The total soluble sugars were determined by the Antrona method, according to Clegg (1956) and the samples reading was done with a spectrophotometer at a wavelength of 620 nm. The results were expressed in mg.g^{-1} of seeds dry weight.

To verify the normal data distribution, the Shapiro-Wilk test (5%) was used; for the homogeneity of variance, the Cochran and Barlett (5%) test was used; and the germination percentage was transformed into sine arc of $(x / 100)^{0.5}$. Next, the data were submitted to analysis of variance by the F test. For analysis of seed physiological quality and initial content of biochemical components, the Tukey test at 5% error probability was used and for the analysis of reserves mobilization over time, regression analysis was used. Analyses were performed via SAS® (SAS, 2009) computer program.

Results and Discussion

Comparing the vigor of the two varieties of seeds from non-desiccated plants (Table 1) it was possible to observe, by the accelerated aging test, that seeds of cultivar NA 5909 RG showed superior vigor, with 89% of normal seedlings and 19 cm in length, compared to seeds of cultivar Benso1RR, which showed 79% of normal seedlings and 16 cm in length (Table 1). Desiccation did not affect the seed vigor of cultivar Benso1RR compared to its control (Table 1).

Table 1. Physiological quality of soybean seeds of two cultivars from plants desiccated with glufosinate ammonium (GLA) or not (TST) in reproductive stage R7.1.

	NA 5909 RG TST High vigor	NA 5909 RG GLA Low Vigor	BENSO1RR TST High vigor	BENSO1RR GLA High vigor	CV (%)
Germination (%)	92 a	79 b	94 a	89 a	3.58
AA (%)	89 a	81 b	79 b	82 b	3.35
SL (cm)	19.18 a	17 b	17.06 b	18.44 ab	2.10

On the row, the means followed by the same letter do not differ at 5% by Tukey test.
SL – seedling length; AA – accelerated aging.

The seeds from desiccated plants of cultivar NA 5909 RG were negatively impacted by the application of glufosinate ammonium, with low vigor, lower germination percentage (79%) and lower seedling length (17 cm) compared to control. Similar results have already been observed previously in the literature, indicating the low herbicide efficiency in obtaining seeds with high physiological quality, promoting low percentage of germination and low vigor (Guimarães et al., 2012; Lacerda et al., 2005a; Pinto et al., 2014).

During the soaking process, it was observed that the application of glufosinate ammonium had changed the pattern of water absorption of cultivar Benso1RR within 48 h, extending phase II (Figure 1), allowing for more time to repair the seeds cellular structures and allocation of their reserves, which may have contributed to maintaining the cultivar vigor on the application of desiccant. This same change in soaking pattern was not observed for seeds of cultivar NA 5909 RG from desiccated plants, which presented low vigor.

Within 24 hours of soaking, all treatments showed radicle protrusion, with water content in the seeds close to 60%, coinciding with Mengarda et al. (2015) and Castro et al. (2004) who have reported that at the end of phase II during germination, the seeds water content is close to 60%, starting phase III, when root protrusion starts. Han et al. (2013) have defined that phase II is between 12 h and 24 h of soaking in soybean seeds.

Also in 24 hours of soaking, desiccated seeds from plants of cultivar NA 5909 RG showed 28% of root protrusion and their control showed 68%. For cultivar Benso1RR from desiccated plants there were 42% of root protrusion and its control 67% (Figure 2). This delay in the root protrusion process could be an advantage from the point of view of tolerance to some sort of stress that might occur in the field, as only few seeds would be affected, since not all would issue sprouts during stress.

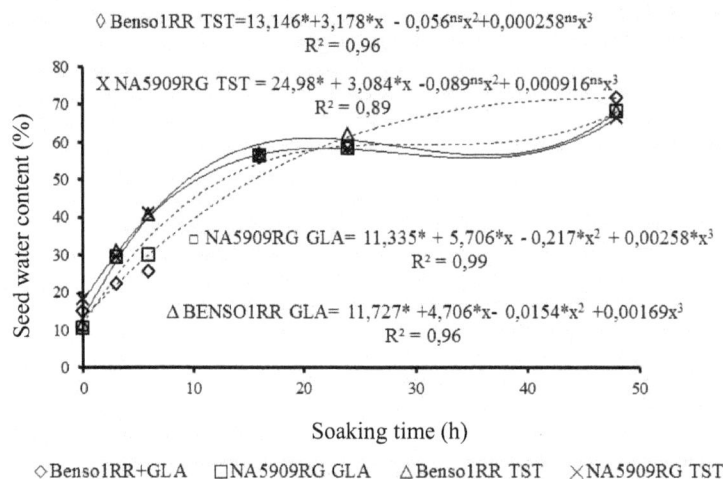

Figure 1. Soybean seeds soaking three-phase pattern for cultivars NA 5909 RG and Benso1RR from plants that were desiccated with glufosinate ammonium (GLA) and non-desiccated (TST).

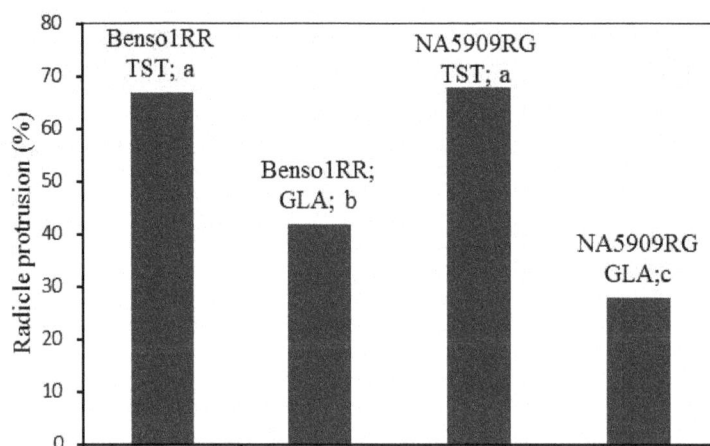

Figure 2. Radicle protrusion from soybean seeds of cultivars NA 5909 RG and Benso1RR from desiccated plants with glufosinate ammonium (GLA) or not (TST). Means followed by the same letter do not differ at 5% by Tukey test.

In 48 h of soaking, seeds from desiccated plants reached almost the same percentage (100%) of root protrusion compared to their controls (Figure 2), indicating the start of phase III. Similar data for soybeans and beans seeds have been observed, and phase III occurs at 40 hours of soaking for soybeans seeds and at 30 h for beans seeds (Teulat-Merah et al., 2011; Cheng et al., 2009).

Regarding the biochemical components of dry seeds reserves, it was observed that the initial phytate and soluble protein content did not change with the application of glufosinate ammonium, when compared with their controls, but there was a reduction of the initial levels of soluble sugar for cultivars NA 5909 RG and Benso1RR (Table 2) regarding their controls. A similar result was observed by Lacerda et al. (2003b), who have not observed differences in total protein content when comparing seeds from plants desiccated with glufosinate ammonium compared to control in different reproductive stages of soybeans. However, they have noted significant differences in the ether extract (lipid) between seeds from desiccated plants in relation to their controls, and, the later the product was applied, the more the ether extract content approached the control.

Phytate is the main storage form of phosphorus in seeds, and can be used as energy reserve, besides being a major precursor of polysaccharides constituting cell walls (Scott, 1991). The initial phytate content of seeds from desiccated plants of cultivars NA 5909 RG and Benso1RR were similar when compared with seeds from non-desiccated plants, ranging from 1.27 to 1.97 $mg.g^{-1}$ of seeds. The levels observed by Moreira et al. (2012) ranged between 1.131 and 2.43 $g.100\ g^{-1}$.

Table 2.　Initial content of biochemical components in soybean seeds from plants desiccated with glufosinate ammonium (GLA) or not (TST).

Treatments	Soluble protein (mg.g^{-1})	Soluble sugar (mg.g^{-1})	Phytate (mg.g^{-1})
NA 5909 RG TST	101.54 a	142.3098 a	1.58 a
NA 5909 RG + GLA	104.26 a	100.8929 ab	1.57 a
Benso1RR TST	110.11 a	94.69 b	1.27 a
Benso1RR + GLA	109.67 a	81.65 b	1.98 a
CV (%)	3.96	16.73	13.76

In the column, means followed by the same letter do not differ at 5% by Tukey test.

The initial content of soluble protein for cultivar NA 5909 RG was 104.26 mg.g^{-1}, being similar for cultivar Benso1RR, with a content of 110.11 mg.g^{-1}, being close to the value found by Stajner et al. (2007) who have observed 95 mg.g^{-1} of soluble protein in soybean seeds, and the authors have not observed any changes when different γ radiation dosages were applied to the seeds, and no difference was observed in the soluble protein contents of this study among the seeds from desiccated plants regarding their control (Table 1).

The application of desiccant promoted the reduction of the initial soluble sugar content in seeds for the two cultivars in relation to their controls (Table 2). For cultivar NA 5909 RG from desiccated plants, 100.89 mg.g^{-1} were observed and their control 142.30 mg.g^{-1}. For cultivar Benso1RR from desiccated plants, 81.65 mg.g^{-1} were observed and their control 94.69 mg.g^{-1}. The values found in this study are similar to those found by Obendorf et al. (2009) who observed between 100 and 150 mg.g^{-1} of total soluble carbohydrate in seeds in physiological maturity.

The high vigor for seeds of cultivar NA 5909 RG in relation to cultivar Benso1RR from non-desiccated plants may be associated with higher initial content of soluble sugar observed for NA 5909 RG in relation to cultivar Benso1RR, and the same was observed by Henning et al. (2010) who claim that soybean seeds with high initial sugar and soluble protein content show high vigor. However, seeds from desiccated plants of cultivar Benso1RR had lower initial content of soluble sugar for dry seeds, showing that the initial content of the reserve components was not the determining factor for the maintenance of the vigor, since there was no difference in vigor between desiccated plant seeds regarding their control for this cultivar.

Seeds germination is a physiological and important process in the development of seedlings, which involves complex processes and the use of their reserves for the formation and establishment of new seedlings (Ishibashi et al., 2013; Kim et al., 2011). During the soaking period in this study, there was a mobilization of soluble sugars and decreases in protein content, indicating the use of these components to form new

seedlings. However, there were no significant changes to the phytate content over time. According to Martinez et al. (2011), 48 hours after soaking is a very short period to observe phytate hydrolysis during soybeans seed germination.

The soluble protein content decreased during the soaking time (Figure 3), more sharply for cultivar Benso1RR, with and without desiccation, and for cultivar NA 5909 RG without desiccation, not interfering in vigor. However, for seeds of cultivar NA 5909 RG from desiccated plants, there was a less pronounced mobilization of soluble protein, reflecting at the end of five days in low germination percentages with smaller seedlings and low vigor when compared to control plants seeds.

The lower mobilization of soluble protein over time can be reflected in reduced synthesis of amino acids. According to Han et al. (2013), the increased amino acid content during the germination of soybean seeds is related to protein degradation, which can be translocated to the embryonic axis growing points, thus contributing in forming living tissues and participating in reactions of the respiratory chain.

With 24 hours of soaking, the cultivars from plants control of cultivar NA 5909 RG had lower soluble sugar content (291.99 mg.g^{-1}) in relation to cultivar Benso1RR (379.23 mg.g^{-1}) to issue nearly the same percentage of rootlets, 67% e 68%, respectively. For seeds from desiccated plants, the differences between cultivars were high, the soluble sugar content for cultivar Benso1RR was 279.27 mg.g^{-1} with 42% of root protrusion compared with cultivar NA 5409RG, with soluble sugar content 146.87 mg.g^{-1} and 28% of root protrusion.

In 48 h of soaking, when all treatments showed 100% root protrusion, there was greater mobilization of soluble sugar of cultivar Benso1RR in relation to cultivar NA 5909 RG, both for seeds from control plants as for seeds from desiccated plants (Figure 4). However, when comparing seeds of both cultivars from control plants, cultivar NA 5909 RG (241.44 mg.g^{-1}) was more efficient in the use of soluble sugar, because with less mobilization (Figure 4) it showed high vigor in relation to cultivar Benso1RR (365 mg.g^{-1}).

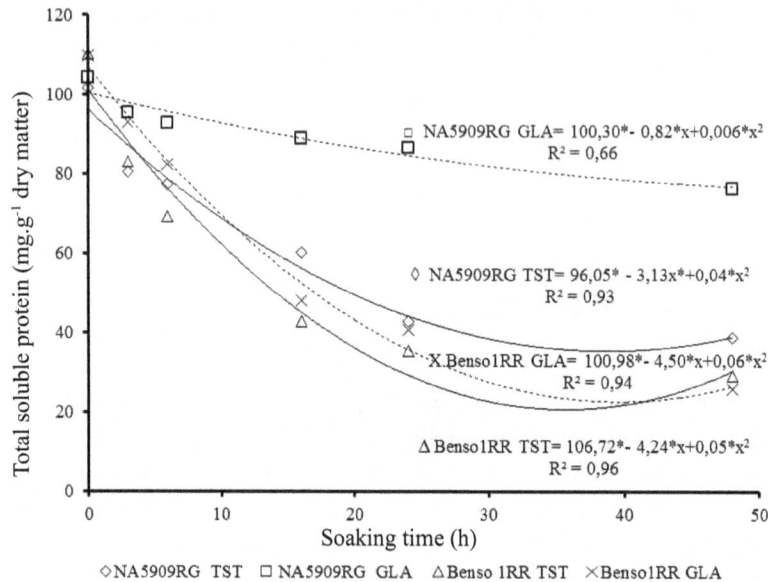

Figure 3. Mobilization of soluble proteins of soybean seeds during germination of cultivars NA 5909 RG and Benso1RR from plants desiccated with glufosinate ammonium (GLA) and non-desiccated (TST). *Significant value ($p < 0.05$) – Tukey test.

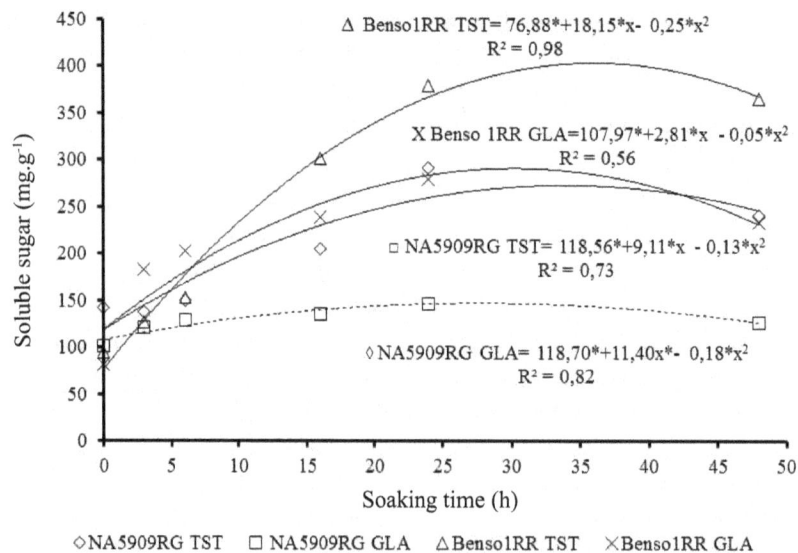

Figure 4. Mobilization of soluble sugar of soybean seeds during seed germination of cultivars NA 5909 RG and Benso1RR from desiccated plants (GLA) and non-desiccated (TST). *Significant value ($p < 0.05$) – Tukey test.

For seeds from desiccated plants (Figure 4), an opposite behavior was observed, in this case the increased mobilization of soluble sugar of cultivar Benso1RR (233.62 mg.g⁻¹) during soaking because it may have been the determining factor for maintaining vigor and germination percentage on the application of glufosinate ammonium on physiological maturity in relation to cultivar NA 5909 RG (126.98 mg.g⁻¹), and not the higher levels of soluble sugar in the dry seed of cultivar NA 5909 RG in relation to seeds of cultivar Benso1RR.

One possible explanation for glufosinate ammonium having influenced the seeds quality is the presence of the herbicide molecule in the seeds. As this herbicide has the ability to translocate at a short distance, one hypothesis is that it could have crossed the pod and come into contact with the seed, or the product has the ability to change the permeability of the pods membranes, favoring translocation of water and the herbicide molecule itself into the pod, contacting the seed. According to Oliveira et al. (2014), greater absorption of water by soybean pods harvested from reproductive stages R7 and R8 takes place, when compared with stages R4, R5 and R6, influencing loss of seed vigor.

Thus, more specific study is being carried out, seeking to understand how the herbicide molecule applied in plants may be affecting the seeds quality and which enzymatic and non-enzymatic mechanisms may be involved in maintaining these seeds vigor, such as, for example, the activity of antioxidant enzymes, non-enzymatic compounds and cellular changes.

Conclusions

It is concluded that the use of glufosinate ammonium has negatively interfered in the mobilization of reserves for NA 5909 RG, reflecting low vigor and low germination percentage. For cultivar Benso1RR the application of desiccant changed the pattern of water absorption, prolonging phase II of soaking, allowing more time for cellular organization of seeds and mobilization of their reserves, contributing to the maintenance of their vigor.

The use of glufosinate ammonium as a desiccant is not recommended, since for cultivar NA 5909 RG there was a reduction of seed vigor, and for cultivar Benso1RR there was no effective difference between seeds from desiccated plants regarding their control.

Acknowledgments

To FUMDES program for granting the doctoral scholarship and to CNPq Notice 14, for the financial support granted to project 4818212011-6 and FAPESC (Fundação de Amparo a Pesquisa e Inovação do Estado de Santa Catarina) and CNPq by productivity grant for second author.

References

AZEVEDO, R.A.; ALAS, R.M.; SMITH, R.J.; LEA, P.J. Response of antioxidant enzymes to transfer from elevated carbon dioxide to air and ozone fumigation, in the leaves and roots of wild-type and a catalase-deficient mutant of barley. *Physiologia Plantarum*, v.104, p.280-292, 1998.

BRADFORD, M.M. A rapid and sensitive method for the quantification of microgram quantities of protein utilizing the principle of protein-dye binding. *Analytical Biochemistry*, v.72, p.248-254, 1976.

BRASIL. Ministério da Agricultura, Pecuária e Abastecimento. *Regras para análise de sementes*. Ministério da Agricultura, Pecuária e Abastecimento. Secretaria de Defesa Agropecuária. Brasília: MAPA/ACS, 2009. 395p. http://www.agricultura.gov.br/arq_editor/file/2946_regras_analise__sementes.pdf

CARVALHO, L.B. Dinâmica fisiológica. In: CARVALHO, L.B. *Herbicidas*,1. ed. Lages-SC, 2013. p.21-52.

CASTRO, R.D.; BRADFORD, K.J.; HILHORST, H.W.M. Embebição e reativação do metabolismo. In: FERREIRA, A.G.; BORGHETTI, F. (Orgs.). *Germinação do básico ao aplicado*. Porto Alegre: Artmed, 2004. p.149-162.

CHENG, L.B.; LI, S.Y.; HE, G.Y. Isolation and expression profile analysis of genes relevant to chilling stress during seed imbibition in soybean [*Glycine max* (L.) Merr.]. *Agricultural Sciences in China*, v.8, n.5, p.521-528, 2009. http://www.sciencedirect.com/science/article/pii/S1671292708602424

CLEGG, K.M. The application of the anthrone reagent to the estimation of starch in cereals. *Journal of the Science of Food and Agricultural*, v.3, p.40-44, 1956.

GARCIA, J.S.; GRATÃO, P.L.; AZEVEDO, R.A.; ARRUDA, M. Metal contamination effects on sunflower (*Helianthus annuus* L.) growth and protein expression in leaves during development. *Journal Agriculture Food Chemistry*, v.54, p.8623–8630, 2006. http://www.scielo.br/scielo.php?script=sci_nlinks&ref=000068&pid=S0103-9016200800005000150004&lng=pt

GUIMARÃES, V.F.; HOLLMANN, M.J.; FIOREZE, S.L.; ECHER, M.M.; RODRIGUES-COSTA, A.C.P.; ANDREOTTI, M. Produtividade e qualidade de sementes de soja em função de estádios de dessecação e herbicidas. *Planta Daninha*, v.30, n.3, p.567-573, 2012. http://www.scielo.br/scielo.php?pid=S0100-83582012000300012&script=sci_arttext

HAN, C.; YIN, X.; HE, D.; YANG, P. Analysis of proteome profile in germinating soybean seed, and its comparison with rice showing the styles of reserves mobilization in different crops. *Plos One*, v.8, n.2, p.1-9, 2013. http://www.ncbi.nlm.nih.gov/pubmed/23460823

HENNING, F.A.; MERTZ, L.M.; JACOB, E.A.; MACHADO, R.D.; FISS, G.; ZIMMER, P.D. Composição química e mobilização de reservas em sementes de soja de alto e baixo vigor. *Revista Bragantia*, v.69, n.3, p.727-734, 2010. http://www.scielo.br/pdf/brag/v69n3/26.pdf

ISHIBASHI, Y.; KODA, Y.; ZHENG, S.H.; YUASA, T.; IWAYA-INOUE, M. Regulation of soybean seed germination through ethylene production in response to reactive oxygen species. *Annals of Botany*, v.1, p.1-8, 2013. http://www.ncbi.nlm.nih.gov/pubmed/23131300

KIM, H.T.; CHOI, U.K.; RYU, H.S.; LEE, S.J.; KWON, O.S. Mobilization of storage proteins in soybean seed (*Glycine max* L.) during germination and seedling growth. *Biochimica et Biophysica Acta*, v.1814, n.9, p.1178-1187, 2011. http://www.ncbi.nlm.nih.gov/pubmed/21616178

LACERDA, A.L.S.; LAZARINI, E.; SÁ, M.E.; VALÉRIO FILHO, W.V. Efeitos da dessecação de plantas de soja no potencial fisiológico e sanitário de sementes. *Revista Bragantia*, v.64, n.3, p.447-457, 2005a. http://www.scielo.br/pdf/brag/v64n3/26439.pdf

LACERDA, A.L.S.; LAZARINI, E.; SÁ, M.E.; VALÉRIO FILHO, W.V. Armazenamento de sementes de soja dessecadas e avaliação da qualidade fisiológica, bioquímica e sanitária. *Revista Brasileira de Sementes*, v.25, n.2, p.97-105, 2003b. http://www.scielo.br/scielo.php?pid=S0101-31222003000400014&script=sci_arttext

LATTA, M.; ESKIN, M. A simple and rapid method for phytate determination. *Journal Agricultural Food Chemistry*, v.28, p.313-315, 1980.

MARCOS-FILHO, J. Teste de envelhecimento acelerado. In: KRZYZANOWSKI, F.C.; VIEIRA, R.D.; FRANÇA-NETO, J.B. (Ed.). *Vigor de sementes*: conceitos e testes. ABRATES, 1999. cap.3, p.3.1-3.24.

MA, F.; CHOLEWA, E.; MOHAMED, T.; PETERSON, C.A.; JZEN, M.G. Cracks in the palisade cuticle of soybean seed coats correlate with their permeability to water. *Annals of Botany*, v.94, p.213–228, 2004. http://aob.oxfordjournals.org/content/94/2/213.full

MARTINEZ, A.P.C.; MARTINEZ, P.C.C.; SOUZA, M.C.; CANNIATTI BRAZACA, S.G. Alterações químicas em grãos de soja com a germinação. *Ciência Tecnologia de Alimentos*, v.31, n.1, p.23-30, 2011. http://www.scielo.br/pdf/cta/v31n1/04.pdf

MENGARDA, L.H.G.; LOPES, J.C.A.; RODRIGO, S.A.; ZANOTTI, R.F.; MANHONE, P.R. Alternating temperature and accelerated aging in mobilization of reserves during germination of *Carica papaya* L. seeds. *Journal of Seed Science*, v.37, n.1, p.16-25, 2015. http://www.scielo.br/scielo.php?script=sci_abstract&pid=S0101312220120004000016&lng=en&nrm=iso&tlng=en

MOREIRA, A.A.; MANDARINO, J.M.G.; NEVES, S.R.D.; LEITE, R.S.; OLIVEIRA, M.A. Teor de ácido fítico em cultivares de soja cultivados em diferentes regiões do Paraná e São Paulo. *Alimentos e Nutrição*, v.23, n.3, p.393-398, 2012. http://serv-bib.fcfar.unesp.br/seer/index.php/alimentos/article/viewFile/393/2047

NAKAGAWA, J. Testes de vigor baseados no desempenho das plântulas. In: KRZYZANOWSKI, F.C.; VIEIRA, R.D.; FRANÇA-NETO, J.B. (Ed.). *Vigor de sementes*: conceitos e testes. Londrina: ABRATES, 1999. p.2.1-2.24.

OBENDORF, R.L.; ZIMMERMAN, A.D.; ZHANG, Q.I.; CASTILLO, A.; KOSINA, S.M.; BRYANT, E.G.; SENSENING, E.M.; WU, J.; SHNEBLY, S.R. Accumulation of soluble carbohydrates during seed development and maturation of low-raffinose, low-stachyose soybean. *Crop Science*, v. 49, p.329 – 341, 2009. https://www.crops.org/publications/cs/abstracts/49/1/329

OLIVEIRA, C.M.G.; KRYZANOWSKI, F.C.; OLIVEIRA, M.C.N.; FRANÇA-NETO, J.B.; HENNING, A.A. Relationship between pod permeability and seed quality in soybean. *Journal of Seed Science*, v.36, n.3, p.273-281, 2014. http://www.scielo.br/pdf/jss/v36n3/aop0214.pdf

PEREIRA, W.A.; PEREIRA, S.M.A.; DIAS, D.C.F.S. Dynamics of reserves of soybean seeds during the development of seedlings of different commercial cultivars. *Journal of Seed Science,* v.37, n.1, p.63-69, 2015. http://www.scielo.br/pdf/jss/v37n1/2317-1537-jss-37-01-00063.pdf

PINTO, M.A.B.; BASSO, C.J.; KULCZYNSKI, S.M.; BELLE, C. Productivity and physiological quality of seeds with burn down herbicides at the preharvest of bean crops. *Journal of Seed Science*, v.36, n.4, 384-391, 2014. http://www.scielo.br/pdf/jss/v36n4/aop0814.pdf

RITCHIE, S.; HANWAY, J.J.; THOMPSON, H.E. How a soybean plant develops. Ames: Iowa State University of Science and Technology, 1982. 20p.

SAS. SAS Institute Inc® 2009. Cary, NC, USA, Lic. UDESC: SAS Institute Inc, 2009.

SCOTT, J.J. Alkaline phytase activity in nonionic detergent extracts of legume seeds. *Plant Physiology*, v.95, n.1, p.1298-1301, 1991.

STAJNER, D.; MILOŠEVIĆ, M.B.; POPOVIĆ, M. Irradiation effects on phenolic content, lipid and protein oxidation and scavenger ability of soybean seeds. *International Journal of Molecular Sciences*, v.8, p.618-627, 2007. http://www.mdpi.org/ijms/papers/i8070618.pdf

TEULAT-MERAH, B.; MORERE-LE PAVEN, M.C.; RICOULT, C.; AUBRY, C.; PELTIER, D. cDNA-AFLP profiling in the embryo axes during common bean germination. *Biologia Plantarum*, v.55, n.3, p.437-447, 2011. http://link.springer.com/article/10.1007/s10535-011-0108-5

WEITBRECHT, K.; MÜLLER, K.; LEUBNER-METZGER, G. First off the mark: early seed germination. *Journal of Experimental Botany*, v.62, n.10, p.3289-3309, 2011. http://www.ncbi.nlm.nih.gov/pubmed/21430292

Morphoanatomy of fruit, seed and seedling of *Ormosia paraensis* Ducke

Breno Marques da Silva e Silva[1*], Camila de Oliveira e Silva[1],
Fabíola Vitti Môro[2], Roberval Daiton Vieira[3]

ABSTRACT – *Ormosia paraensis* Ducke, known as "tento", has seeds that are used to make handcrafts and wood that is worked on by furniture makers. For forest identification and seeds technology, the information about the morphoanatomy of their fruits, seeds and seedlings is scarce. Therefore, the purpose of this study was to morphoanatomically describe the fruit, seeds and the post-seminal development of "tento". For the morphoanatomical description, the evaluations were examined by optical and electron microscopy scanning. The fruit is a nutant legume, brown to black, dehiscent and with one or two seeds of lateral placentation, being the epicarp slim, the mesocarp woody and the endocarp spongy, measuring about 4.4 cm, 3.9 cm and 2.0 cm in length, width and thickness, respectively. The seeds are bitegmic, exalbuminous and rounded with average dimensions: length of 12.36 mm, width of 9.68 mm and thickness of 8.03 mm. The "tento" seedlings have simple and alternate leaves, with tap roots and cylindrical stem, being its germination hypogeal cryptocotyledonary.

Index terms: morphology, post-seminal development, germination, Fabaceae.

Morfoanatomia do fruto, da semente e da plântula de *Ormosia paraensis* Ducke

RESUMO – *Ormosia paraensis* Ducke, conhecida popularmente por tento, possui sementes ornamentais amplamente usadas na confecção de biojóias, assim como, sua madeira é usada na indústria moveleira. Para a identificação florestal e tecnologia de sementes, as informações sobre a morfoanatomia de seus frutos, sementes e plântulas são escassas. Desta forma, o objetivo no presente trabalho foi descrever morfoanatomicamente o fruto, a semente e a plântula de tento. Para a descrição morfoanatômica de tento, os frutos, as sementes e as plântulas foram avaliados por meio de microscopia óptica e eletrônica de varredura. O fruto de tento é um legume nucóide, pseudo-septado, castanho a preto, deiscente, portando uma ou duas sementes, de placentação lateral, sendo o epicarpo delgado, mesocarpo lenhoso e endocarpo esponjoso, medindo cerca de 4,4; 3,9 e 2,0 cm de comprimento, largura e espessura, respectivamente. As sementes são bitegumentadas, exalbuminosas, arredondadas, com dimensões médias: comprimento de 12,36 mm, largura de 9,68 mm e espessura de 8,03 mm. As plântulas de tento apresentam folhas simples e alternas, com raiz pivotante e caule cilíndrico, sendo a germinação do tipo hipógea criptocotiledonar.

Termos para indexação: morfologia, desenvolvimento pós-seminal, germinação, Fabaceae.

Introduction

The Fabaceae family has a worldwide distribution, including about 650 genera and 18,000 species, divided into four subfamilies: Caesalpinioideae, Faboideae (Papilionoideae), Mimosoideae and Cercideae with approximately 200 genera and about 1,500 species in Brazil (Souza and Lorenzi, 2012).

In the Amazon rain forest, the Fabaceae's economic importance is marked by many timber, herbal (antifungal and anti-inflammatory), dyes, oil, fruit, insecticides, repellents, and ornamental species and are used to make handicrafts, among others (Ferreira et al., 2005; Chevreuil et al., 2011).

Ormosia paraensis Ducke, popularly known in Brazilian Portuguese as tento, tenteiro or olho-de-cabra, is a tree species widely distributed in the dense ombrophilous forests and ancient capoeira shrubs in the Amazon rain forest. By virtue of the straight shaft without branches and dense wood, tento is of interest to the wood industry. Moreover, the potential for nodulation and nitrogen fixation, characteristic of legumes, is more attractive for use in reforestation, afforestation and replanting of degraded areas with tenteiro (Carneiro et al., 1998; Lorenzi, 2002).

The characterization of Fabaceae's morphological

---
[1]Universidade do Estado do Amapá, 68906-970 – Macapá, AP, Brasil.

[2]Departamento de Biologia Aplicada a Agropecuária, UNESP, 14884-900 – Jaboticabal, SP, Brasil.

[3]Departamento de Produção Vegetal, UNESP, 14884-900 – Jaboticabal, SP, Brasil.

*Corresponding author <silvabms@hotmail.com.br>

and anatomical structures of fruits, seeds and seedlings is the goal of several studies aiming to solve taxonomic problems, subsidize species identification, and understand ecological and physiological strategies, among others (Melo-Pinna et al., 1999; Barroso et al., 2007; Silva and Môro, 2008).

Seeds internal and external morphology, combined with seedlings observations, allow the identification of structures, offering, in the laboratory, subsidies for the correct interpretation of germination, identification and certification tests of physiological quality. Therefore, it can assist seed technology studies, and, in the nursery, contribute to the recognition of the species and methods suitability for producing seedlings for various purposes (Andrade et al., 2003; Rego et al., 2007). However, for most Amazonian legume species, basic information about the fruits, seeds and seedlings morphology and anatomy are sparse, and for tento, fruit and seed morphology is partially dealt with by Ducke (1925). This way, to aid in the identification in the field, for the benefit of ecological information and the seeds physiology interpretation, the aim of this study was to describe morphologically and anatomically the fruit, seed and post-seminal development of *Ormosia paraensis* Ducke.

Material and Methods

The fruits were harvested from ten tento matrices trees (*Ormosia paraensis* Ducke) located at Parque Natural Municipal "Arivaldo Gomes Barreto" (00°02'21"S and 51°05'35"W), in the Brazilian city of Macapá, AP. Then the fruits were sent for morphological and anatomical analysis at Plant Morphology Laboratory at Department of Biology Applied to Agriculture (DBAA) and at Electron Microscopy Laboratory at Jaboticabal UNESP, Campus Jaboticabal, Jaboticabal, SP, Brazil.

For biometry evaluation, measurements were performed at random for length, width and thickness on 100 fruits and 200 seeds, with the aid of a caliper and a millimeter ruler. Thereafter the fruits and seeds dry matter was determined by drying four replicates of 10 fruits and 10 seeds in a greenhouse at 70 °C during 72 hours (Benincasa, 2003).

The post-seminal development phases were characterized by sowing 100 seeds without chiseling with no. 8 wood sandpaper, germinated in trays (35 cm x 20 cm x 12 cm) in washed and sterilized sand, moistened with Maxin XL 0.1% aqueous solution, maintained in a germination chamber at 30 °C and with a photoperiod of 8 hours. Every day seedlings descriptions were performed in sequential stages of development, demonstrating the development of the primary root, the emergence of secondary roots, the growth beginning

of the first leaf and the conspicuous apical bud, and the expansion of the first eophylls.

The seeds were fixed in FAA (formalin-aceto-alcohol) and then preserved in ethanol 70% (Johansen, 1940). For the cuts, the seed integument softening was carried out in glycerin 50% at 100 °C and then the seeds were soaked for about four hours.

The transverse and longitudinal sections were done by freehand and in a rotary microtome. In this case, the material was previously dehydrated and embedded in paraffin as described by Johansen (1940) and Sass (1958). Subsequently, double staining of the tento seeds sections was done in astra blue and basic fuchsin, according to Kraus et al. (1998).

For the permeability assay, tento seeds chiseled with sandpaper and non-chiseled were dipped in a methylene blue 1% aqueous solution for 0, 1, 3, 6, 12, 24, 48 and 72 hours. Subsequently, these seeds were longitudinally sectioned and examined with a stereoscopic microscope to determine the limit penetration of the dye which has molecular weight close to that of water (Melo-Pinna et al., 1999).

Histochemical tests were conducted for starch and cellulose, evidenced with acid phloroglucinol; lipid substances with Sudan IV; mucilage and pectic substances with methylene blue (Johansen, 1940; Sass, 1958).

To observe the tento seedlings leaves surfaces and seeds, they were dried for observation under a scanning electron microscope (SEM), following the methodology described by Santos (1996). The fruits, seeds and post-seminal development phases illustrations were done using a stereo microscope with a camera lucida coupled.

The morphological and anatomical descriptions of fruits, seeds and seedlings were performed following the criteria and terminologies adopted by Esau (1993), Pezzato-da-Glória and Carmello-Guerreiro (2006), Barroso et al. (2004; 2007) and Damião-Filho and Môro (2005).

Results and Discussion

Tento's (*Ormosia paraensis* Ducke) fruits are nucoid vegetables, pseudo-septate, brown to black, dehiscent, carrying one or two seeds, of side placentation, and the epicarp is thin, the mesocarp is woody and the endocarp is spongy, measuring about 4.4 cm, 3.9 cm and 2.0 cm in length, width and thickness, respectively (Figures 1A-B). Likewise, Barroso et al. (2004; 2007) have described the *Ormosia* Jacks. fruits as vegetable nucoids.

Tento's vegetable nucoids are circular with conspicuous sutures, a slightly peaked end and dorsoventrally prominent due to the central presence of seeds, but they are laterally linked to the fruit through the funiculus, and parietal or side placentation (Figures 1A-B). In fruits with more than one seed, the fruit is constricted between the seeds, but they are false septa, as observed in different legumes (Barroso et al., 2007; Silva and Môro, 2008). According to Ducke (1925), tento's vegetables have one to three bicolor seeds.

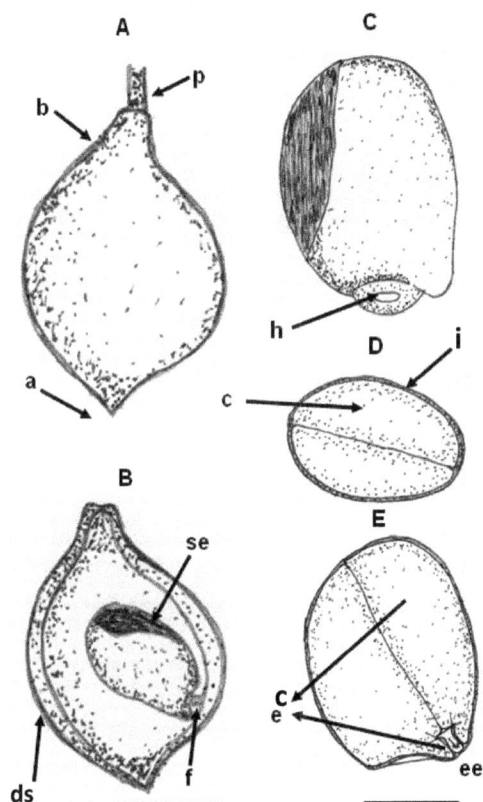

Figure 1. A. Fruit external view. B. Fruit internal view. (Scale: 1 cm). C. Seed external view. D. Seed cross cut. E. Seed longitudinal cut (Scale: 0.5 cm) of tento (*Ormosia paraensis* Ducke). Caption: a – apex, b – base, p – peduncle, ds – dorsal suture, se – seed and f – funiculus, i – integument, c – cotyledon, h – hilum, ea – embryonic axis and e – embryo. Bar = 0.5 cm.

Tento's seeds are stenospermic, bitegmic, exalbuminous, and rounded (Figures 1C-D-E; Figures 2A-B; Figures 3C-D), with average dimensions: length 12.36 mm, width 9.68 mm, thickness 8.03 mm, average volume 0.463 mL.seed^{-1}, dry matter of 0.45 g.seed^{-1} and density 0.98 g.mL^{-1}. Ducke (1925) has similarly noted that tento's seeds have measure 12-13 mm. The integument is formed by a red and black outer layer, glabrous and thin, with a heterochrome (white), elliptical and small hilum relative to the seed, with a side bulge indicating the presence of the embryonic axis (Figures 1C-E; Figure 3C). The tegmen is white and purple and firmly adhered to the embryo. The two cotyledons are free, pale yellow, starchy, massive and plane-convex. The embryo is invaginated

(papilionaceous) with a differentiated and small embryonic axis in relation to the seeds (Figures 1D-E; Figure 3C).

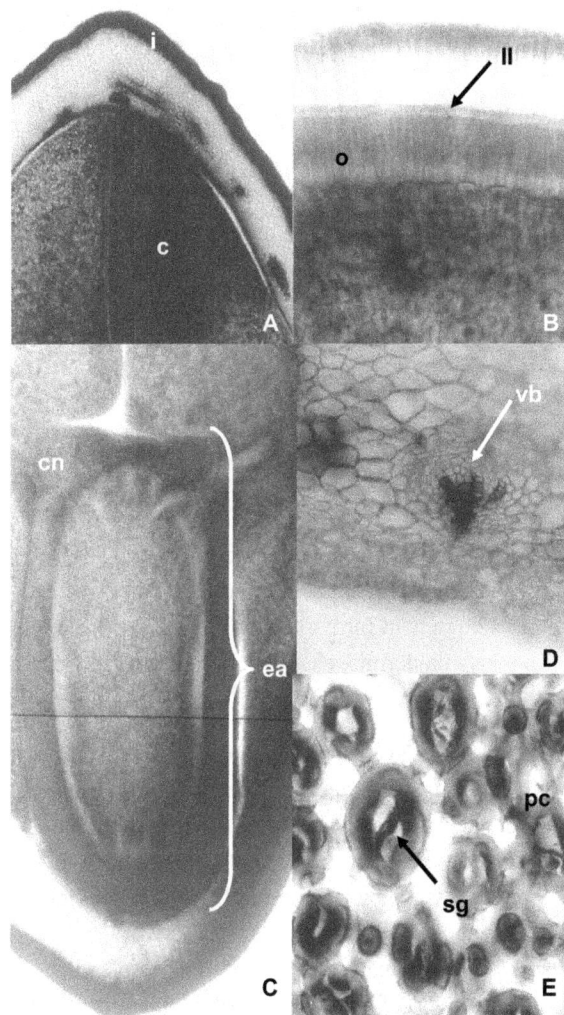

Figure 2. Tento seeds cut (*Ormosia paraensis* Ducke). A. Integument and cotyledon (12.5x). B. Integument (lucid line, osteosclereids and palisade layer) (50x) (Transversal). C. Embryonic axis (12.5x) (Longitudinal). D. Vascular bundle (cotyledon) (50x). E. Parenchyma cells (cotyledon) (100x) (Transversal). Caption: i – integument, c – cotyledon, ea – embryonic axis, cn – cotyledon node, ll – lucid line, o – osteosclereids, sg – starch grains, pc – parenchyma cells, vb – vascular bundle.

The tento's seeds integument has four distinct layers: cuticle (coating with hydrophobic substances), skin (with a compact palisade layer, consisting in radially elongated macrosclereids, unevenly thickened cell walls), hypodermis (hourglass cells, or pillar cells or osteoesclereids) and parenchyma cells (Figures 2A-B/E; Figures 3A-D).

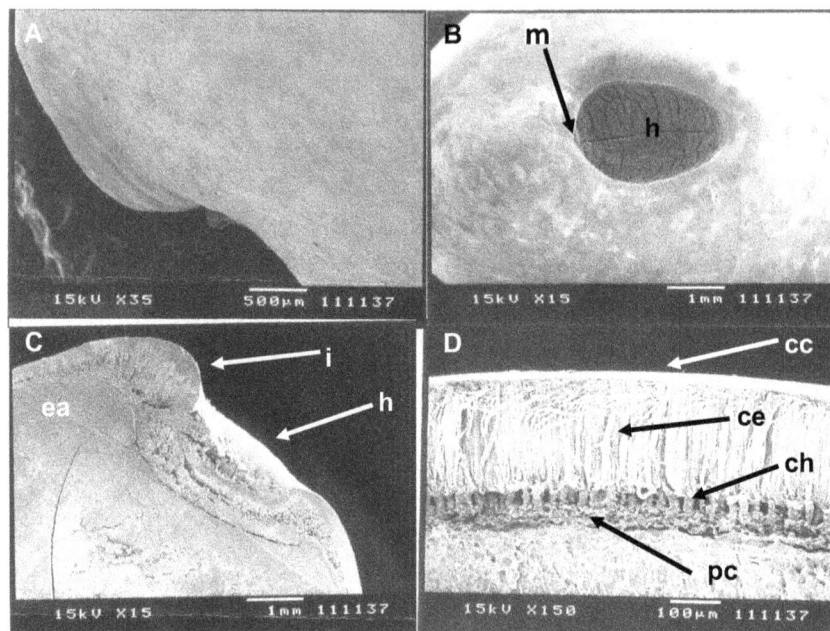

Figure 3. Tento seeds scanning electron micrograph (*Ormosia paraensis* Ducke – Fabaceae). A. Integument (outer surface view). B. Hilum and micropyle (outer surface view). C. Hilum, micropyle and embryonic axis (longitudinal cut). D. Integument (cross-section). Caption: i – integument, h – hilum, m – micropyle, ea – embryonic axis, cc – cuticle, ce – epidermis, ch – hypodermis and pc – parenchyma cells.

According to Carvalho and Nakagawa (2012) and Baskin and Baskin (2014), the chemical composition, arrangement and intercellular substances of the palisade layer influence the absorption of water by the seed. Lewis and Yamamoto (1990) have reported that lignin is a natural polymer present in the seeds outer layer.

The sclerenchyma cells, called macrosclereids, form the palisade layer of the tento's seeds integument, which continuously appears throughout the outer layer, except for the hilum area, where two palisade layers can be observed (Figure 2B and Figure 3C). In this case, the innermost palisade layer comes from the funiculus, while the outermost layer comes from the external integument of the ovule (Esau, 1993). The presence of discontinuity in the integument (pleurogram) was not observed in tento's seeds (Figure 2A and Figure 3D).

The embryo is formed by two cotyledons consisting mainly in parenchyma cells with starch grains and permeated by characteristic vascular bundles of the eudicotyledons, responsible for the translocation of nutrients for growth of the embryonic axis, and the fully differential embryonic axis, adnate through node that is cotyledonary to the cotyledons (Figures 2 C-D-E). Likewise, Corner (1951), Barroso et al. (2007) and Silva and Môro (2008) describe embryos of various legumes seeds.

The hypodermis consists in a uniform layer of cells, continuous across the outer layer, except for the hilum area, where it is absent. It is formed by sclerenchyma cells with cell walls of uneven thickness, called osteoesclereids, with the presence of large intercellular spaces (Figures 3C-D), as observed by Corner (1951) in legume seeds.

The lucid line, a refractive line that runs across the macrosclereids through the integument, is located just above the middle portion of these cells (Figure 2B; Figure 3D). According to Melo-Pinna et al. (1999), the lucid line is usually found in legumes seed integument.

The parenchyma occurs below the hypodermic layer of the seed integument, formed by thickened cell walls of (Figures 3 C-D) of cellulosic nature. Such cells are slightly more elongated at the hilum-chalaza direction, have a thin cell wall and absent protoplasm (Esau, 1993), and are placed in numerically variable strata, according to the seed area.

The tento's seeds germination starts from four days after sowing with primary, cylindrical and white to cream root protrusion. At six days after sowing, there is growth and formation of absorbent hair in the primary root (Figures 4A-B). The development of secondary roots is accompanied by meager hypocotyl elongation. Emergence is marked by strong growth of the epicotyl, cylindrical, green and with thin and hyaline trichomes, and consequent output of sand plumule in about 16 days after sowing (Figure 4C). The epicotyl becomes erect, extends, and at 24 days the formation of protophilus, simple and opposite, of conduplicate pre-foliation, is observed. Concurrently, the cotyledons remain below the sand and covered by the integument due to unimpressive development and growth of the hypocotyl (Figures 4C-D).

Thus, tento seeds germination was characterized as hypogeal-cryptocotyledonary. Similarly, seed germination of *Ormosia arborea* Ducke is classified as crypto-hypogeal-reserve (Rodrigues and Tozzi, 2007).

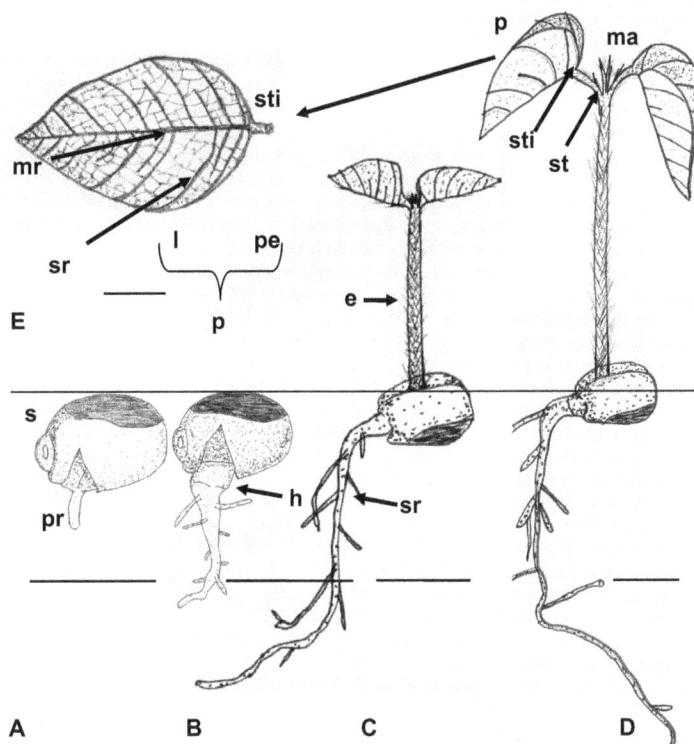

Figure 4. A-D. Post-seminal development. E. Tento seedling protophilus (*Ormosia paraensis* Ducke – Fabaceae). Caption: s – substrate, pr – primary root, sr – secondary root, h – hypocotyl, e – epicotyl, st – stipule, sti – stipels, ma – apical meristem, p – protophilus, mr – midrib, sr – secondary rib, pe – petiole and l – limbo. Bar = 1 cm.

The tento's protophilus are simple and opposite, with slightly rough leaf blades, and thin and hyaline trichomes, mainly distributed in the midrib, with stipules at the base of the petiole and stipels at the base of the limbo (Figures 4D-E and 5A-B). Likewise, *Ormosia arborea* Ducke seedlings show stipules and stipels, but with alternate protophilus (Rodrigues and Tozzi, 2007).

The tento's seedling shows the taproot system with axial, cylindrical, and sub-woody primary root, thicker at the base and yellowish, tapered and whitish at the apex, with well developed, branched and irregularly distributed secondary roots (Figure 4D). There is the presence of nodules in the roots, which confirms the statement by Raven et al. (2007) that many members of the Fabaceae family are able to fix nitrogen when associated with bacteria of the genus *Rhizobium*; for this reason they colonize little fertile soils.

The tento's seedlings leaves are simple, petiolate, with pulvinus, lanceolate, with acuminate apex, smooth margin, reticulate venation, foliaceous limbo, which initially presents

clear red color and turns green with development, with the midribs and secondary ribs being evident and the tertiary ones less evident (Figure 4E, Figures 5C-F). The petiole is green, cylindrical, thin, with trichomes, has two stipules and two lanceolate stipels and pulvinus at the base (Figures 4D-E; Figures 5A-B). According to Rodrigues and Machado (2006), the presence of pulvinus is striking in the Fabaceae family.

The protophilus adaxial face is darker than the abaxial and glabrous, and the adaxial midribs are immersed in mesophyll (Figures 5C-E-I), while the abaxial side has trichomes and protruding midribs. The trichomes are concentrated mainly in the midribs and on the edges of the leaf blade (Figures 5D-F-H).

The leaves are hypostomatic and the stomata are paracytic, with epicuticular wax on both surfaces of the blade (Figures 5I-J). Probably the presence of stomata on the leaf abaxial surface is related to the control of water loss, because the incidence of light energy is lower on this surface (Larcher, 2006).

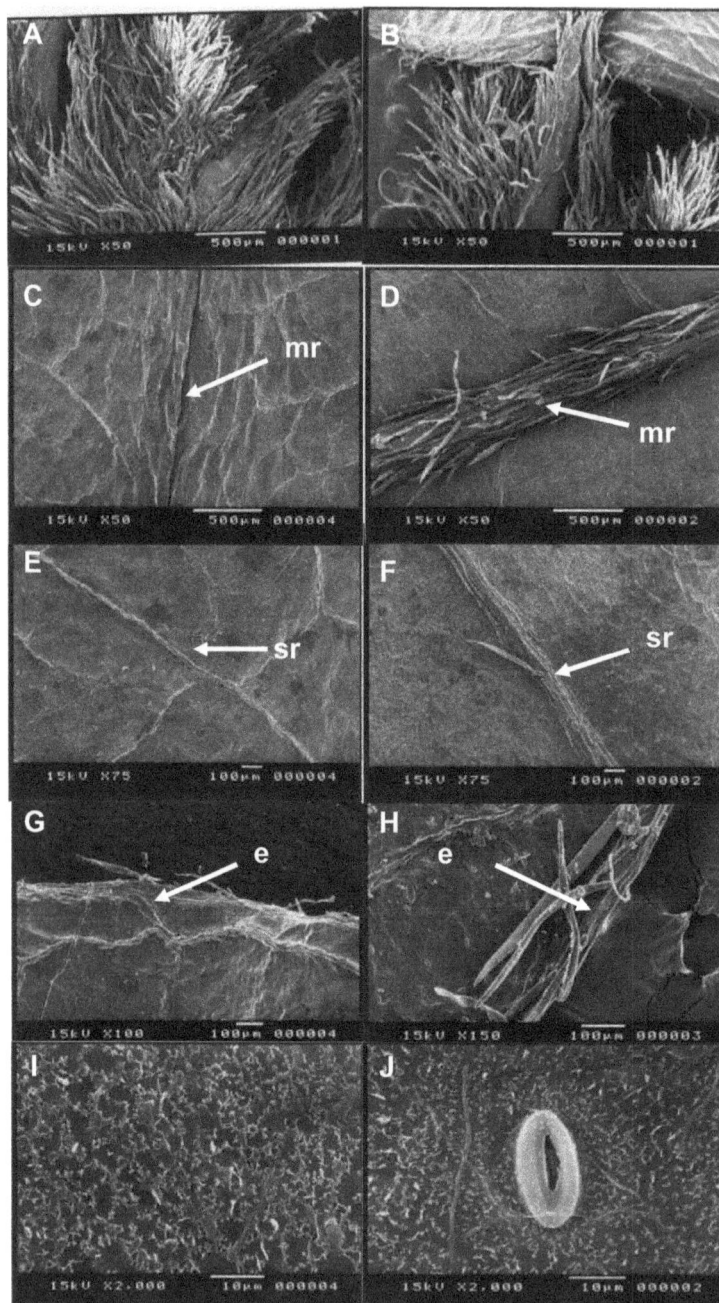

Figure 5. A. Stipules. B. Stipels. C. Adaxial surface – midrib. D. Abaxial surface – midrib. E. Adaxial surface – secondary
rib. F. Abaxial surface – secondary rib. G. Adaxial surface – edge. H. Abaxial surface – edge. I. Adaxial surface
– epicuticular wax. J. Abaxial surface – epicuticular wax and stomata on the surface of tento seedlings leaf blades
(*Ormosia paraensis* Ducke). Caption: mr – midrib, sr – secondary rib, e – edge.

Conclusions

The characters of fruits, seeds and seedlings are suitable
for the test interpretation of germination and vigor, as well as
species recognition in the field.

Acknowledgments

To Jaboticabal UNESP (Universidade Estadual Paulista
"Júlio de Mesquita Filho") for the infrastructure and equipment
availability. To CAPES for the doctorate scholarship granted
to the first author. To CNPq for the Productivity in Research

scholarship granted to the last two authors.

References

ANDRADE, A.C.S.; CUNHA, R.; SOUZA, A.F.; REIS, R.B.; ALMEIDA, K.J. Physiological and morphological aspects of seed viability of a neotropical savannah tree, *Eugenia dysenterica* DC. *Seed Science and Technology*, v.31, p.125-137, 2003. http://webcache.googleusercontent.com/search?q=cache:KOhZaWxDCawJ:https://sigaa.ufrn.br/sigaa/verProducao%3FidProducao%3D885331%26key%3D2a9ba33f1090f3e9f1f80d703decf45e+&cd=1&hl=pt-BR&ct=clnk&gl=br

BARROSO, G.M.; PEIXOTO, A.L.; ICHASO, C.L.F.;GUIMARÃES, E.F.; COSTA, C.G. *Sistemática de angiospermas do Brasil*. 2. ed. Viçosa: Editora UFV, 2007. 309 p.

BARROSO, G.M; MORIN, M.P.; PEIXOTO, A.L.; ICHASO, C.L.F. *Frutos e sementes*: morfologia aplicada à sistemática de dicotiledôneas. Viçosa: Editora UFV, 2004. 443 p.

BASKIN, C.C., BASKIN, J.M. *Seeds*: ecology, biogeography and evolution of dormancy and germination, 2ed. San Diego, CA, USA: Academic/Elsevier, 2014. 205p.

BENINCASA, M.M.P. *Análise de crescimento de plantas*: noções básicas. Jaboticabal: FUNEP, 2003. 41 p.

CARNEIRO, M.A.C.; SIQUEIRA, J.O.; MOREIRA, F.M.S.; CARVALHO, D.; BOTELHO, S.A.; JUNIOR, O.J.S. Micorrizaarbuscular em espécies arbóreas e arbustivas de ocorrência no sudeste do Brasil. *Cerne*, v.4, n.1, p.129-145, 1998. http://www.dcf.ufla.br/cerne/artigos/13-02-20093894v4_n1_artigo%2009.pdf

CARVALHO, N.M.; NAKAGAWA, J. *Sementes*: ciência, tecnologia e produção. 5 ed. Jaboticabal: FUNEP, 2012. 588 p.

CHEVREUIL, L.R.; GONÇALVES, J.F.C.; SCHIMPL, F.C.; SOUZA, C.S.C.R.; SOUZA, L.A.G.; PANDO, S.C. Prospecção de inibidores de serinoproteinases em folhas de leguminosas arbóreas da floresta Amazônica. *Acta Amazonica*, v.41, n.1, p.163-170, 2011. http://www.scielo.br/pdf/aa/v41n1/a19v41n1.pdf

CORNER, E. J. H. The leguminous seed. *Phytomorphology*, v.1, p.117-150. 1951.

DAMIÃO-FILHO, C.F.; MÔRO, F.V. *Morfologia vegetal*. 2. ed. Jaboticabal: FUNEP, 2005. 172 p.

DUCKE, W.A. Plantes nouvelles ou peu connues de la région amazonienne (IIIe partie). *Archivos do Jardim Botânico do Rio de Janeiro*, v.4, p.62-63, 1925. http://www.biodiversitylibrary.org/page/31246559#page/484/mode/1up

ESAU, K. *Anatomia das plantas com sementes*. São Paulo: Edgard Blucher, 1993. 293 p.

FERREIRA, L.V.; VENTICINQUE, E.; ALMEIDA, S.S. O desmatamento na Amazônia e a importância das áreas protegidas. *Estudos Avançados*, v.19, n.53, p.1-10, 2005. http://www.scielo.br/pdf/ea/v19n53/24086.pdf

JOHANSEN, D.L. *Plant microtechnique*. New York: McGraw-Hill, 1940. 222 p.

KRAUS, J.E., SOUSA, H.C., REZENDE, M.H., CASTRO, N.M., VECCHI, C.; LUQUE, R. Astra blue and basic fucsin double staining of plant materials. *Biotechnie & Histochemistry*, v.73, p.235-243, 1998.http://informahealthcare.com/doi/abs/10.3109/10520299809141117?journalCode=bih

LARCHER, W. *Ecofisiologia vegetal*. São Carlos: RIMA Artes e Textos, 2006. 532 p.

LEWIS, N.G.; YAMAMOTO, E. Lignin: occurrence, biogenesis and biodegradation. *Annual Review of Plant Physiology Plant Molecular Biology*, v.41, p.455-496, 1990. http://www.annualreviews.org/doi/pdf/10.1146/annurev.pp.41.060190.002323

LORENZI, H. Árvores brasileiras: manual de identificação e cultivo de plantas arbóreas nativas do Brasil. Nova Odessa: Instituto Plantarum, v.2, 2002. 368 p.

MELO-PINNA, G.F.; NEIVA, M.S.M.; BARBOSA, D.C.A. Estrutura do tegumento seminal de quatro espécies de Leguminosae (Caesalpinioideae), ocorrentes numa área de caatinga (PE – Brasil). *Revista Brasileira de Botânica*, v.22, n.3, p.375-379, 1999.http://www.scielo.br/pdf/rbb/v22n3/22(3)a04.pdf

PEZZATO-DA-GLÓRIA, B.; CARMELLO-GUERREIRO, S.M. (Editoras) *Anatomia Vegetal*. 2ª ed. Viçosa: Ed. UFV, 2006. 438 p.

RAVEN, P.H.; EVERT, R.F.; EICHHORN, S.E. *Biologia Vegetal*. São Paulo: Editora Guanabara Koogan S. A., 2007. 856 p.

REGO, S.S.; SILVA, A.J.C.; BRONDANI, G.E.; GRISI, F.A.; NOGUEIRA, A.C.; KUNIYOSHI, Y.S. Caracterização morfológica do fruto, semente e germinação de *Duranta vestita* Cham. (Verbenaceae). *Revista Brasileira de Biociências*, v.5, p. 474-476, 2007. http://www.ufrgs.br/seerbio/ojs/index.php/rbb/article/viewFile/480/383

RODRIGUES, T.M.; MACHADO, S.R.Anatomia comparada do pulvino primário de leguminosas com diferentes velocidades de movimento foliar. *Revista Brasileira de Botânica*, v.29, n.4, p. 709-720, 2006. http://www.scielo.br/pdf/rbb/v29n4/19.pdf

RODRIGUES, R.S.; TOZZI, A.M.G.A. Morfologia de plântulas de cinco leguminosas genistóides arbóreas do Brasil (Leguminosae-Papilionoideae). *Acta Botanica Brasilica*, v.21, n.3, p.599-607, 2007. http://www.scielo.br/pdf/abb/v21n3/a07v21n3.pdf

SANTOS, J.M. Microscopia de varredura aplicada às ciências biológicas. Jaboticabal: FUNEP, 1996. 56 p.

SASS, J.E. *Elements of botanical microtechnique*. New York: McGraw-Hill Book Company, 1958. 222 p.

SILVA, B.M.S.; MÔRO, F.V. Aspectos morfológicos do fruto, da semente e desenvolvimento pós-seminal de faveira (*Clitoria fairchildiana* R. A. Howard. – Fabaceae). *Revista Brasileira de Sementes*, v.30, n.3, p.195-201, 2008. http://www.scielo.br/pdf/rbs/v30n3/26.pdf

SOUZA, V.C.; LORENZI, H. Botânica sistemática: guia ilustrado para identificação das famílias de angiospermas da flora brasileira, baseado em APGIII. Ed. 3. Nova Odessa: Instituto Plantarum de Estudos da Flora Ltda. 2012. 640 p.

Biometric description of fruits and seeds, germination and imbibition pattern of desert rose [*Adenium obesum* (Forssk.), Roem. & Schult.]

Ronan Carlos Colombo[1]*, Vanessa Favetta[1], Lilian Yukari Yamamoto[1],
Guilherme Augusto Cito Alves[1], Julia Abati[1], Lúcia Sadayo Assari
Takahashi[1], Ricardo Tadeu de Faria[1]

ABSTRACT – This study has aimed to carry out the description of fruits and seeds and germination process of desert rose, from two years of observations. The fruits and seeds were characterized based on length, diameter and number of seeds per fruit. The seeds internal structure and germinating process were also described. Germination test was performed at 25 and 30 °C temperatures, determining the germination percentage and germination speed index. Seeds harvested in 2013 were stored for 12 months and submitted to a new germination test. Parallel to these tests, the seeds imbibition curve was determined in substrates over and between sheets of paper at 15, 20, 25 and 30 °C temperatures. The fruits and seeds had similar lengths and diameters in both years of observations. The seeds can be stored for up to 12 months without loss in viability and temperatures of 25 and 30 °C are suitable for performing the germination test of this species. The water absorption curve of desert rose seeds follows a triphasic pattern of soaking.

Index terms: *Adenium obesum* (Forssk.), Roem. & Schult., Apocynaceae, seeds storage.

Descrição biométrica de frutos e sementes, germinação e padrão de embebição de rosa do deserto [*Adenium obesum* (Forssk.), Roem. & Schult.]

RESUMO – Objetivou-se nesse estudo realizar a descrição de frutos e sementes e do processo de germinação de rosa do deserto, a partir de dois anos de observações. Os frutos e as sementes foram caracterizados com base no comprimento, diâmetro e número de sementes por fruto; também, descreveu-se a estrutura interna das sementes e o processo de germinação. Realizaram-se testes de germinação às temperaturas de 25 e 30 °C, determinando-se a porcentagem e índice de velocidade de germinação. As sementes colhidas em 2013 foram armazenadas por 12 meses e submetidas a novo teste de germinação. Paralelo a esses testes determinou-se a curva de embebição das sementes em substrato sobre e entre papel às temperaturas de 15, 20, 25 e 30 °C. Os frutos e as sementes apresentaram comprimentos e diâmetros semelhantes nos dois anos de observações. As sementes podem ser armazenadas por até 12 meses sem perdas na viabilidade e as temperaturas de 25 e 30 °C são adequadas para realizar o teste de germinação dessa espécie. A curva de absorção de água das sementes de rosa do deserto segue um padrão trifásico de embebição.

Termos para indexação: *Adenium obesum* (Forssk.), Roem. & Schult., Apocynaceae, armazenamento de sementes.

Introduction

The *Adenium* genus belongs to the botanical family Apocynaceae, which includes many tropical ornamental species such as *Catharanthus* spp., *Beaumontia* spp., *Carissa* spp., *Allamanda* spp., *Mandevilla* spp., *Nerium* spp. and *Plumeria* spp.

Adenium obesum [(Forssk.), Roem. & Schult.], commonly known as desert rose, is found in sub-Saharan Africa from Sudan to Kenya and from west of Senegal to the south of Natal and Swaziland. In areas where it is native, dry and cold winters are sufficient to induce a dormant period, including the loss of leaves. The flowers of this species are tubular and the color ranges from deep purple-red through pink and white. However, commercial cultivars have different nuances of color, shape and size, and some feature an attractive fragrance. The flowers are produced in clusters (corymbs) at the apex of the branches during most of the year, although in some cultivars flowering is more restricted (Brown, 2012; Dimmitt et al., 2009; McBride et al., 2014).

In addition to the ornamental value, Adamu et al. (2005)

[1]Universidade Estadual de Londrina, Caixa Postal 1011, 86057-970 - Londrina, PR, Brasil.
*Corresponding author <ronancolombo@yahoo.com.br>

report the antimicrobial potential of extracts prepared from the desert rose suberin, so that the species is considered a broad spectrum in controlling the microorganisms tested. There are also scientific studies related to the cytotoxic compounds produced by this plant in the fight against cancer cells (Nakamura et al., 2000) and the influenza virus (Kiyohara et al., 2012).

The propagation of the species occurs mainly by seeds, whereas plants from seeds present more developed caudex and main root when compared to the ones propagated by cuttings. However, there are few studies describing the seeds and their germination process, as well as other elements involved in this process, such as the ideal temperature for germination and the water absorption curve.

In the process of germination, the water regulates the tissues rehydration, with the consequent intensification of respiration, and digestion and translocation metabolic activities, which are essential to supply energy and nutrients for growth resumption of the embryonic axis (Kikuchi et al., 2006; Ataíde et al., 2014).

Determining the water absorption curve is also related to integument impermeability studies, determining the treatments duration with plant growth regulators, osmotic conditioning and pre-hydration (Carvalho and Nakagawa, 2012). Still, it is interesting to understand this process for determining the temperature and humidity conditions when willing to store the seeds for an indeterminate period of time, without, however, affecting their physiological quality.

As in the germination process, humidity and temperature can trigger physiological processes during the storage phase and thereby accelerate the seeds deterioration or promote the growth of pathogens (Pontes et al., 2006; Medina et al., 2009).

Thus, this work aimed to do a descriptive study of fruits and seeds and the *A. obesum* germination process from two years of observations.

Material and Methods

Plant material

Fruits of *A. obesum* were collected between the months of February and March 2013 and March and April 2014 from 20 plants grown in a greenhouse at University State of Londrina when they started the spontaneous opening. The fruits were transferred to the Seed Analysis Laboratory and packaged in plastic trays until completing the natural opening and releasing the seeds.

Characterization of fruits and seeds

The biometrical description of the fruit (2013 and 2014)

was done from a sample of 13 fruits for each year. Therefore, the length and equatorial diameter of each fruit at harvest were measured; the number of seeds per fruit was also determined.

For biometrical description of seeds (2013 and 2014) determining the thousand seed weight from eight samples of 100 seeds was held; and water content, determined by the oven method at 105 °C (Brasil, 2009), using two samples of 20 seeds. The seeds length and diameter were measured in a sample of 200 seeds with a digital caliper.

For internal morphological observations, the seeds were previously soaked in distilled water for three hours to soften and moisturize the tissues. After this period, longitudinal sectioning was done with a metallic blade and then they were observed through a Motic® SMZ-168 magnifying glass.

Germination test and water absorption curve

Freshly harvested seeds in 2013 were submitted to the germination test at 25 °C. Simultaneously, the seeds water absorption curves were determined at temperatures of 15, 20, 25 and 30 °C in two substrates: over and between sheets of paper.

The remaining seeds were stored in a glass bottle with a polyethylene cap at a temperature of 25 °C and relative humidity of 65% ± 5 for a period of 12 months and submitted to seed germination test under the same conditions in the year 2014.

In 2014, a fresh lot of seeds was harvested and they were submitted to a germination test at temperatures 25 and 30 °C. Similarly, the seeds water absorption curves were determined at 25 and 30 °C temperatures over and between sheets of paper, based on the results obtained in 2013.

For the germination tests, four replications of 50 seeds were employed, which were placed on towell paper moistened with distilled water in the amount of 2.5 times the dry paper mass (Brasil, 2009), stored in crystal polystyrene boxes (Gerbox®) in a B.O.D. (Biochemical Oxygen Demand) incubator under constant light. The variables evaluated were germination percentage and germination speed index (GSI) (Maguire, 1962), starting the counts from the third (2013) and fourth days (2014) after installing the tests.

The imbibition curve was performed with four replications of ten seeds for each treatment. The seeds were laid out over or between sheets of paper, as described for the germination test. Initially, the seeds were weighed every hour during the first eight hours of imbibition by means of an analytical precision scale (0.0001 g). After this time, the seeds were weighed at intervals of 24 hours, marked from the beginning of the experiment, ending the weighing 24 hours after the primary root protrusion of 50% of seeds. At every weighing the seeds were removed from the Gerbox® boxes and placed over paper to absorb external moisture, weighed and then returned to the

Gerbox® and B.O.D.

Statistical analysis

The measures relating to the length and diameter of fruits and seeds of years 2013 and 2014 are followed by descriptive statistics and grouped in a diagram of frequency distribution. The averages obtained for the germination percentage in both lots were submitted to analysis of variance and compared based on the *p-value*. For the soaking curves in both years, a regression model was adjusted by testing the polynomial regression of first, second and third degrees.

Results and Discussion

Characterization of fruits and seeds

The desert rose fruits are classified as follicles, dehiscent, having a break line in the longitudinal direction, and striking when the fruits are in an advanced ripening stage. Their length and diameter can vary depending on factors such as pollination efficiency, environmental conditions and nutritional status of the plants. Table 1 shows desert rose fruits length and diameter values measured in years 2013 and 2014. The fruits average length was between 18.7 and 20.0 cm and the diameter between 12.9 and 13.3 mm.

The number of seeds per fruit is also related to the same conditions affecting the development of the fruit. In 2013 there was an average of 75 seeds per fruit and 84 seeds in 2014 (Table 1).

The seeds have a cylindrical shape, a brown staining integument and a brown-gold pappus (bristles) on both ends, which helps the dispersion by wind (Figure 1A). Desert rose seeds are similar to the the ones in oleander (*Nerium oleander*), but in this species bristles are on only one end thereof.

Table 1.   Biometric characterization of *Adenium obesum* fruits harvested in 2013 and 2014.

	Length (cm)		Diameter (mm)		Number of seeds	
	2013	2014	2013	2014	2013	2014
Minimum	15.1	16.7	11.1	11.3	28.0	55.0
Maximum	25.1	23.4	15.2	15.4	117.0	118.0
Average	18.7 ± 2.9	20.0 ± 1.7	13.3 ± 1.2	12.9 ± 1.3	74.7 ± 27.9	84.4 ± 19.6

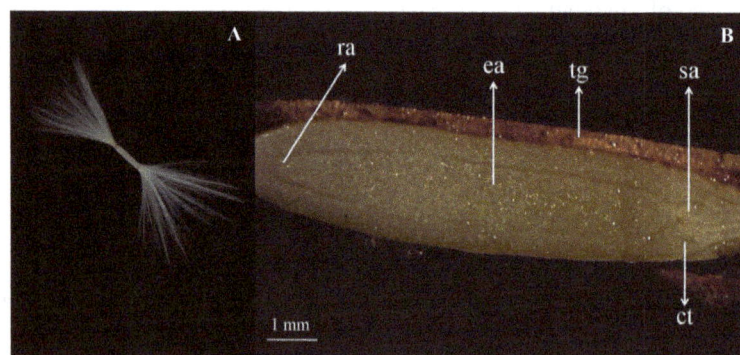

Figure 1. Morphological characterization of *Adenium obesum* seeds, external (A) and internal (B) aspects.
Caption: (B): tg – integument; ea – embryonic axis; ra – root apex; sa – stem apex; ct – cotyledons.

For seeds length and diameter, it can be seen in Table 2 and Figure 2 that in 2013 the predominant classes of length and diameter were 9.1 to 10.0 mm and 1.81 to 2.0 mm, respectively. In 2014, the highest frequency for diameter was observed in the same class of 2013; however, the predominant length was between 11.1-12.0 mm.

For the thousand seed weight were obtained the average values of 15.8 g (2013) and 15.6 g (2014). Water content had an average of 7.1% in 2013 and for this same lot reduction of the water content to 6.5% was found after a 12-month storage. However, for the seeds harvested in 2014 the water content was 5.6%, which explains the reduction in the thousand seed weight in that year.

Table 2.   Biometric characterization of *Adenium obesum* seeds harvested in 2013 and 2014.

	Length (mm)		Diameter (mm)	
	2013	2014	2013	2014
Minimum	7.7	7.9	1.5	1.4
Maximum	13.1	14.4	2.7	2.6
Average	9.8 ± 1.0	10.7 ± 1.3	1.9 ± 0.2	2.0 ± 0.2

As for the seeds internal structure (Figure 1B), it was found that when they are hydrated, the embryo is easily observed: it is cylindrical, has white coloring and occupies almost the whole kernel space. The cotyledonary reserve tissue is also white colored and has a firm consistency.

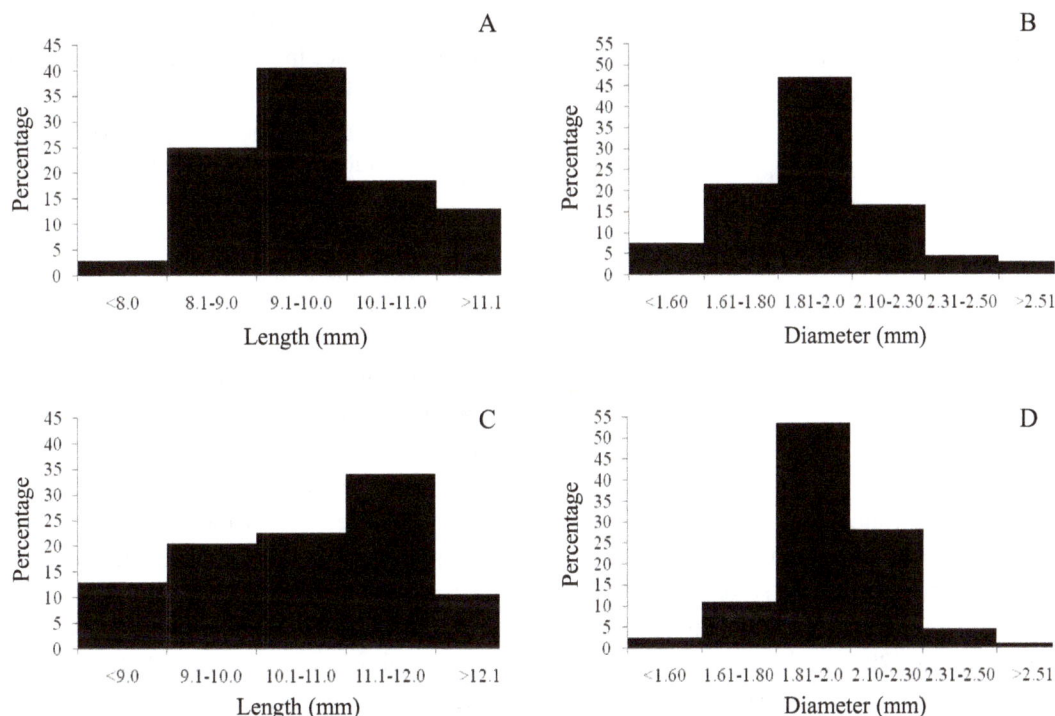

Figure 2. Frequency distribution for seed length and diameter of *Adenium obesum* seeds collected in years 2013 (A and B) and 2014 (C and D).

Germination test and water absorption curve

Parallel to the germination test, the description of the desert rose germination process was carried out at 25 °C temperature, over paper. 96 hours after seeding, seed integument disruption and primary root protrusion were observed. After 120 hours, there was greening of the base stem and an average increase of 3 mm of the radicle. After 168 hours, there was a total disruption of the integument and exposure of the cotyledonary leaves (Figure 3) due to water absorption and increasing volume of seed cotyledons, causing integument disruption and emergence of the root hypocotyl axis and other internal structures of the seed (Borges et al., 2009).

As for the germination tests, it was observed at harvest and 12 months after it that the average values for this variable were 90 and 91%, respectively (Table 3), with little influence of storage at 25 °C on germination percentage. Thus, it turns out that storing seeds of this species at 25 °C for up to 12 months does not alter its germinability ($p > 0.7$). Regarding the germination speed index (GSI), the value observed (7.4) remained constant in the two periods in which the germination test was conducted.

Figure 3. Germination process characterization of *Adenium obesum*. (A) Primary root protrusion, (B) primary root elongation, (C) issuance of root hairs and (D) exposure of cotyledonary leaves.

For the lot of seeds harvested in 2014, the germination test was conducted at temperatures of 25 and 30 °C. This year, however, there were higher germination percentages compared to 2013. At the temperature of 25 °C, the average germination percentage was 96%, and at 30 °C it was 98%; however, there was no statistically significant difference between them (Table 3). Carpenter and Boucher (1992), in a study of Madagascar periwinkle (or rosy periwinkle) (*Catharanthus roseus*) 'Grape cooler' seeds, found that maximum germination occurs at temperatures between 25 and 35 °C, and temperatures below 25 °C are unfavorable to this process.

Table 3.   Germination (%) and germination speed index (GSI) of *Adenium obesum* seeds harvested in 2013 and 2014.

	2013 A*	2013 B	2014	
	25 °C	25 °C	25 °C	30 °C
Germination (%)	90	91	96	98
F	0.09[ns]		1.04[ns]	
p-value	0.7		0.3	
CV (%)	4.7		2.9	
GSI	7.4	7.4	8.2	10.7

*2013 A: freshly harvested seeds; 2013 B: seeds stored for 12 months. [ns] non significant F-value.

Thus, to evaluate the quality of a given seed lot in a laboratory, there is a need for a germination pattern for each species because each one presents seeds with different characteristics regarding their germinative and physiological behaviors (Wielewicki et al., 2006).

On the other hand, GSI increased due to the germination temperature increase, from 8.2 to 25 °C to 10.7 to 30 °C, so that the temperature of 30 °C accelerated the seeds germination process (Table 3). Similar results are described by Gordin et al. (2012), wherein the *Guizotia abyssinica* seeds GSI increased as the germination temperature increased from 25 to 30 °C.

As the temperature and kinetic energy are increased, water viscosity is reduced, thus favoring imbibition and the seeds metabolism components reactions rate (Marcos-Filho, 2015), such as changing the structure of proteins and nucleic acids, besides modifying the cell membranes fluidity (Rodrigues et al., 2010; Zinn et al., 2010).

For the seeds water absorption in 2013 (Figure 4), an increase in mass was observed due to imbibition, from 7.24% after the first hour at 15 °C over paper up to 25.46% at 25 °C between sheets of paper. For this same seeds lot after 12 months of storage there is, in the first hour of imbibition, an increase from 7.33% to 10.40% in mass of

seeds kept over paper (25 °C) and between sheets of paper (30 °C), respectively (Figures 5A and F). Similar values are observed for seeds harvested in 2014 and submitted to the same conditions (Figures 5C and H).

This rapid water absorption by the seeds occurs due to the large difference in water potential between the seeds and the substrate, which characterizes phase I of the germination process (Bewley and Black, 1994). After five hours from the start of imbibition, seeds harvested in 2013 and kept between sheets of paper at temperatures of 20, 25 and 30 °C reached water contents between 35 and 50% (Figures 4 D, F and H). For imbibition curves reached in 2014, it has also been found, after the fifth hour, that the seeds had already absorbed 35 to 42% of water in all conditions tested, except for those submitted to 25 °C over paper (Figure 5).

For Carvalho and Nakagawa (2012), cotyledonary seeds finish phase I of the germination process once they reach water contents between 35 and 40%, and from these levels phase II would be started. However, the duration of each phase depends on properties inherent to seeds and environmental conditions present (Bewley and Black, 1994).

In the experiments conducted in 2013 and 2014, the primary root protrusion occurred between 48 and 72 hours in seeds kept at a temperature of 30 °C in both substrates (over paper and between sheets of paper); with that, the end of phase II of the germination process is characterized. However, in 2013, the seeds maintained at temperatures of 15 and 20 °C in both substrates began the process of deterioration after 120 hours of incubation under these conditions, and only some seeds issued the primary root. Thus, the importance of temperature to trigger the germination process is evident.

With respect to substrates over paper and between sheets of paper, there was no direct influence of these on the pattern of seeds water absorption. Although not being entirely surrounded by moisture (as it occurs between sheets of paper), seeds absorb water in a similar manner when exposed over paper; it is likely that this is due to the high permeability of the integument. Gordin et al. (2012) have found no differences either in the pattern of *Guizotia abyssinica* seeds water absorption in the substrates over paper and between sheets of paper.

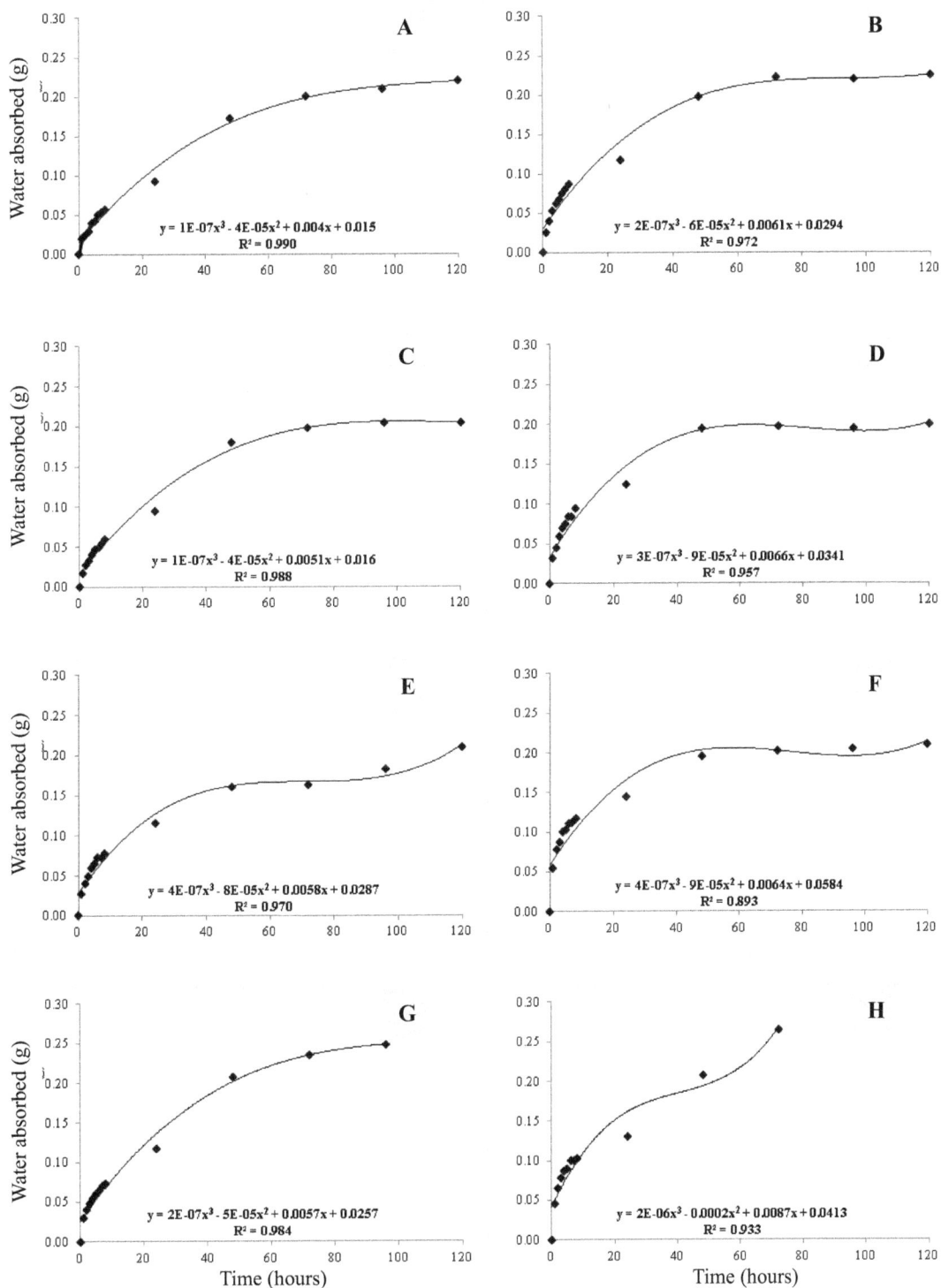

Figure 4. Water absorption curves in *Adenium obesum* seeds harvested in 2013 in a substrate over paper and between sheets of paper, at temperatures of 15, 20, 25 and 30 °C.

Substrates: over paper (A, C, E and G); between sheets of paper (B, D, F and H). Temperatures: 15 °C (A and B), 20 °C (C and D), 25 °C (E and F) and 30 °C (G and H).

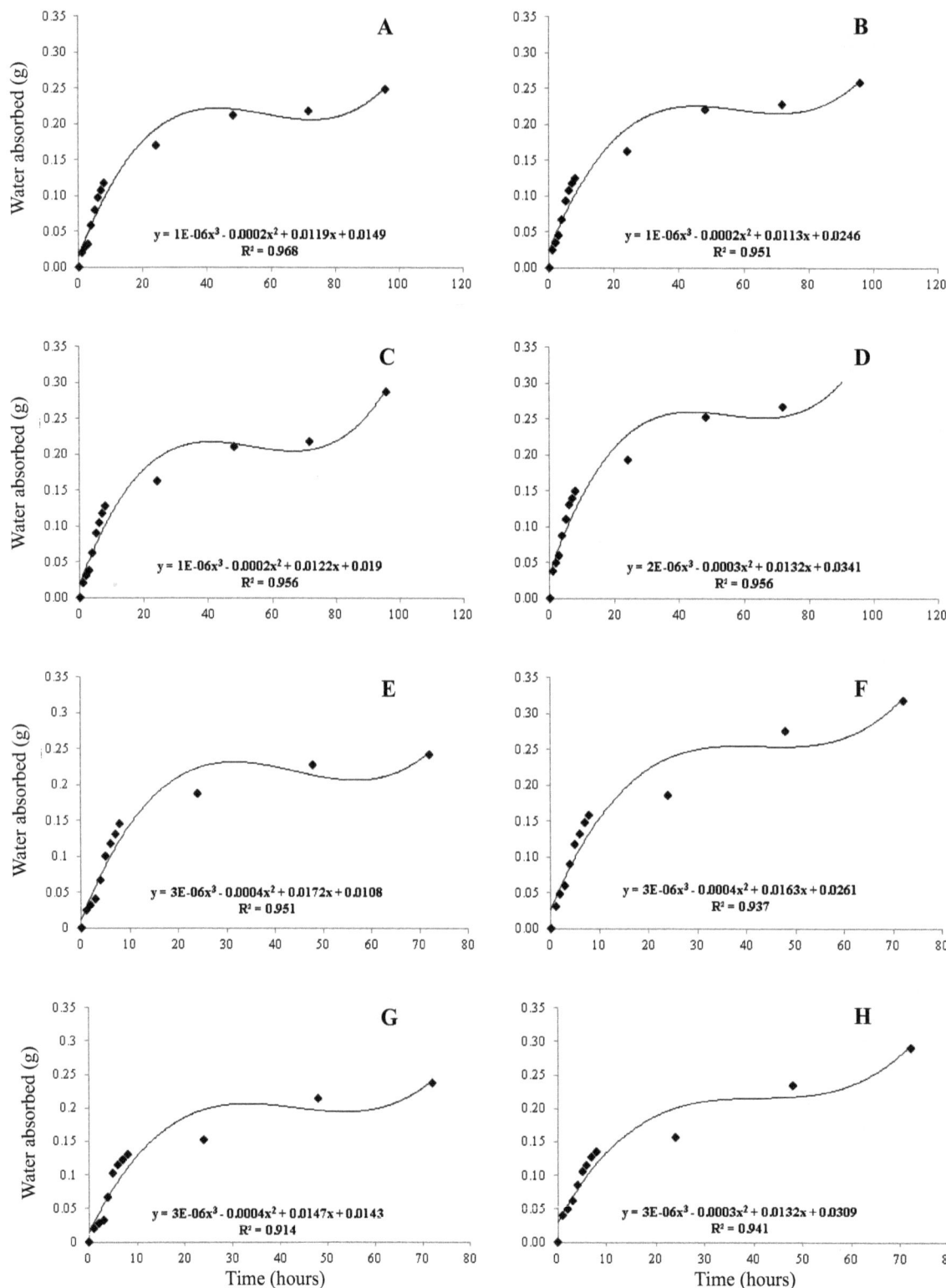

Figure 5. Water absorption curves in *Adenium obesum* seeds harvested in 2013 and stored for 12 months and harvested in 2014 in a substrate over paper and between sheets of paper, at temperatures of 25 and 30 °C.

Seeds stored for 12 months: (A, B, E and F); seeds harvested in 2014: (C, D, G and H); 25 °C, over paper: (A and C); 25 °C, between sheets of paper: (B and D); 30 °C, over paper: (E and G); 30 °C, between sheets of paper: (F and H).

Conclusions

Desert rose seeds can be stored at 25 °C for up to 12 months without loss of viability.

Temperatures of 25 and 30 °C are effective for the desert rose germination test.

The water absorption curve of desert rose seeds follows the triphasic pattern of imbibition, and the seeds reach germination phase I after five hours of imbibiton.

References

ADAMU, H.M.; ABAYEH, O.J.; AGHO, M.O.; ABDULLAHI, A.L. An ethnobotanical survey of Bauchi State herbal plants and their antimicrobial activity. *Journal of Ethnopharmacology*, v.99, p.1-4, 2005. http://www.academia.edu/7550317/An_ethnobotanical_survey_of_Bauchi_State_herbal_plants_and_their_antimicrobial_activity

ATAÍDE, G.M.; BORGES, E.E.L.; FLORES, A.V.; CASTRO, R.V.O. Avaliação preliminar da embebição de sementes de jacarandá-da-bahia. *Pesquisa Florestal Brasileira*, v.34, n.78, p.133-139, 2014. http://pfb.cnpf.embrapa.br/pfb/index.php/pfb/article/viewFile/520/362

BEWLEY, J.D.; BLACK, M. *Seeds*: physiology of development and germination. New York: Plenum Press, 1994. 445p.

BORGES, R.C.F.; COLLAÇO JUNIOR, J.C.; SCARPARO, B.; NEVES, M.B.; CONEGLIAN, A. Caracterização da curva de embebição de sementes de pinhão-manso. *Revista Científica Eletrônica de Engenharia Florestal*, v.8, n.13, p.1-8, 2009. http://faef.revista.inf.br/site/e/engenharia-florestal-13-edicao-fevereiro-de-2009.html#tab270

BRASIL. Ministério da Agricultura, Pecuária e Abastecimento. *Regras para análises de sementes*. Ministério da Agricultura, Pecuária e Abastecimento. Secretaria de Defesa Agropecuária. Brasília: MAPA/ACS, 2009. 395p. http://www.agricultura.gov.br/arq_editor/file/2946_regras_analise__sementes.pdf

BROWN, S.H. *Adenium obesum*. Horticulture Agent Lee County Extension, Fort Myers, Florida, 2012. 8 p. http://lee.ifas.ufl.edu/Hort/GardenPubsAZ/Dessert_Rose_Adenium_obesum.pdf

CARPENTER, W.J.; BOUCHER, J.F. Germination and storage of vinca seed is influenced by light, temperature, and relative humidity. *HortScience*, v.27, n.9, p.993-996, 1992. http://hortsci.ashspublications.org/content/27/9/993.full.pdf

CARVALHO, N.M.; NAKAGAWA, J. *Sementes*: ciência, tecnologia e produção. 5 ed. Jaboticabal: FUNEP, 2012. 590p.

DIMMITT, M.; JOSEPH, G.; PALZKILL, D. *Adenium*: Sculptural Elegance, Floral Extravagance. 1ed. Tucson: Scathingly Brilliant Idea. 2009. 152p.

GORDIN, C.R.B.; MARQUES, R.F.; MASETTO, T.E.; SCALON, S.P.Q. Germinação, biometria de sementes e morfologia de plântulas de *Guizotia abyssinica* Cass. *Revista Brasileira de Sementes*, v.34, n.4, p. 619-627, 2012. http://www.scielo.br/pdf/rbs/v34n4/13.pdf

KIKUCHI, K.; KOIZUMI, M.; ISHIDA, N.; HIROMI, K. Water uptake by dry beans observed by micro-magnetic resonance imaging. *Annals of Botany*, v.98, p.545-553, 2006. http://www.ncbi.nlm.nih.gov/pmc/articles/PMC3292055/

KIYOHARA, H.; ICHINO, C.; KAWAMURA, Y.; NAGAI, T.; SATO, N.; YAMADA, H.; SALAMA, M.M.; ABDEL-SATTAN, E. In vitro anti-influenza virus activity of a cardiotonic glycoside from *Adenium obesum* (Forssk.). *Phytomedicine: International Journal of Phytotherapy and Phytopharmacology*, v.19, n.2, p. 111-114, 2012. http://www.sciencedirect.com/science/article/pii/S0944711311002650

MARCOS-FILHO, J. *Fisiologia de sementes de plantas cultivadas*. Londrina: ABRATES, 2015, 660p.

MAGUIRE, J.D. Speed of germination-aid in selection and evolution for seedling emergence and vigor. *Crop Science*, v.2, n.2, p.176-177, 1962. https://dl.sciencesocieties.org/publications/cs/abstracts/2/2/CS0020020176

McBRIDE, K.M.; HENNY, R.J.; CHEN, J.; MELLICH, T.A. Effect of light intensity and nutrition level on growth and flowering of *Adenium obesum* 'Red' and 'Ice Pink'. *HortScience*, v.49, n.4, p.430-433, 2014. http://www.researchgate.net/profile/Jianjun_Chen8/publication/262005470_Effect_of_light_intensity_and_nutritional_level_on_growth_and_flowering_of_Adenium_obesum_Red_and_Ice_Pink/links/0c9605363e5973e183000000.pdf

MEDINA, P.F.; TANAKA, M.A.S.; PARISI, J.J.D. Sobrevivência de fungos associados ao potencial fisiológico de sementes de triticale (*X. triticosecale* Wittmack) durante o armazenamento. *Revista Brasileira de Sementes*, v.31, n.4, p. 17-26, 2009. http://www.scielo.br/pdf/rbs/v31n4/02.pdf

NAKAMURA, M.; ISHIBASHI, M.; OKUYAMA, E.; KOYANO, T.; KOWITHAYAKORN, T.; HAYASHI, C.M.; KOMIYAMAD, K. Cyto-toxic pregnanes from leaves of *Adenium obesum*. *Natural Medicines*, v.54, n.3, p.158-159, 2000. http://ci.nii.ac.jp/els/110008732044.pdf?id=ART0009809011&type=pdf&lang=en&host=cinii&order_no=&ppv_type=0&lang_sw=&no=1437697895&cp

PONTES, C.A.; CORTE, V.B.; BORGES, E.E.L.; SILVA, A.G.; BORGES, R.C.G. Influência da temperatura de armazenamento na qualidade das sementes de *Caesalpinia peltophoroides* Benth. (sibipiruna). *Revista Árvore*, v.30, n.1, p.43-48, 2006. http://www.scielo.br/pdf/rarv/v30n1/28507.pdf

RODRIGUES, A.P.D.C.; LAURA, V.A.; PEREIRA, S.R.; SOUZA, A.L.; FREITAS, M.E. Temperatura de germinação em sementes de estilosantes. *Revista Brasileira de Sementes*, v.32, n.4, p. 166-173, 2010. http://www.scielo.br/pdf/rbs/v32n4/19.pdf

ZINN, K.E.; TUNC-OZDEMIR, M.; HARPER, J.F. Temperature stress and plant sexual reproduction: uncovering the weakest links. *Journal of Experimental Botany*, v.61, n.7, p. 1959-1968, 2010. http://jxb.oxfordjournals.org/content/61/7/1959.full.pdf+html

WIELEWICKI, A.P.; LEONHARDT, C.; SCHLINDWEIN, G.; MEDEIROS, A.C.S. Proposta de padrões de germinação e teor de água para sementes de algumas espécies florestais presentes na região sul do Brasil. *Revista Brasileira de Sementes*, v.28, n.3, p. 191-197, 2006. http://www.scielo.br/pdf/rbs/v28n3/27.pdf

Seed germination of Brazilian guava (*Psidium guineense* Swartz.)

Márcia Adriana Carvalho dos Santos[1], Manoel Abílio de Queiróz[1],

Jaciara de Souza Bispo[1], Bárbara França Dantas[2]*

ABSTRACT- Brazilian guava (*Psidium guineense* Swartz.) is a plant species native from Brazil and present in all Brazilian biomes. This species occurs in the Caatinga biome as a wild fruit that has broad utility and can generate income, as well as provide material for breeding. The objective of this work was to study seed germination of guava accessions collected in two municipalities in Bahia. Seeds imbibition curve was studied. Also, seed germination and seedlings initial growth were evaluated for 44 days on different temperatures (15, 20, 25, 30 and 35 °C), fruit maturation, drying, pre-soaking and priming. Seeds imbibition curve was triphasic in which lag-phase begun after 30 hours imbibition and lasted 256 hours. An effect of guava genotype was observed in seeds physiological quality. Temperatures of 20 and 25 °C were ideal for guava seeds germination. Seeds pre-soaking and priming techniques are feasible to reduce germination time and increase seedling performance.

Index terms: genetic resources, Myrtaceae, vigor.

Germinação de sementes de araçá (*Psidium guineense* Swartz.)

RESUMO- O araçá (*Psidium guineense* Swartz) é uma espécie nativa do Brasil e presente em todos os biomas brasileiros. Ocorre na Caatinga de forma silvestre e é uma fruteira que apresenta ampla utilidade, podendo gerar renda além de fornecer material para o melhoramento genético. O objetivo deste trabalho foi estudar o processo germinativo de sementes de quatro acessos de araçá coletados em dois municípios da Bahia. Foi estudada a curva de embebição das sementes, bem como o efeito de diferentes temperaturas (15, 20, 25, 30 e 35°C), estágio de maturação, secagem, pré-embebição e condicionamento osmótico na germinação, vigor e crescimento inicial de plântulas dos acessos de araçá. A curva de embebição das sementes se mostrou trifásica, na qual a fase-lag iniciou após 30 h de embebição e durou 256 h. Foi observado um efeito dos genótipos na qualidade fisiológica das sementes de araçá. As temperaturas de 20 e 25 °C foram ideais para a germinação de sementes de araçá. As técnicas de pré-embebição e condicionamento osmótico das sementes reduziram o tempo de germinação e aumentaram o desempenho das sementes.

Termos para indexação: recursos genéticos, Myrtaceae, vigor.

Introduction

Numerous species of genus *Psidium* that produce edible fruits can be found in all regions of Brazil, such as common guava (*Psidium* sp) and guava-pear (*Psidium acutangulum* DC) (Giacobbo et al., 2008).

Psidium guineense Swartz, commonly named araçá, Brazilian guava, or simply guava is a wild plant from Myrtaceae family, which occurs throughout Brazil. Its shrub or tree has twisted stem, smooth bark with evergreen leathery leaves and are well adapted to climate and soil of northeastern Brazil, where it grows spontaneously in many places (Bezerra et al., 2006). These guavas are classified as berry type fruits, with yellow, red

or purple peel and whitish pulp, with many seeds (Santos et al., 2004). Fruits are rich in minerals and functional elements, such as vitamins and phenolic compounds (Caldeira et al., 2004). Evaluation studies of guava indicate good prospects to be introduced as functional food, which is a strong interest of food industry (Degáspari and Waszczynskyj, 2004).The breeding programs seek agronomically important traits, such as, more productive plants and low vulnerability to pests and diseases, coupled with products rich in nutrients. Many studies have evaluated the effectiveness of guava rootstock associated with different varieties of guava (*Psidium guajava* L.), in order to reduce or mitigate the damage caused by nematodes (Miranda et al, 2012; Souza et al, 2014).

[1]Departamento de Tecnologia e Ciências Sociais, Universidade do Estado da Bahia, Caixa Postal 171, 48905-680 - Juazeiro, BA, Brasil.

[2]Embrapa Semiárido, Caixa Postal 23 , CEP 56302-970- Petrolina, PE, Brasil.
*Corresponding author < barbara.dantas@embrapa.br>

Few literature on *P. guineense* seeds germination state that alternate temperatures from 20-30 ºC, or 25 ºC constant temperatures are the most appropriate to access seed quality (Mugnol et al., 2014). However, it is not agreed whether seeds should undergo any pre-treatment for higher germination percenteges and rates in accessions harvested in the Cerrado biome of Brazilian midwest region (Mugnol et al., 2014; Masseto et al., 2014). Although it is common scientific knowledge that different mother plants produce seed with different physiological responses (Turesson, 1922; Andersson and Milberg, 1998; Galloway, 2005), literature have not displayed these differences regarding seeds behaviour of different accessions of Brazilian guava in different temperatures, fruit maturation stage, desiccation tolerance and pre-germinative treatments.

Due to its potential for different uses, as well as the lack of studies regarding Brazilian guava seed propagation, this study aimed to evaluate germination process of seeds native from the Caatinga biome of Brazilian northeastern semiarid region.

Material and Methods

Four independent essays were performed in order to characterize guava seeds germination and its optimal conditions. The essays used seeds of four different guava accessions in the cities of Jacobina and Campo Formoso, Bahia State, Brazil. Accessions and geographic coordinates of occurrence were: Y52 (11°15`S e 040°31`W), Y53 (11°15`S e 040°31`W) and Y95 (11°11`S e 040°27`W) harvested in Jacobina- BA; Y85 (10°29`S e 040° 17`W) harvested in Campo Formoso - BA. The experimental design of all essays was totally randomized.

Fruits of each accession were collected manually. For processing of seeds, mucilage was removed in running tap water over a sieve. Then seeds were placed to dry in shade. After extraction, shriveled, withered and damaged seeds were eliminated and the remaining good seeds were packed in paper bags and placed in a cold chamber at 10 °C and 40% relative humidity. Accessions Y85 and Y95 were stored for one year and accessions Y52 and Y53 for few days prior to essays.

Seeds imbibition curve: eight replications of 20 guava seeds of Y85 accession, were initially weighted on a precision scale (0.001g). After weighting, seeds were distributed on blotting paper moistened with 13 mL of distilled water and fungicide (Captan, 1 mg. g^{-1}) solution in germination boxes (gerboxes) and incubated in BOD (Biochemical Oxygen Demand) germination chamber at 25 °C. Soaked seeds were weighted after 3, 6, 24, 26, 30, 48, 50, 54, 72, 78, 96, 98, 102, 120 hours and every 24 hours until 408 hours of imbibition,

evaluating water gain. Data was plotted into a non-fitted scattered curve with average mean error bars.

Optimum temperature: in a 5x2 factorial scheme (temperature x acessions), four replicates with 20 seeds of Y85 and Y95 accessions were placed in gerboxes on blotting paper moistened with 13 mL of distilled water and fungicide solution. Gerboxes were placed in plastic bags to prevent evaporation of water and incubated at different temperatures, which were 15, 20, 25 30 and 35 °C, in BOD germination chambers. Germination evaluation was performed daily for 44 days, considering germinated those seeds with 2 mm emission radicle length.

Fruit post-harvest ripening and seed drying: an essay was performed in a 2x2x2 factorial with two accessions (Y52 and Y53), in two maturation stages (ripe and overripe) and seed drying (not dried and dried seeds), with 80 seeds per treatment and four replications.

Guava fruits of Y52 and Y53 accessions were harvested at physiological maturity (ripe) in Jacobina - BA. Seeds were immediately extracted from half of the fruits of each accession and the other half was kept in a laboratory environment (25±4 °C, RH 60%) for eight days to overripe before seed extraction. Thus, after extracting seeds of each accession (Y52 and Y53) and each maturation stage (ripe and overripe), they were divided into two lots. A lot, without seed drying, was immediately submitted to germination test. The remaining seeds were placed to dry in shade for 24 hours, in lab environment with room temperature ranging from 25 to 35 °C and relative humidity ranging from 30-40%. For germination evaluation seeds were distributed on blotting paper moistened with 13 mL of distilled water and fungicide (Captan, 1 mg. g^{-1}) solution in gerboxes and incubated in BOD germination chambers at 25 °C. Germinated seeds were daily counted during 44 days, considering germinated those seeds with a 2 mm radicle length.

Pre-soaking and seed priming: 160 seeds of each accession (Y52, Y53, Y85 and Y95) were distributed in gerboxes on two sheets of blotting paper moistened with 13 mL of polyethylene glycol (PEG 6000) solution with osmotic potential -1.0 MPa (Villela et al., 1991). Gerboxes were maintained in BOD germination chamber at 20 °C for eight days. After this period, seeds were washed in running tap water, dried superficially and the half of the seeds (pre-soaked seeds) was immediately submitted to germination test. The other half was placed to shade dry for eight days (primed seeds) and submitted to germination test after that period, along with seeds that have not undergone any pre-treatment (control). The experiment was arranged in a 4x3 factorial scheme with four accessions (Y52, Y53, Y85 and Y95) x three pre-treatments (pre-soaking, priming and control). For germination evaluation seeds were distributed on blotting paper moistened with 13 mL of distilled water and fungicide solution in gerboxes and incubated in BOD germination chambers at 20 °C. Germination evaluation was performed daily for 44 days, considering germinated those seeds with 2 mm emission radicle length.

In the three later essays, germination data accessed were used to estimate germination percentage (%), germination speed index (Maguire, 1962), mean germination time (days⁻¹, Labouriau, 1983), germination speed (seedling.days⁻¹, Labouriau, 1983). Seedlings were evaluated for length (cm), fresh and dry matter accumulation (mg) after 44 days. Data were submitted to ANAVA and the averages compared by Tukey's test at 5 % probability.

Results and Discussion

Guava seeds imbibition curve showed a triphasic pattern (Figure 1). Phase I (FI) was completed in 30 hours (left arrow in Figure 1), due to rapid water absorption. The second phase (FII) or lag phase, in which seeds slowly uptake water and does not display any embryo growth (Bewley et al., 2013), lasted about 250 hours (time elapsed between arrows in Figure 1). The third phase of imbibition (FIII) began after that. After approximately 280 hours (right arrow in Figure 1) 10% of seedlings had root protrusion and a sharp increase in seeds water uptake begun. At this stage, embryo starts axis its growth and seedlings newly formed cells require large volumes of water (Dantas et al., 2008a; 2008b; Borges et al., 2009; Smiderle et al., 2013). These results showed guava seeds do not have dormancy, because only non-dormant and viable seeds reach phase III of imbibition curve (Bewley et al., 2013). The imbibition

curve was evaluated during 408 hours in which the Y85 accession seed obtained about 70% radicle protrusion (Figure 1).

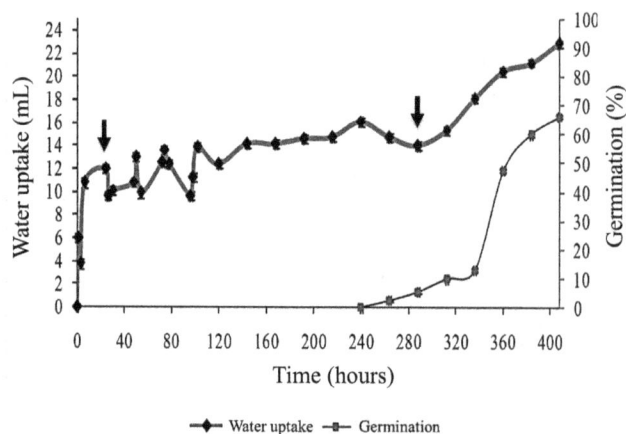

Figure 1. Imbibition and germination curves of guava (*Psidium guineense*), accession Y85. Vertical bars indicate the standard error of the mean.

Guava accessions seeds subjected to different temperatures showed a significant interaction for the accessed variables. Seeds of Y85 accession showed a higher physiological quality than Y95 accession (Table 1). Germination percentage showed no statistical difference among higher temperatures (20, 25 and 30°C) for Y85 accession and at temperatures 20 and 25 °C for Y95 accession. For both accessions of guava (Y85 and Y95), mean germination time (MGT), mean germination speed (MGS) and germination speed index (GSI) showed best results at 20 and 25 °C. At 15 °C, Y95 accession seeds showed no germination, whilst Y85 seeds showed 35% germination, but not seedling development. None of the seeds germinated at 35 ºC.

Table1. Physiological quality of guava (*Psidium guineense*) seeds submitted to different temperatures.

Variables	Accessions	Temperatures (°C)				CV (%)
		15	20	25	30	
Germination	Y85	35.0 Ab	88.33 Aa	85.00 Aa	71.66 Aa	19.5
(%)	Y95	0.00 Bb	70.0 Ba	56.56 Ba	10.00 Bb	
MGT	Y85	42.25 Ab	28.37 Aab	26.0 Aa	29.23 Bab	24.57
(days)	Y95	-	33.0 Aa	32.6 Aa	10.0 Ab	
MGS	Y85	0.02 Ab	0.037 Aa	0.04 Aa	0.033 Aab	27.08
(seedlings.days⁻¹)	Y95	0.0 Bb	0.03 Aa	0.03 Aa	0.01 Bb	
GSI	Y85	0.17 Ac	0.63 Aab	0.68 Aa	0.50 Ab	19.8
(seedlings.days⁻¹)	Y95	0.0 Bb	0.43 Ba	0.35 Ba	0.07 Bb	
Seedling length	Y85	0.0 Ac	2.82 Ab	3.89 Aa	2.82 Ab	9.75
(cm)	Y95	0.0 Ac	1.68 Bb	2.18 Ba	1.68 Bb	
Fresh matter	Y85	0.0 Ab	45.0 Aa	61.33 Aa	45.0 Aa	25.77
(mg)	Y95	0.0 Ab	34.67 Aa	49.33 Aa	34.67 Aa	
Dry matter	Y85	0.0 Ab	14.0 Aa	14.0 Aa	14.0 Aa	16.90
(mg)	Y95	0.0 Ac	7.67 Ab	11.66 Aa	7.67 Bb	

MGT (mean germination time), MGS (mean germination speed, GSI (germination speed index). Means followed by same capital letters in the column and lower case letters in the line, do not differ statistically at 5%.

Variations in air temperature and rainfall during development and maturation of seeds can provide various physiological responses in mature seeds, such as thermal requirement and basal temperature for germination (Lamarca et al., 2013). Seedlings from both accessions grown at 25 °C showed increased length in comparison to other temperatures. Accession Y85, which showed higher seedling fitness than Y95, was not influenced by temperature regarding dry matter. Seedlings fresh matter, however, was not altered by temperatures ranging from 20 to 30 °C.

Seeds germinate when environmental conditions (moisture, oxygen, temperature and light) indicate a temporal or spatial window for seedling emergence, development and survival (Long et al., 2014). Usually seeds germinate more efficiently, with higher percentages and speed, in temperatures similar to those environments were they were produced. Native tree species from Cerrado and Atlantic Forest biomes usually germinate well at constant temperatures of 25 °C, however for those that occur in the Amazon biome the ideal temperature for the germination test is 30 °C (Brancalion et al., 2010). Altough Caatinga native seeds germinate better in temperatures closer to 30 °C (Oliveira et al., 2014a), some species that are widely spread in different biomes seem to have lower optimum temperatures for seed germination. Myracrodruon urundeuva seeds, as an exemple, has optimum germination temperature around 20 °C (Oliveira et al., 2014b).

Seed vigor can be defined as all characteristics of a lot that determine potential for a uniform and rapid emergence and development in a wide range of environmental conditions (Rajjou et al., 2012). Therefore, seeds with higher vigor, as shown by Y85 accession, are able to tolerate abiotic stresses better than those with lower vigor (Dantas et al., 2007), such as Y95 accession (Table 1). Alves et al. (2005) concluded that the origin of Mimosa caesalpiniifolia Benth. accessions strongly influenced seed germination. Thus the areas where the fruits of both guava accessions were harvested, despite having very similar climatic and geographical conditions, may have influenced seeds physiological quality.

On the other hand, changes in germination responses among different seeds of same species may be also due to seed maturity stage when it was harvested or dispersed by motherplants (Lamarca et al., 2013), as well as post-harvest processing and conservation (Pessoa et al., 2010; Resende et al., 2012).

Regarding fruit post-harvest ripening and seed drying essay, there was no interaction between stage of maturation, seed drying and accession (Table 1). Regarding accessions response, there was statistical difference only for germination percentage, germination speed index and seedling dry weight (Table 2). Accession Y53 presented the best averages for germination and seedling dry weight, while accession Y52 showed higher average only for GSI (Table 2). This advantage could be related to low germination percentage obtained by this accession.

Table 2. Fruit post-harvest ripening and seed drying effect on physiological quality of guava (*Psidium guineense*).

Accession	Germination (%)		MGT (days)		MGS (seedlings.days^{-1})		GSI (seedlings.days^{-1})	
	Y53	Y52	Y53	Y52	Y53	Y52	Y53	Y52
	56.0 a	36 b	30.19 a	30.85 a	0.030 a	0.032 a	0.35 b	0.47 a
Maturation stage	overripe	ripe	overripe	ripe	overripe	ripe	overripe	Ripe
	42.5 a	49.5 a	30.92 a	30.12 a	0.031 a	0.031 a	0.35 b	0.42 a
Seed drying	not dried	dried	not dried	dried	not dried	dried	not dried	dried
	46.0 a	35.5 a	30.52 a	33.51 a	0.031 a	0.04 a	0.39 a	0.27 a
CV (%)	23.56		7.39		11.09		23.61	
	Seedling length (cm)		Fresh matter (mg)		Dry matter (mg)			
Accession	Y53	Y52	Y53	Y52	Y53	Y52		
	1.75 a	1.68 a	5.45 a	6.19 a	1.27 a	0.97 b		
Maturation stage	overripe	ripe	overripe	ripe	overripe	ripe		
	1.71 a	1.73 a	5.45 a	6.85 a	1.10 a	1.5 a		
Seed drying	not dried	dried	not dried	dried	not dried	dried		
	2.06 a	1.38 b	5.38 a	6.92 a	1.31 a	0.93 b		
CV (%)	15.73		50.18		35.3			

MGT (mean germination time). MGS (mean germination speed). GSI (germination speed index). Means followed by same capital letters in the column and lower case letters in the line do not differ statistically at 5%.

The effect of fruits maturation stage from which the seeds were extracted was significant only for GSI, on the other hand, seed drying showed no positive effect on seed physiological quality

(Table 2). Among 1034 species of 23 different genera of Myrtaceae family which occur in Brazil, only 90 species occur in Caatinga (Forzza et al., 2012), showing this family has few species that can withstand harsh environmental conditions of this biome. Furthermore, seeds of many Brazilian tropical tree species in Myrtaceae family show high water content at shedding and have been considered to be sensitive to desiccation. However, among some species of this family, there are different desiccation sensitivity levels based on water content (Delgado and Barbedo, 2007; Masetto et al., 2008; Delgado and Barbedo, 2012). The seedlings development (length and dry matter) was better without seed drying (Table 2), thus, accessions Y52 e Y53 showed a mild sensitivity to dessication,

although they do not show recalcitrant characteristics.

Priming and pre-soaking increased seed overall physiological quality of the guava accessions studied in this work, which can be observed in germination percentage, MGT and seedling length (Table 3). This is related to the fact that during these treatments occurs reserve mobilization, activation of DNA and RNA synthesis, ATP production, and repair of damage in membrane system suffered during storage (Bewley et al., 2013) improving seed vigor, increasing germination percentage and uniformity.

Table 3. Physiological quality of guava (*Psidium guineense*) seeds accessions subjected to pre-soaking and priming.

Pre-treatment	Germination (%)			
	Y53	Y52	Y85	Y95
Pre-soaking	76 aA	6.66 bB	81.33 aA	76 aAB
Priming	65.33 bA	16 cB	92.0 aA	86.67 aA
Control	38.66 cB	33.33 cA	88.33 aA	70.0 bB
CV (%)	12.07			
	Mean germination time (days)			
	Y53	Y52	Y85	Y95
Pre-soaking	27.77 cA	26.43 bcA	19.27 aA	21.1 abA
Priming	31.73 bAB	29.27 bAB	20.27 aA	21.6 aA
Control	34.84 bB	32.33 abB	28.37 aB	33.0 abB
CV (%)	7.97			
	Mean germination speed (seedlings.days^{-1})			
	Y53	Y52	Y85	Y95
Pre-soaking	0.03 bA	0.03 bA	0.05 aA	0.05 aA
Priming	0.03 aA	0.03 aA	0.03 aB	0.03 aB
Control	0.04 bA	0.04 bA	0.05 aA	0.05 aA
CV (%)	9.78			
	Germination speed index (seedlings.days^{-1})			
	Y53	Y52	Y85	Y95
Pre-soaking	0.73 bA	0.033 cB	1.10 aA	0.90 abA
Priming	0.53 bA	0.13 cAB	1.17 aA	1.10 aA
Control	0.28 bB	0.26 bA	0.63 aB	0.43 abB
CV (%)	17.46			
	Seedling length (cm)			
	Y53	Y52	Y85	Y95
Pre-soaking	2.78 aA	3.00 aA	3.68 aA	2.48 aA
Priming	2.02 bAB	1.95 bB	3.16 aAB	2.22 bAB
Control	1.3 bB	1.75 bB	2.82 aB	1.59 bB
CV (%)	16.00			
	Seedling fresh matter (mg)			
	Y53	Y52	Y85	Y95
Pre-soaking	3.07 aB	3.13 aB	5.36 aA	2.85 aA
Priming	4.3 aB	3.4 aB	2.37 aA	1.73 aA
Control	8.1 Aa	8.1 aA	4.5 bA	2.33 bA
CV (%)	37.53			
	Seedling dry matter (mg)			
	Y53	Y52	Y85	Y95
Pre-soaking	1.53 aA	1.39aA	1.44 aA	0.92 aA
Priming	1.48 aA	1.17 aA	1.22 aA	0.74aB
Control	1.31 aA	1.09 aA	1.40 aA	0.60 aB
CV (%)	24.86			

Means followed by same capital letters in the column and lower case letters in the line, do not differ statistically at 5%.

Seed priming showed a large increase in germination percentage in Y53 and Y95 accession when compared to control seeds, however there was no statistical difference between the pre-soaking and priming treatments for accession Y53 and among all treatments for accession Y85 (Table 3). Pre-soaking and priming treatments interfered negatively in the germination percentage of accession Y52. Possibly the seeds from this accession are in the deterioration process, resulting in poor germination, as demonstrated in control treatment results (Table 3).

Germination time is an important parameter in seedling establishment in field (Carvalho and Nakagawa, 2012). Tarquis and Bradford (1992) observed priming and prehydration treatments had little effect on seed germination of various species, other than *Psidium*, but significantly reduced mean time to germination by up to 61% relative to untreated seeds. In this work we found up to 57% reduction in MGT, which means 12 days anticipation in germination in Y95 accession (Table 3).

Pre-soaking also influenced length of seedlings (Table 3). Accessions with higher (Y85) and lower (Y52) seed vigor showed less sensitivity to priming or pre-soaking treatments (Table 3). According to Masetto et al. (2014) priming in *Psidium guinea* seeds showed an increase in germination, root growth and reduction in the average time for germination. Differences in the effects produced by priming seeds with different vigor levels, reinforces the statement that responses of seeds submitted to osmopriming are more intense for lots which have begun deterioration process, however are not yet utterly deteriorated (Powell, 1998).

In numerous species, growing conditions of a parent plant, both wild and cultivated, may affect germination of its seeds (Fenner, 1991).Variations in environmental conditions during seed development may result in variations plant and seed performance. Among environmental traits, temperature changes during seed maturation plays a dominant role in both plant and seed performance, whereas light signaling (light intensity and photoperiod) has more impact on plant traits (He et al., 2014). Asides parent environment, some studies have shown there is wide variability for germination and vigor of seeds of different accessions, cultivars and progenies of many fruit species (Cardoso et al., 2009; Nerling et al., 2013; Negreiros et al., 2015), such as observed for guava accessions (Tables 1-4). Santos et al. (2014) described genetic diversity for fruit characteristics among guava accessions harvested in the same municipality and in different municipalities in the semiarid region of Bahia.

For guava, it is likely that an interaction between parental environment, genotype and initial seed physiological quality affects plant and seed performance, as occurs with *Arabidopsis* (He at al., 2014).

Conclusions

Brazilian guava accessions collected in semiarid Bahia showed different germination behavior for all studied traits.

Temperatures of 20 and 25 °C were ideal for germination test of guava seeds in laboratory.

Pre-soaking and priming in guava seeds are recommended to reduce germination time and increase seed physiological quality.

References

ALVES, E.U.; BRUNO, R.L.A.; OLIVEIRA, A.P.O.; ALVES, A.U.; ALVES, A.U.; PAULA, R.C. Influência do tamanho e da procedência de sementes de *Mimosa caesalpiniifolia* Benth. sobre a germinação e vigor. *Revista Árvore*, v.29, n.6, p.877-885, 2005. http://www.scielo.br/pdf/rarv/v29n6/a06v29n6.pdf

ANDERSSON, L.; MILBERG, P. Variation in seed dormancy among mother plants, populations and years of seed collection. *Seed Science Research*, v.8, n.1, p.29-38, 1998. https://people.ifm.liu.se/permi/PlantEcology/SeedSciRes98.pdf

BEWLEY, J.D.; BRADFORD, K.J.; HILHORST, H.W.M.; NONOGAKI, H. *Seeds:* physiology of development germination and dormancy. New York: Springer, 2013. 392p.

BEZERRA, J.E.F.; LEDERMAN, I.E.; SILVA-JÚNIOR, J.F.; PROENÇA, C.E.B. *Guava*. In: VIEIRA, R.F.; AGOSTINI-COSTA, T.; SILVA, D.B.; FERREIRA, F.R.; SANO, S.M. *Frutas Nativas da Região Centro-Oeste do Brasil*. Brasília: Embrapa Recursos Genéticos e Biotecnologia, 2006. p.42-63.

BORGES, R.C.F.; COLLAÇO-JÚNIOR, J.C.; SCARPARO, B.; NEVES, M.B.; CONEGLIAN, A. Caracterização da curva de embebição de sementes de pinhão -manso. *Revista Científica Eletrônica de Engenharia Florestal*, v.13, p.1-8, 2009. http://www.scielo.br/scielo.php?script=sci_nlinks&ref=000080&pid=S0101-3122201200030001900003&lng=pt

BRANCALION, P.H.S.; NOVEMBRE, A.D.L.C.; RODRIGUES, R.R. Temperatura ótima de germinação de sementes de espécies arbóreas brasileiras. *Revista Brasileira de Sementes*, v.32, n.4, p.15-21, 2010. http://dx.doi.org/10.1590/S0101-31222010000400002

CALDEIRA, S.D.; HIANE, P.A.; RAMOS, M.I.L.; RAMOS-FILHO, M.M. Physical-chemical characterization of guava (*Psidium guineense* SW.) and taruma (*Vitex cymosa* Bert.) of Mato Grosso do Sul State, Brasil. *Boletim do Centro de Pesquisa e Processamento de Alimentos Brasil*, v.22, n.1, p.145-154, 2004.

CARDOSO, D.L.; SILVA, R.D.; PEREIRA, M.G.; VIANA, A.P.; ARAÚJO, E.F. Diversidade genética e parâmetros genéticos relacionados à qualidade fisiológica de sementes em germoplasma de mamoeiro. *Revista Ceres*, v.56, n.5, p.572-579, 2009. http://www.ceres.ufv.br/ceres/revistas/V56N005P04108.pdf

CARVALHO, N. M.; NAKAGAWA, J. *Sementes:* ciência, tecnologia e produção. 5.ed. Jaboticabal: FUNEP, 2012. 590p.

DANTAS, B.F.; CORREIA, J.S.; MARINHO, L.B.; ARAGÃO, C.A. Alterações bioquímicas durante a embebição de sementes de catingueira (*Caesalpinia pyramidalis* Tul.). *Revista Brasileira de Sementes*, v.30, n.1, p.221-227, 2008a. http://www.scielo.br/pdf/rbs/v30n1/a28v30n1.pdf

DANTAS, B.F.; RIBEIRO, L.S.; ARAGÃO, C.A. Germination, initial growth and cotyledon protein content of bean cultivars under salinity stress. *Revista Brasileira de Sementes*, v.29, n.2, p.106-110, 2007. http://www.scielo.br/pdf/rbs/v29n2/v29n2a14.pdf

DANTAS, B.F.; SOARES, F.S.J.; LÚCIO, A.A.; ARAGÃO, C.A. Alterações bioquímicas durante a embebição de sementes de baraúna (*Schinopsis brasiliensis* Engl.). *Revista Brasileira de Sementes*, v.30, n.2, p.214-219, 2008b. http://www.scielo.br/pdf/rbs/v30n2/a27v30n2.pdf

DEGÁSPARI, C.H.; WASZCZYNSKYJ, N. Antioxidants properties of phenolic compounds. *Visão Acadêmica*, v.5, p.33-40, 2004. http://ojs.c3sl.ufpr.br/ojs/index.php/academica/article/viewFile/540/453

DELGADO L.F.; BARBEDO, C.J. Tolerância à dessecação de sementes de espécies de *Eugenia. Pesquisa Agropecuária Brasileira*, v.42, n.2, p.265-272, 2007. http://www.scielo.br/pdf/pab/v42n2/16.pdf

DELGADO, L.F.; BARBEDO, C.J. Water potential and viability of seeds of *Eugenia* (Myrtaceae), a tropical tree species, based upon different levels of drying. *Brazilian Archives of Biology and Technology*, v.55, n.4, p.583-590, 2012. http://www.scielo.br/pdf/babt/v55n4/a14v55n4.pdf

FENNER, M. The effects of the parent environment on seed germinability. *Seed Science Research*, v.1, n.2, p.75-84, 1991. http://dx.doi.org/10.1017/S0960258500000696

FORZZA, R.C.; LEITMAN, P.M.; COSTA, A.F.; CARVALHO-JÚNIOR., A.A.; PEIXOTO, A.L.; WALTER, B.M.T.; BICUDO, C.; ZAPPI, D.; COSTA, D.P.; LLERAS, E.; MARTINELLI, G.; LIMA, H.C.; PRADO, J.; STEHMANN, J.R.; BAUMGRATZ, J.F.A.; PIRANI, J.R.; SYLVESTRE, L.; MAIA, L.C.; LOHMANN, L.G.; QUEIROZ, L.P.; SILVEIRA, M.; COELHO, M.N.; MAMEDE, M.C.; BASTOS, M.N.C.; MORIM, M.P.; BARBOSA, M..; MENEZES, M.; HOPKINS, M.; SECCO, R.; CAVALCANTI, T. B.; SOUZA, V.C. 2012. Introdução. In: *Lista de Espécies da Flora do Brasil*. Jardim Botânico do Rio de Janeiro. http://floradobrasil.jbrj.gov.br/2012/

GALLOWAY, L. F. Maternal effects provide phenotypic adaptation to local environmental conditions. *New Phytologist*, v. 166, n. 1, p.93-100. 2005. http://onlinelibrary.wiley.com/doi/10.1111/j.1469-8137.2004.01314.x/pdf

GIACOBBO, C.L.; ZANUZO, M.; CHIM, J.; FACHINELLO, J.C. Avaliação do teor de vitamina c em diferentes grupos de guava-comum. *Revista Brasileira de Agrociência*, v.14, n.1, p.155-159, 2008. http://periodicos.ufpel.edu.br/ojs2/index.php/CAST/article/viewFile/1899/1732

HE, H.; VIDIGAL, S.D.; SNOEK, L.B.; SCHNABEL, S.; NIJVEEN, H.; HILHORST, H.; BENTSINK, L. Interaction between parental environment and genotype affects plant and seed performance in Arabidopsis. *Journal of Experimental Botany*, v.65, n.22, p.6603-6615, 2014. http://jxb.oxfordjournals.org/content/early/2014/09/18/jxb.eru378

LABOURIAU, L.G. *A germinação de sementes*. Washington: OEA, 1983. 174p.

LAMARCA, E.V.; SILVA, C.V.; BARBEDO, C.J.; Limites térmicos para a germinação em função da origem de sementes de espécies de *Eugenia* (Myrtaceae) nativas do Brasil. *Acta Botanica Brasilica*, v.25, n.2, p.293-300, 2013. http://www.scielo.br/pdf/abb/v25n2/a05v25n2.pdf

LONG, R.L.; GORECKI, M.J.; RENTON, M.; SCOTT, J.K.; COLVILLE, L.; GOGGIN, D.E.; COMMANDER, L.E.; WESCOTT, D.A.; CHERRY, H.; FINCH-SAVAGE, W.E. The ecophysiology of seed persistence: a mechanistic view of the journey of germination or demise. *Biological Reviews*, v. 90, n. 1, p. 31-59, 2014. http://onlinelibrary.wiley.com/doi/10.1111/brv.12095/pdf

MAGUIRE, J.D. Speed of germination—aid in selection and evaluation for seedling emergence and vigor. *Crop Science*, v.2, n.2, p. 176-177, 1962.

MASETTO, T.E.; SILVA NEVES, E.M.; SCALON, S.D.P.Q.; DRESCH, D.M. Drying, storage and osmotic conditioning of *Psidium guineense* Swartz seeds. *American Journal of Plant Sciences*, v.5, n.17, 2014. http://www.scirp.org/journal/PaperInformation.aspx?PaperID=48419

MASETTO, T.E; FARIA, J.M.R.; DAVIDE, A.C.; SILVA, E.A.A. Desiccation tolerance and DNA integrity in *Eugenia pleurantha* O. Berg. (Myrtaceae) seeds. *Revista Brasileira de Sementes*, v.30, n.2, p.51-56, 2008. http://www.scielo.br/pdf/rbs/v30n2/a07v30n2.pdf

MIRANDA, G.B.; SOUZA, R.D.; GOMES, V.M.; FERREIRA, T.D.F.; ALMEIDA, A.M. Avaliação de acessos de *Psidium* spp. quanto à resistência a *Meloidogyne enterolobii*. *Bragantia*, v. 71, n. 1, p. 52-58, 2012. http://www.scielo.br/pdf/brag/v71n1/aop1126.pdf

MUGNOL, D.D.; QUINTÃO, S.S.P.; SILVA, N.E.M.; ELISA, M.T.; MARA, M.R. Effect of pre-treatments on seed germination and seedling growth in *Psidium guineense* Swartz. *Agrociencia*, v. 18, n. 2, p. 33-39, 2014. http://www.scielo.edu.uy/pdf/agro/v18n2/v18n2a04.pdf

NEGREIROS, J.R.D.S.; ALEXANDRE, R.S.; ÁLVARES, V.D.S.; BRUCKNER, C.H.; CRUZ, C.D. Divergência genética entre progênies de maracujazeiro- amarelo com base em características das plântulas. *Revista Brasileira de Fruticultura*, v.30, n.1, p.197-201, 2015. http://www.scielo.br/scielo.php?script=sci_arttext&pid=S0100-29452008000100036&lng=en&tlng=pt. 10.1590/S0100-29452008000100036

NERLING, D.; COELHO, C.M.M.; NODARI, R.O. Genetic diversity for physiological quality seeds from corn (*Zea mays* L.) intervarietal crossbreeds. *Journal of Seed Science*, v.35, n.4, p.449-456, 2013. http://www.scielo.br/pdf/jss/v35n4/06.pdf

OLIVEIRA, G. M.; MATIAS, J. R.; DANTAS, B. F. Temperatura ótima para germinação de sementes nativas da Caatinga. *Informativo ABRATES*, v.24, n.3, p.44-47, 2014a. http://www.abrates.org.br/images/--Informativo/v24_n3/Palestras.pdf

OLIVEIRA, G. M.; RODRIGUES, J. M.; RIBEIRO, R. C.; BARBOSA, L. G.; SILVA, J. E. S. B.; DANTAS, B. F. Germinação de sementes de espécies arbóreas nativas da Caatinga em diferentes temperaturas. *Scientia Plena*, v.10, n.4, p.1-6, 2014b. http://www.scientiaplena.org.br/sp/article/view/1790/954

PESSOA, R.C.; MATSUMOTO, S.N.; MORAIS, O.M.; VALE, R.S.D.; LIMA, J.M. Germinação e maturidade fisiológica de sementes de *Piptadenia viridiflora* (Kunth.) Benth relacionadas a estádios de frutificação e conservação pós-colheita. *Revista Árvore*, v.34, n.4, p.617-625, 2010. http://www.scielo.br/pdf/rarv/v34n4/v34n4a06.pdf

POWELL, A.A. Seed improvement by selection and invigoration. *Scientia Agricola*, v.55, p.126-133, 1998.

RAJJOU, L.; DUVAL, M.; GALLARDO, K.; CATUSSE, J.; BALLY, J.; JOB, C.; JOB, D. Seed germination and vigor. *Annual Review of Plant Biology*, v.63, p.507-533, 2012. http://www.annualreviews.org/doi/pdf/10.1146/annurev-arplant-042811-105550

RESENDE, O.; ALMEIDA, D.P.; COSTA, L.M.; MENDES, U.C.; SALES, J.D.F. Adzuki beans (*Vigna angularis*) seed quality under several drying conditions. *Ciência Tecnologia, Alimentos*, v.32, n.1, p.151-155, 2012. http://www.scielo.br/pdf/cta/v32n1/aop_cta_5007.pdf

SANTOS, C.M.R.; FERREIRA, A.G.; ÁQUILA, M.E.A. Características de frutos e germinação de sementes de seis espécies de Myrtaceae nativas do Rio Grande do Sul. *Ciência Florestal*, v.14, n.2, p.13-20, 2004. http://cascavel. ufsm.br/revistas/ojs-2.2.2/index.php/cienciaflorestal/article/view/1802

SANTOS, M.A.C.; QUEIRÓZ, M.A.; SANTOS, A.S.; SANTOS, L.C.; CARNEIRO, P.C.S. Diversidade genética entre acessos de guava de diferentes municípios do semiárido baiano. *Revista Caatinga*, v.27, p.48-57, 2014. http://periodicos.ufersa.edu.br/revistas/index.php/sistema/article/view/2887/pdf_113

SMIDERLE, O.J.; LIMA, J.M.E.; PAULINO, P.P.S. Curva de absorção de água em sementes de *Jatropha curcas* L. com dois tamanhos. *Revista Agro@mbiente*, v.7, n.2, p.203-208, 2013. http://revista.ufrr.br/index.php/agroambiente/article/view/1056/1150

SOUZA, A.G.; RESENDE, L.V.; LIMA, I.P.; SANTOS, R.M.; CHALFUN, N.N.J. Variabilidade genética de acessos de araçazeiro e goiabeira suscetíveis e resistentes a *Meloidogyne enterolobii*. *Ciência Rural*, v.44, n.5, p.822-829, 2014. http://www.scielo.br/pdf/cr/v44n5/a14514cr6237.pdf

TARQUIS, A.M.; BRADFORD, K.J. Prehydration and priming treatments that advance germination also increase the rate of deterioration of lettuce seeds. *Journal of Experimental Botany*, v.43, n.3, p.307-317, 1992.

TURESSON, G. The genotypical response of the plant species to the habitat. *Hereditas*, v.3, n.3, p.211-350, 1922. http://onlinelibrary.wiley.com/doi/10.1111/j.1601-5223.1922.tb02734.x/abstract

VILLELA, F.A.; DONI FILHO, L.; SEQUEIRA, E.L. Tabela de potencial osmótico em função da concentração de polietileno glicol 6000 e da temperatura. *Pesquisa Agropecuária Brasileira*, v.26, n.11-12, p.1957-1968, 1991. http://seer.sct.embrapa.br/index.php/pab/article/view/3549/882

Effect of sodium nitroprusside (SNP) on the germination of *Senna macranthera* seeds (DC. ex Collad.) H. S. Irwin & Baneby under salt stress

Aparecida Leonir da Silva[1]*, Denise Cunha Fernandes dos Santos Dias[2], Eduardo Euclydes de Lima e Borges[3], Dimas Mendes Ribeiro[1], Laércio Junio da Silva[2]

ABSTRACT – Nitric oxide (NO) has been used as stimulating of the germination process for many species. However, there are few studies evaluating the effect of nitric oxide donor in the regulation of seed germination under salt stress, especially for native forest species. The objective was to evaluate the effects of SNP, an NO donor substance, on germination of *Senna macranthera* seeds under salt stress. The seeds were germinated at different osmotic potentials induced by NaCl solution (0.0, -0.1, -0.2, -0.3, -0.4 and -0.5 MPa). To evaluate the effect of the SNP, potentials -0.3 and -0,4 MPa were selected, applying SNP at different concentrations: 100, 200, 300 and 400 μM. Germination tests were conducted at 25 °C, with photoperiod of 8 hours. Percentage of radicle protrusion, radicle protrusion speed index, percentage of normal seedlings, shoots and roots length and dry matter were evaluated. Salt stress with NaCl is harmful to germination of *S. macranthera* seeds. SNP has the potential to recover germination under salt stress, especially in the concentration of 100 μM.

Index terms: "fedegoso", NaCl, nitric oxide, native species.

Efeito do nitroprussiato de sódio (SNP) na germinação de sementes de *Senna macranthera* (DC. ex Collad.) H. S. Irwin & Baneby sob estresse salino

RESUMO – O óxido nítrico (ON) vem se destacando como estimulador do processo de germinação para muitas espécies. Há poucos estudos que avaliam o efeito do doador de óxido nítrico na regulação da germinação de sementes sob estresse salino, principalmente na germinação de espécies florestais nativas. Objetivou-se avaliar os efeitos do nitroprussiato de sódio (SNP), uma substância doadora de ON, na germinação de sementes de *Senna macranthera* sob estresse salino. As sementes foram colocadas para germinar em diferentes potenciais osmóticos, induzidos por solução de NaCl (0,0; -0,1; -0,2; -0,3; -0,4 e -0,5 MPa). Para testar o efeito do SNP, selecionou-se os potenciais de -0,3 e -0,4 MPa, com aplicação de SNP em diferentes concentrações: 100, 200, 300 e 400 μM. Os ensaios germinativos foram conduzidos em câmara de germinação a 25 °C, com fotoperíodo de 8 horas. Foram avaliados a protrusão radicular, o índice de velocidade de protrusão radicular, a porcentagem de plântulas normais, o comprimento da parte aérea e do sistema radicular, massa seca da parte aérea e do sistema radicular. O estresse salino com NaCl é prejudicial a germinação de sementes de *S. macranthera*. O SNP tem potencial para promover a recuperação da germinação das sementes sob estresse salino, sendo a concentração de 100 μM a mais eficaz.

Termos para indexação: fedegoso, NaCl, óxido nítrico, espécie nativa.

Introduction

Senna macranthera is an arboreal species, belonging to the Fabaceae – Caesalpinioideae family, popularly known as "Fedegoso" (Lorenzi, 2000). The species is widely used in landscaping due to its ornamental characteristics and small size, which is ideal for urban trees. It is a pioneering, fast-growing species, recommended for use in plantations in degraded areas (Lorenzi, 2000).

Demand for seedlings of native forest species has been growing due to the need of reforestation. Most of these species are propagated by seeds; success in the formation of

[1]Departamento de Biologia Vegetal, Universidade Federal de Viçosa, 36570-000 – Viçosa, MG, Brasil.
[2]Departamento de Fitotecnia, Universidade Federal de Viçosa, 36570-000 _ Viçosa, MG, Brasil.
[3]Departamento de Engenharia Florestal, Universidade Federal de Viçosa, 36570-000 – Viçosa, MG, Brasil.
*Corresponding author <aparecidaleonir@gmail.com>

seedlings depends on the knowledge about the regulation of seeds germination and the quality thereof, for each species (Rego et al., 2009).

The seeds are often exposed to various environmental stresses that negatively interfere in germination, vegetative development, plant productivity and, in severe cases, can lead to seedlings death. According to Braga et al. (2009), the time period required for germination is important for the survival of forest species, especially where water availability is limited during certain periods of the year.

Germination is a physiological event that depends on seed quality and environmental conditions such as water supply and oxygen, and temperature and light suitability (Maekawa et al., 2010). Seed hydration is the most important event, because water is the matrix where most biochemical and physiological processes occur and result in primary root protrusion (Moraes et al., 2005). Salt stress affects one of the main processes of the plant life cycle, namely, seeds germination (Cesur and Tabur, 2011; Zheng et al., 2009), leading to a reduced and delayed germination rate (Singh et al., 2012). One of the most widespread methods to determine plants tolerance to salt stress is the observation of germination in saline substrates (Lima and Torres, 2009).

Salinity affects germination, not only due to hindering water absorption, but also to facilitating the entry of toxic amounts of ions in the seeds during imbibition (Simaei et al., 2012). Moreover, salt stress causes changes in seeds physiology due to ionic toxicity, osmotic stress and increased reactive oxygen species (ROS) (Mittler, 2002), leading to gradual lipid peroxidation and antioxidant enzyme inactivation (Tanou et al., 2009). Fan et al. (2013) emphasize the importance of further studies to understand the physiological mechanisms involved on seed germination under salt stress and develop appropriate measures to alleviate the negative effects of salinity on germination.

There are several reports in the literature on the harmful effect of salt stress on seed germination. In seeds of forest species, this effect for species *Schizolobium amazonicum* (Braga et al., 2008), *Enterolobium schomburgkii* (Braga et al., 2009), *Chorisia speciosa* (Fanti and Perez, 2004), *Dimorphandra mollis* (Masetto et al., 2014), *Gliricidia sepium* (Farias et al., 2009) and *Zizyphus joazeiro* (Lima and Torres, 2009) have already been observed.

Among the factors that favor germination under stress conditions, nitric oxide (NO) has stood out as a stimulator of the process in many species. NO is a free radical produced from L-arginine, toxic, inorganic, and colorless gas, with seven nitrogen atoms and eight oxygen atoms, being one of the most important mediators of cellular signaling (Dusse et

al., 2003). One of the limiting factors in the studies on the possible actions of NO in plants is the absence of mutants with a differential production of NO, in addition to the lack of mechanisms to perceive and translate the signals induced by this compound (Ederli et al., 2009). Therefore, ON donor and sequestering substances have been widely used in studies to elucidate their functions. The most used reagents as ON donors are sodium nitroprusside (SNP) and S-Nitroso-N-acetylpenicillamine (SNAP).

Several studies have shown a positive effect of NO on the recovery of seed germination under different types of stress. However, there are few studies that evaluate the effect of NO on seeds germination in salt stress conditions, particularly in native species. It was found in *Lupinus luteu* seeds that SNP was effective in reversing the negative impact of NaCl on germination (Kopyra and Gwóźdź´, 2003); this reversal was also observed in *Cucumis sativus* (Fan et al., 2013) and *Ocimum basilicum* (Saeidnejad et al., 2013) seeds.

The positive effect of NO on the seed germination process is observed in some studies, but there are no reports of the effect of SNP on seed germination of *S. macranthera* under saline conditions. Given the above, this study has aimed to evaluate the effects of SNP on seed germination of *S. macranthera* under salt stress induced by NaCl.

Material and Methods

Senna macranthera seeds were collected in 2012 in the Brazilian city of Viçosa, Minas Gerais. The seeds were stored in cold room at 5 °C and relative humidity of 60%. Before establishing the tests, all seeds were mechanically scarified with sandpaper number 100 on the opposite side to the hilum and then they were treated with fungicide Captan at 0.2%.

Experiment I – Evaluation of root protrusion and seed germination of S. macranthera under salt stress conditions

The seeds were germinated in saline stress conditions, obtained by using saline solution of sodium chloride (NaCl, P.M. 58.44). The following osmotic potentials were tested: -0.1; -0.2; -0.3; -0.4 and -0.5 MPa. The salt concentrations for each osmotic potential were obtained based on the equation by J. H. van't Hoff, cited by Salisbury and Ross (1992): $\psi_{os} = - RTC$, where:

ψ_{os} = osmotic potential (atm);
R = general ideal gas constant (0.082 atm. 1. mol^{-10}. k^{-1});
T = temperature (°K); and
C = concentration (mol/1) (No. of moles/l).

For the germination test, four replications of 50 seeds

were placed on paper towel rolls moistened with water (control) or saline solution at a ratio of 2.5 times the dry paper weight. Subsequently, the rolls were placed in plastic bags. The germination test was conducted in a germinator at 25 °C, with a photoperiod of 8 hours.

The numbers of seeds with protruded primary radicle and germination percentage (normal seedlings) were daily assessed. Seeds whose primary root had at least 2 mm in length were considered protruded. The percentage of normal seedlings was detemined following the criteria established by Rules for Seed Testing (Brasil, 2009). On the seventh day after sowing, the percentage of root protrusion and germination was calculated. With daily data, radicle protrusion speed index (RPSI) was calculated.

At the end of the germination test, shoot length (CPA) and root system length (RSL) of seedlings were evaluated, with the aid of a digital caliper. The results were expressed in millimeters per seedling. Dry matter of shoot (DMS) and root dry matter (RDM) were determined by the oven method with forced air circulation, set at 70 °C, where the seedlings remained until constant weight.

Experiment II – Nitric oxide effect on root protrusion and germination of S. macranthera seeds under salt stress conditions

The seeds were germinated under salt stress conditions in osmotic potentials -0.3 and -0.4 MPa of NaCl (defined according to the results of the previous experiment). For each potential, sodium nitroprusside solution (SNP) were applied in concentrations of 100, 200, 300 and 400 µM.

For the germination test, four replications of 50 seeds were germinated in accordance with the procedures adopted in the previous experiment. The substrate was moistened with a mixture of NaCl and SNP solutions, in the proportion of 2.5 times the weight of the dry paper. Percentage of primary root protrusion, RPSI and germination percentage were determined.

Statistical analysis

Experiment I was conducted in a completely randomized design with four replications. The data were submitted to analysis of variance (ANOVA) and expressed as mean ± standard deviation. Experiment II was conducted in a completely randomized design with four replications in a 2 x 4 factorial arrangement, i.e., two osmotic potentials (-0.3 and -0.4 MPa) and four concentrations of SNP (100, 200, 300 and 400 µM). Root protrusion data were transformed by the arcsine function and then they were subjected to analysis of variance. The results obtained for the salt concentrations were expressed as mean ± standard deviation.

Results and Discussion

S. macranthera seeds present integument hardness, which corresponds to the characteristic primary dormancy mechanism of the species (Lorenzi, 2000). In this study, the control treatment seeds (water) presented high root protrusion (87%), showing the efficiency of the mechanical scarification procedure indicated by Lemos Filho et al. (1997) in overcoming the species dormancy.

There was a gradual decrease in root protrusion and seed vigor by reducing the osmotic potential (Figure 1), where the seeds protrusion percentage under potential 0.0 MPa was 87% to 36% in potential -0.5 MPa. A more drastic effect of salinity was observed, with a significant reduction in radicle protrusion speed index, germination, shoot length and root system length with increased salt stress (Figures 1B, 1C, 1D and 1E). Although some seeds still emit radicle, there was no formation of seedling in saline potentials -0.4 and -0.5 MPa. There was a slight increase in the aerial part dry matter in potentials -0.3 and -0.4 MPa and root remained stable at concentrations of -0.1; -0.2 and -0.3 MPa in relation to control (Figures 1F and 1G).

In seeds of other forest species, harmful effect of salt stress has also been found, such as for *Schizolobium amazonicum* (Braga et al., 2008) and *Enterolobium schomburgkii* (Braga et al., 2009), especially in reducing germination percentage, with significant decreases from potentials at -0.2 MPa in the solution of NaCl. However, in studies conducted by Fanti and Perez (2004) significant decreases in germination percentage to the osmotic potential of -0.4 MPa in *Chorisia speciosa* seeds were not observed, but from -0.6 MPa significant reductions in viability were recorded. For *Gliricidia sepium* (Farias et al., 2009) germination was significantly affected only in osmotic potentials above -1.0 MPa. In *Zizyphus joazeiro* all salt concentrations used (-0.3; -0.6 and -0.9 MPa) significantly decreased germination (Lima and Torres, 2009).

According to the results found in this study, *S. macranthera* seeds showed greater decrease in root protrusion from osmotic potential -0.4 MPa; for other species, this potential also proved harmful, such as for *Senna occidentalis* (Norsworthy and Oliveira, 2005), *Senna obtusifolia* (Pereira et al., 2014) and *Dimorphandra mollis* (Masetto et al., 2014).

It is noticed that the seeds of different species support different osmotic salt stresses on germination, and the seedlings are the most sensitive. One of the factors responsible for the reduction in germination may have been an excess of ions Na^+ and Cl^-, since they cause decreased protoplasmic swelling (Ferreira and Borghetti, 2004). Another factor may have been the excess of soluble salts, resulting in reduced water potential, i.e., the water absorption capacity of the seeds is reduced. This reduction of water potential and the salts toxic effects initially interfere with the process of water uptake by seeds, influencing vigor, affecting speed and thus the germination time of these seeds in non-toxic levels of salinity (Cavalcante and Perez, 1995).

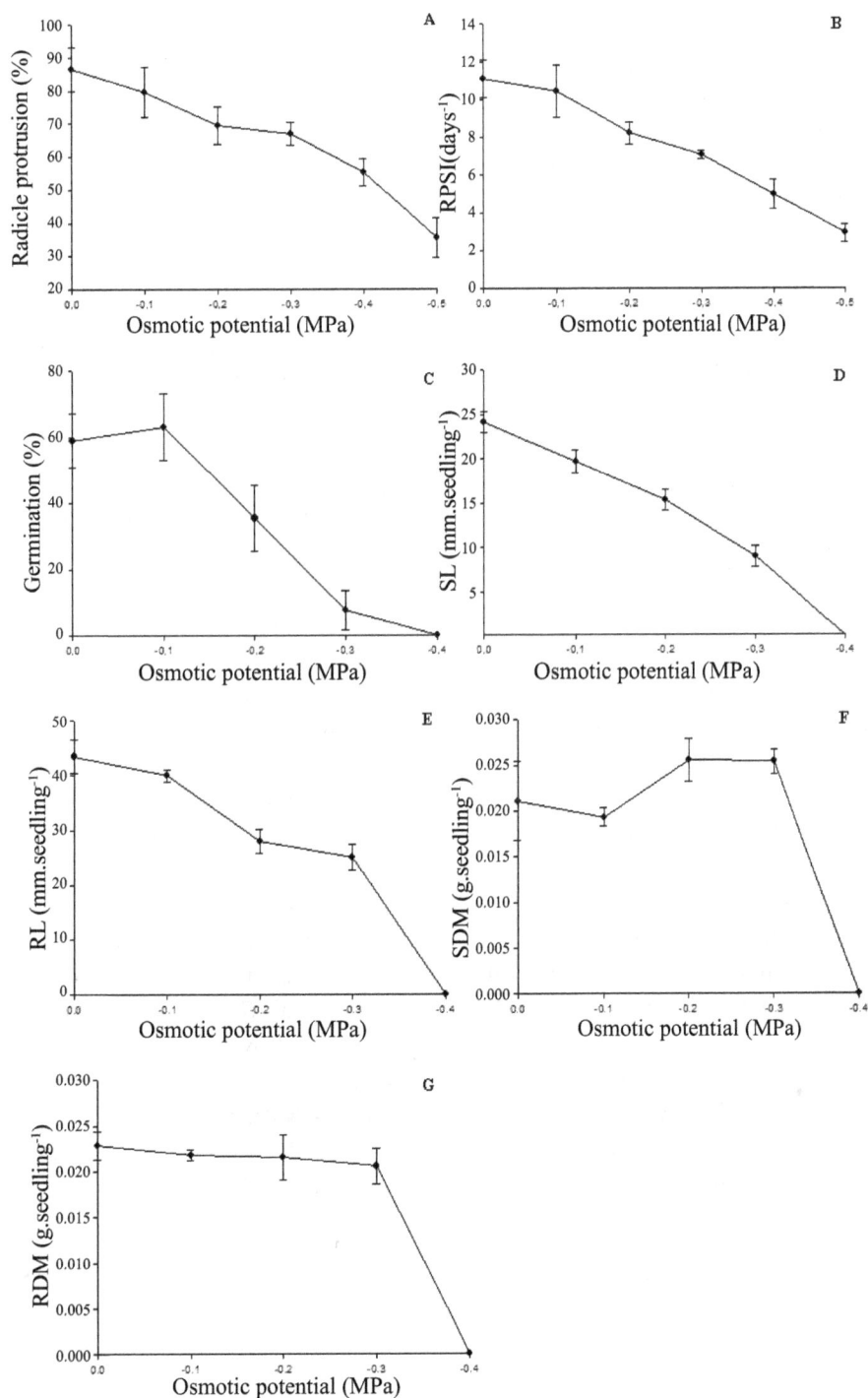

Figure 1. A: Percentage of root protrusion, B: root protrusion speed index (RPSI), C: percentage of germination, D: shoot length (SL), E: root length (RL), F: dry matter of shoot (DMS) and G: root dry matter (RDM), of *Senna macranthera* seeds, scarified and germinated in different osmotic potentials with NaCl (0.0; -0.1; -0.2; -0.3; -0.4 and -0.5 MPa) at 25 °C.

Bewley et al. (2013) have pointed out that the seeds moderate resistance to salt stress is useful when trying to use saline soils in dry regions, since 6% of terrestrial continents are made up of saline soils. Knowledge of species able to withstand these conditions can aid in the proper recommendation of species for planting, especially for native species, for which there is limited information (Rego et al., 2011).

Species tolerant to salt stress are classified as halophytes or glycophytes due to their tolerance to salt stress, both respond similarly to salt stress, and the percentage and germination rate are inversely proportional to increased salinity, varying only the salt tolerance limit (Jeller and Perez, 2001). Halophyte species are highly tolerant, germinating in the environment with up to 8% of NaCl (Ungar, 1978), and the ones that are little tolerant have their germination inhibited at 1.5% of NaCl. Most glycophytes do not germinate in medium with concentrations higher than 1.5% of NaCl. In this study, S. macranthera may be included among the little salt-tolerant glycophytes because it does not have a high tolerance limit (up to -0.4 MPa). Similar results were found for Senna spectabilis, also included among the little salt-tolerant glycophytes, with a limit of -1.6 MPa (Jeller and Perez, 2001).

From the results of radicle protrusion percentage obtained in the first assay (Figure 1A), two salt concentrations were selected, which caused moderate stress, with root protrusion around 50% to evaluate SNP effect on the recovery of the seeds physiological potential. The selected concentrations were -0.3 and -0.4 MPa, due to promoting moderate stress levels in radicle protrusion of S. macranthera seeds (67 and 55%, respectively).

Thus, there was an increase in root protrusion and shoot length of the seedlings, and decrease in radicle protrusion speed index for all concentrations of SNP tested (100, 200, 300 and 400 µM) in the potential of -0.3 MPa (Figures 2 A and B). There was an increase of 18% in root protrusion and of 225% in germination, compared to the control (without application of SNP) at a concentration of 100 µM of SNP (Figure 2). Although there is a small reduction in dry matter of shoot and root system compared to the control, this was not significant (Figures 2 F and G).

There was an increase in seeds root protrusion under salt stress -0.4 MPa with the application of SNP at concentrations of 100 (15%), 200 (10%) and 300 µM (17%), compared to pure salt stress (Figure 3). The radicle protrusion speed index decreased in all concentrations (100, 200, 300 and 400 µM) of SNP, compared to the control (-0.4 MPa). Thus, it is possible to observe the limitation of SNP effect in reversing the inhibitory effect of salt stress in seedling production.

There are no reports in the literature about the germination behavior of native forest species with nitric oxide donor application in the reversal of salt stress. However, in Lupinus luteus seeds from the Fabaceae family, SNP also proved effective in reversing the decline of germination caused by NaCl (Kopyra and Gwóźdź´, 2003). For Cucumis sativus, the recovery of salt stresses was more effective at the concentration of 50 µM of SNP; however, there was a decrease in germination at the highest concentration of SNP (400 µM) (Fan et al., 2013). In Triticum aestivum seeds there was a significant increase in germination after seven days of seed germination with SNP under salt stress with NaCl (Zheng et al., 2009).

However, in Ocimum basilicum seeds there was no increase in seed germination with nitric oxide donor, probably because it relieves salt stress through changes in physiological properties and the antioxidant system (Saeidnejad et al., 2013).

According to Correa-Aragunde et al. (2004) and Lombardo et al. (2006), nitric oxide is involved in the regulation of root morphology. Correa-Aragunde et al. (2004) have found that nitric oxide affects the growth of Lycopersicon esculentum roots in a dose-dependent manner, wherein SNP at low concentrations stimulates root growth, while high concentrations have an inhibitory effect on growth.

In general, S. macranthera seeds are sensitive to salt stress induced by NaCl, with reduction of root protrusion, germination and seed vigor, especially in osmotic potential below -0.4 MPa. SNP was efficient in the recovery of the physiological potential of S. macranthera seeds at the concentration of 100 µM.

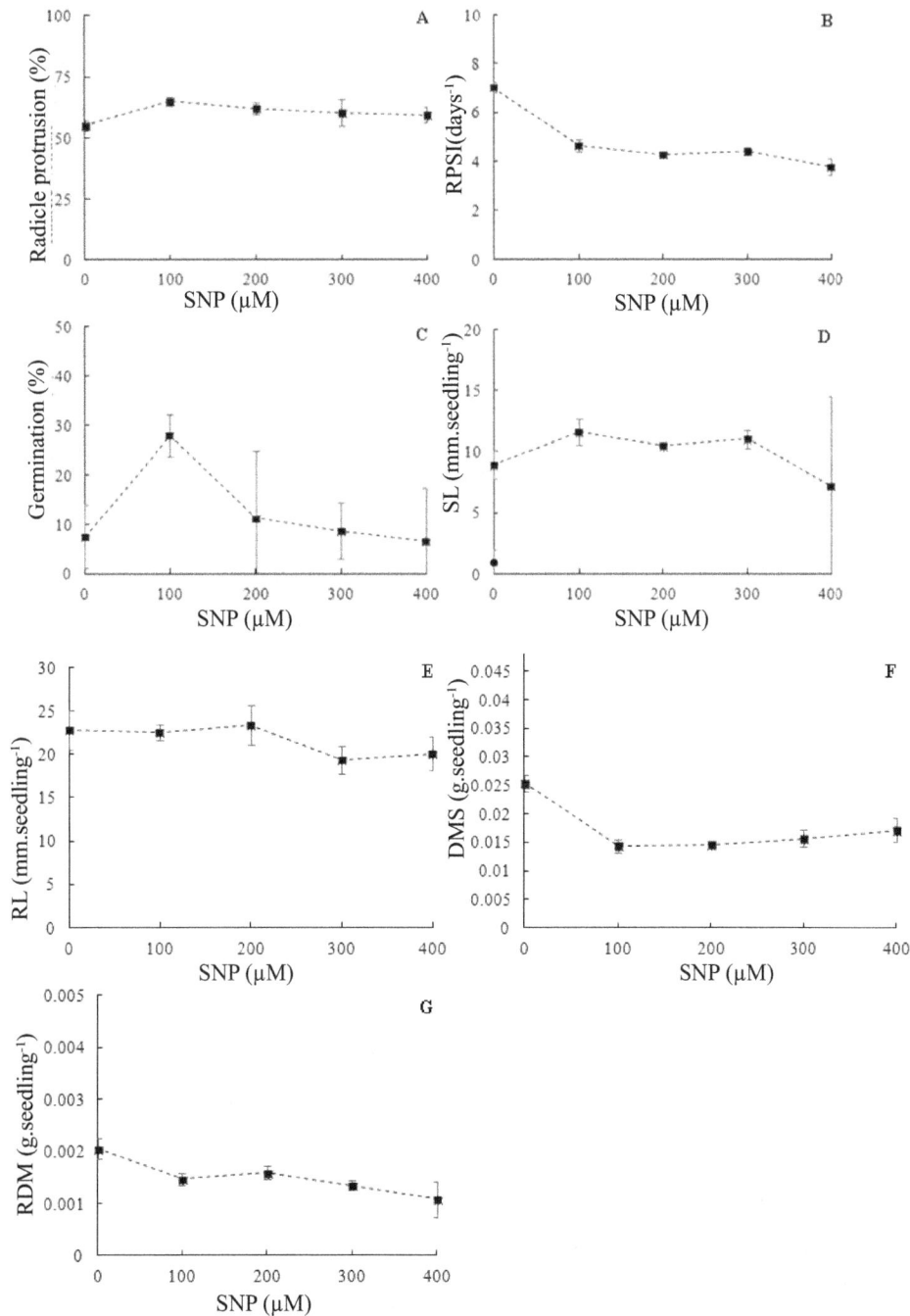

Figure 2. A: Percentage of root protrusion, B: radicle protrusion speed index (RPSI), C: percentage of germination, D: shoot length (SL), E: root length (RL), F: dry matter of shoot (DMS) and G: root dry matter (RDM), of *Senna macranthera* seeds, scarified and germinated in osmotic potential with NaCl -0.3 MPa together with different concentrations of SNP (100, 200, 300 and 400 μM) at 25 °C.

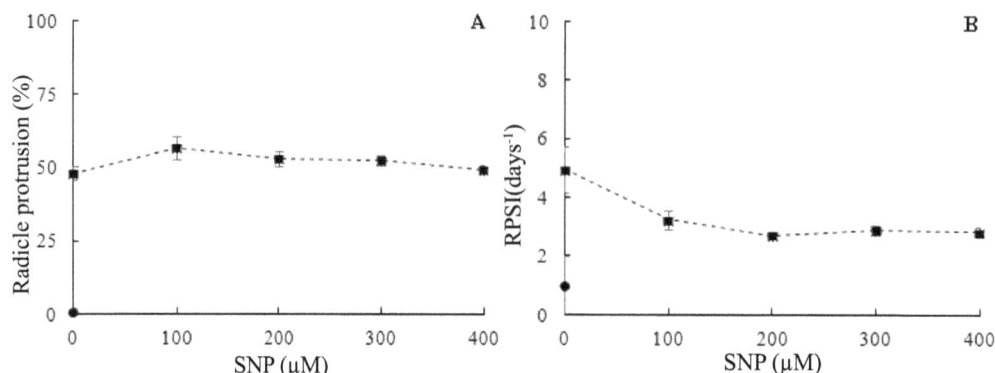

Figure 3. A: Percentage of root protrusion, B: radicle protrusion speed index (RPSI) of *Senna macranthera* seeds, scarified and germinated in osmotic potential with NaCl -0.4 MPa together with concentrations of SNP (100, 200, 300 and 400 μM) at 25 °C.

Conclusions

Salt stress is harmful to germination of *S. macranthera* seeds. SNP has the potential to promote germination recovery under salt stress more effectively when applied at a concentration of 100 μM.

Acknowledgments

We thank CAPES (Coordenação de Aperfeiçoamento de Pessoal de Nível Superior) for supporting and granting a scholarship.

References

BEWLEY, J.D.; BRADFORD, K.J.; HILHORST, H.W.M.; NONOGAKI, H. *Seeds:* Physiology of development, germination and dormancy. New York: Plenum, 2013. 392p.

BRAGA, L.F.; SOUSA, M.P.; CESARO, A.S.; LIMA, G.P.P.; GONÇALVES, A.N. Germinação de sementes de pinho-cuiabano sob deficiência hídrica com diferentes agentes osmóticos. *Scientia Forestalis*, v.36, n.78, p.157-163, 2008. http://www.ipef.br/publicacoes/scientia/nr78/cap08.pdf

BRAGA, L.F.; SOUSA, M.P.; ALMEIDA, T.A. Germinação de sementes de *Enterolobium schomburgkii* (Benth.) Benth. submetidas a estresse salino e aplicação de poliamina. *Revista Brasileira de Plantas Medicinais*, v.11, n.1, p.63-70, 2009. http://www.scielo.br/scielo.php?script=sci_arttext&pid=S1516-05722009000100011

BRASIL. Ministério da Agricultura, Pecuária e Abastecimento. *Regras para análise de sementes*. Ministério da Agricultura, Pecuária e Abastecimento. Secretaria de Defesa Agropecuária. Brasília: MAPA/ACS, 2009. 395p. http://www.agricultura.gov.br/arq_editor/file/2946_regras_analise__sementes.pdf

CAVALCANTE, A.M.B.; PEREZ, S.C.J.G.A. Efeitos dos estresses hídrico e salino sobre a germinação de sementes de *Leucaena leucocephala* (Lam.) de Witt. *Pesquisa Agropecuária Brasileira*, v.30, p.281-289, 1995. http://www.alice.cnptia.embrapa.br/bitstream/doc/104187/1/pab17fev95.pdf

CESUR, A.; TABUR, S. Chromotoxic effects of exogenous hydrogen peroxide (H_2O_2) in barley seeds exposed to salt stress. *Acta Physiology Plant*, v.33, p.705-709, 2011. http://link.springer.com/article/10.1007%2Fs11738-010-0594-7#page-1

CORREA-ARAGUNDE, N.; GRAZIANO, M.; LAMATTINA, L. Nitric oxide plays a central role in determining lateral root development in tomato. *Planta*, v.218, p.900-905, 2004. http://www.ncbi.nlm.nih.gov/pubmed/14716561

DUSSE, L.M.S.; VIEIRA, L.M.; CARVALHO, M.G. Revisão sobre óxido nítrico. *Jornal de Patologia e Medicina Laboratorial*, v.39, n.4, p.343-350, 2003. http://www.scielo.br/scielo.php?script=sci_arttext&pid=S1676-24442003000400012

EDERLI, L.; REALE, L.; MADEO, L.; FERRANTI, F.; GEHRING, C.; FORNACIARI, M.; ROMANO, B.; PASQUALINI, S. NO release by nitric oxide donors in vitro and in planta. *Plant Physiology and Biochemistry*, v.47, p.42-48, 2009. http://www.ncbi.nlm.nih.gov/pubmed/18990582

FAN, H.F.; DU, C.X.; DING, L.; XU, Y.L. Effects of nitric oxide on the germination of cucumber seeds and antioxidant enzymes under salinity stress. *Acta Physiology Plant*, v.35, p.2707-2719, 2013. http://download.springer.com/static/pdf/171/art%253A10.1007%252Fs11738-013-1303-0.pdf?originUrl=http%3A%2F%2Flink.springer.com%2Farticle%2F10.1007%2Fs11738-013-1303-0&token2=exp=1450179762~acl=%2Fstatic%2Fpdf%2F171%2Fart%25253A10.1007%25252Fs11738-013-1303-0.pdf%3ForiginUrl%3Dhttp%253A%252F%252Flink.springer.com%252Farticle%252F10.1007%252Fs11738-013-1303-0*~hmac=ac97ec09bae92db87ec4bcff025c51c74688767d1196dee4060882822baded03

FANTI, S.C.; PEREZ, S.C.J.G.A. Processo germinativo de sementes de paineira sob estresses hídrico e salino. *Pesquisa Agropecuária Brasileira*, v.39, n.9, p.903-909, 2004. http://www.scielo.br/scielo.php?pid=s0100-204x2004000900010&script=sci_arttext

FARIAS, S.G.G.; FREIRE, A.L.O.; SANTOS, D.R.; BAKKE, I.A.; SILVA, R.B. Efeitos dos estresses hídrico e salino na germinação de sementes de Gliricidia [*Gliricidia sepium* (JACQ.) STEUD.]. *Revista Caatinga*, v.22, n.4, p.152-157, 2009. http://periodicos.ufersa.edu.br/revistas/index.php/sistema/article/view/1329

FERREIRA, A.G.; BORGUETTI, F. *Germinação*: do básico ao aplicado. Porto Alegre: Artmed, 2004. 323p.

Effect of sodium nitroprusside (SNP) on the germination of Senna macranthera seeds (DC. ex Collad.) H. S. Irwin...

39

JELLER, H.; PEREZ, S.C.J.G.A. Efeitos dos estresses hídrico e salino e da ação de giberelina em sementes de *Senna spectabilis*. *Ciência Florestal*, v.11, n.1, p.93-104, 2001. http://coral.ufsm.br/cienciaflorestal/artigos/v11n1/art9v11n1.pdf

KOPYRA, M.; GWÓŹDŹ', E.A. Nitric oxide stimulates seed germination and counteracts the inhibitory effect of heavy metals and salinity on root growth of *Lupinus luteus*. *Plant Physiology Biochemistry*, v.41, p. 1011-1017, 2003. http://www.sciencedirect.com/science/article/pii/S098194280300175X

LEMOS FILHO, J.P.; GUERRA, S.T.; LOVATO, M.B.; SCOTTI, M.R.M.M.L. Germinação de sementes de *Senna macranthera*, *Senna multijuga* e *Stryphnodendron polyphyllum*. *Pesquisa Agropecuária Brasileira*, v.32, n.4, p.357-361, 1997. http://seer.sct.embrapa.br/index.php/pab/article/view/4653

LIMA, B.G.; TORRES, S.B. Estresses hídrico e salino na germinação de sementes de *Zizyphus joazeiro* Mart. (Rhamnaceae). *Revista Caatinga*, v.22, n.4, p.93-99, 2009. http://www.redalyc.org/pdf/2371/237117843016.pdf

LOMBARDO, M.C.; GRAZIANO, M.; POLACCO, J.C.; LAMATTINA, L. Nitric oxide functions as a positive regulator of root hair development. *Plant Signaling & Behavior*, v.1, p.28-33, 2006. http://www.ncbi.nlm.nih.gov/pubmed/19521473

LORENZI, H. Árvores Brasileiras: Manual de Identificação e Cultivo de Plantas Arbóreas Nativas do Brasil. 3. ed. Nova Odessa, SP: Instituto Plantarum de Estudos da Flora Ltda, 2000. 352p.

MAEKAWA, L.; ALBUQUERQUE, M.C.F.; COELHO, M.F.B. Germinação de sementes de *Aristolochia esperanzae* O. Kuntze em diferentes temperaturas e condições de luminosidade. *Revista Brasileira de Plantas Medicinais*, v.12, n.1, p.23-30, 2010. http://www.scielo.br/pdf/rbpm/v12n1/v12n1a05.pdf

MASETTO, T.E.; SCALON, S.P.Q.; REZENDE, R.K.S.; OBA, G.C.; GAMBATTI, M.; PATRÍCIO, V.S. Germinação de sementes de *Dimorphandra mollis* Benth.: efeito de salinidade e condicionamento osmótico. *Revista Brasileira de Biociências*, v. 12, n.3, p.127-131, 2014. http://www.ufrgs.br/seerbio/ojs/index.php/rbb/article/view/2736

MITTLER, R. Oxidative stress, antioxidants and stress tolerance. *Trends in Plant Science*, v.7, n. 9, p.405-410, 2002. http://www.sciencedirect.com/science/article/pii/S1360138502023129

MORAES, G.A.F.; MENEZES, N.L.; PASQUALLI, L.L. Bean seed performance under different osmotic potentials. *Ciência Rural*, v.35 n.4, p.776-780, 2005. http://www.scielo.br/scielo.php?script=sci_arttext&pid=S0103-84782005000400004

NORSWORTHY, J.K.; OLIVEIRA, M.J. *Coffee senna* (*Cassia occidentalis*) germination and emergence is affected by environmental factors and seedling depth. *Weed Science*, v.53, n.5, p. 657-662, 2005. http://www.jstor.org/stable/4047034?seq=1#page_scan_tab_contents

PEREIRA, M.R.R.; MARTINS, C.C.; MARTINS, D.; SILVA, R.J.N. Estresse hídrico induzido por soluções de PEG e de NaCl na germinação de sementes de nabiça e fedegoso. *Bioscience Journal*, v.30, n. 3, p. 687-696, 2014. http://www.seer.ufu.br/index.php/biosciencejournal/article/view/18049

REGO, S.S.; NOGUEIRA, A.C.; KUNIYOSHI, Y.S.; SANTOS, A.F. Germinação de sementes de *Blepharocalyx salicifolius* (H.B.K.) Berg. em diferentes substratos e condições de temperaturas, luz e umidade. *Revista Brasileira de Sementes*, v. 31, n.2, p. 212-220, 2009. http://www.scielo.br/scielo.php?script=sci_arttext&pid=S0101-31222009000200025

REGO, S.S.; FERREIRA, M.M.; NOGUEIRA, A.C.; GROSSI, F.; SOUSA, R.K.; BRONDANI, G.E.; ARAUJO, M.A.; SILVA, A.L.L. Estresse hídrico e salino na germinação de sementes de *Anadenanthera colubrina* (Veloso) Brenan. *Journal of Biotechnology and Biodiversity*, v.2, n.4, p.37-42, 2011. http://revista.uft.edu.br/index.php/JBB/article/view/212/146

SAEIDNEJAD, A.H.; PASANDI-POUR, A.; PAKGOHAR, N.; FARAHBAKHSH, H. Effects of exogenous nitric oxide on germination and physiological properties of basil under salinity stress. *Journal of Medicinal Plants and By-products*, v.1, p.103-113, 2013. http://www.academia.edu/9493044/Effects_of_Exogenous_Nitric_Oxide_on_Germination_and_Physiological_Properties_of_Basil_under_Salinity_Stress

SALISBURY, F.B.; ROSS, C.W. *Plant physiology*. 4.ed. Belmont: Wadsworth Publishing Company, 1992. 682p.

SIMAEI, M.; KHAVARI-NEJAD, R.A.; BERNARD, F. Exogenous application of salicylic acid and nitric oxide on the ionic contents and enzymatic activities in NaCl-stressed soybean plants. *American Journal of Plant Sciences*, v.3, n.10, p.1495-1503, 2012. http://www.scirp.org/journal/PaperInformation.aspx?PaperID=24169

SINGH, J.; SASTRY, E.V.D.; SINGH, V. Effect of salinity on tomato (*Lycopersicon esculentum* Mill.) during seed germination stage. *Physiology Molecular Biology Plants*, v.18, n.1, p.45–50, 2012. http://www.ncbi.nlm.nih.gov/pmc/articles/PMC3550529/

TANOU, G.; MOLASSIOTIS, A.; DIAMANTIDIS, G. Induction of reactive oxygen species and necrotic death-like destruction in strawberry leaves by salinity. *Environmental and Experimental Botany*, v.65, n.2-3, p.270-281, 2009. http://www.sciencedirect.com/science/article/pii/S009884720800107X

UNGAR, I.A. Halophyte seed germination. *Botany Review*, v.44, p.233-264, 1978. http://www.jstor.org/stable/pdf/4353933.pdf?acceptTC=true

ZHENG, C.; JIANG, D.; LIU, F.; DAI, T.; LIU, W.; JING, Q.; CAO, W. Exogenous nitric oxide improves seed germination in wheat against mitochondrial oxidative damage induced by high salinity. *Environmental and Experimental Botany*, v.67, p.222-227, 2009. http://www.researchgate.net/publication/260163276_Exogenous_nitric_oxide_improves_seed_germination_in_wheat_against_mitochondrial_oxidative_damage_induced_by_high_salinity._Environ_Exp_Bot

6

Viability of Brazilwood seeds (*Caesalpinia echinata* Lam.) stored at room temperature in controlled atmospheres

Nestor Martini Neto[1], Claudio José Barbedo[1*]

ABSTRACT – Seed storage at room temperature is an important and low-cost tool for *ex situ* conservation. However, the high rates of seed deterioration could reduce the potential for storage in this condition. Therefore, the knowledge of the suitable water content for this type of storage plays a critical role. This study aimed to assess the time required to stabilize the relative humidity (RH) in sealed flasks with saturated salt solutions, with or without the introduction of seeds of *Caesalpinia echinata*, as well as to assess the viability of these seeds stored in environments with different hygroscopic equilibrium. The results showed that 2 and about 12 days are needed to stabilize the RH, respectively, without or with the seeds. The amount of saturated salt solutions in this airtight environment influences both the speed to equilibrate the RH and the final values of the RH. Seeds of *Caesalpinia echinata* tolerate drying up to 5% water content (wet basis); however, the viability of these seeds at room temperature is maintained for short periods even at low water content.

Index terms: storage, hygroscopic equilibrium, desiccation tolerance.

Viabilidade de sementes de pau-brasil (*Caesalpinia echinata* Lam.) mantidas em temperatura ambiente com atmosferas controladas

RESUMO – O armazenamento de sementes em temperatura ambiente é uma importante ferramenta como estratégia de conservação *ex situ*. No entanto, as altas taxas de deterioração diminuem o potencial de armazenamento. Diante disso, a busca do teor de água ótimo para este tipo de armazenamento assume papel fundamental. Este trabalho teve por objetivo avaliar o período necessário para equilíbrio da umidade relativa do ar (UR) em frascos herméticos a partir da introdução de soluções salinas saturadas e de sementes de *Caesalpinia echinata*, bem como avaliar a capacidade de manutenção da viabilidade dessas sementes em ambientes com diferentes equilíbrios higroscópicos. Os resultados permitiram verificar que são necessários dois dias para equilíbrio da UR a partir da inclusão de soluções salinas saturadas em ambiente hermético sem a presença de sementes, e cerca de 12 dias com a presença das sementes; que a quantidade de soluções salinas presente no ambiente hermético influencia a velocidade de equilíbrio da UR e o equilíbrio ao final do período e as sementes de *Caesalpinia echinata* suportam secagem até teores de água de 5% (base úmida), mas mesmo com baixo teor de água estas sementes conservam-se por curtos períodos quando em temperatura ambiente.

Termos para indexação: armazenamento, equilíbrio higroscópico, tolerância à dessecação.

Introduction

Ex situ conservation of germplasm not only aims to exploit the economic potential of plant material, but also the preservation of endangered species, especially those with shortages of natural remnants (Pilatti et al., 2011; Barbedo et al., 2013). Among the forms of conservation, seed storage is considered safe and economically viable because, depending on the species and environmental conditions, the seeds can maintain their viability for decades, centuries or even millennia (Marcos-Filho, 2005; Ellis and Hong, 2006;

Sallon et al., 2008; Barbedo et al., 2013). However, the lack of basic knowledge about the behavior of seeds in many forest species, mainly the ones native from Brazil, hampers the use of this conservation strategy (Pilatti et al., 2011). It is known that the conservation of *Caesalpinia echinata* Lam., Brazilwood, for example, requires sub-zero temperatures (Hellmann et al., 2006), but the factors that lead to loss the viability of these seeds in less than 60 days when stored at or close to the ambient temperature are not known (Barbedo et al., 2002). This is a species that, in addition to its historical importance to Brazil and being one of the most important

[1]Instituto de Botânica, Núcleo de Pesquisa em Sementes, Caixa Postal, 68041, 04301012 – São Paulo, SP, Brasil.
*Corresponding author< claudio.barbedo@pesquisador.cnpq.br>

flora currently at risk of extinction and of its economic and pharmacological potential (Pilatti et al., 2011; Lamarca and Barbedo, 2012; Cruz-Silva et al., 2013; Gomes et al., 2014), produces seeds of great scientific interest because they are tolerant to desiccation, but with low storage capacity.

The seeds keep a constant exchange of moisture with the environment and therefore the water has great influence on their physiology. One of the major goals in seed storage studies is to control the movement of water between the seed and the environment, especially due to its participation in the deterioration processes of the seeds. The amount and energy of water in the seed are intrinsically related to the deterioration speed, activating the respiratory metabolism when in high amounts, causing deleterious reactions during excessive drying (Marcos-Filho, 2005; Barbedo et al., 2013). Seeds release or absorb water from their surrounding air, depending on the difference in vapor pressure of water in the seed and air. When the water pressure of the seed surface equals the ambient air vapor pressure, the hygroscopic equilibrium is obtained (Marcos-Filho, 2005; Carvalho and Nakagawa, 2012). Such dynamics can in many cases be controlled. For example, saturated salt solutions produce constant water vapor pressure at constant temperature (Vertucci and Roos, 1990; Sun, 2002) and are used in research aimed at analyzing the relationship between relative humidity (RH), temperature and moisture content equilibrium of the seeds (Choudhury et al., 2011; Bazin et al., 2011).

This relationship is obtained from water sorption isotherms, an important tool in the study of water relations in seeds that help find the ideal water content for the storage of each type of seed (Vertucci and Roos, 1990; Sun, 2002). Studies of this kind are primarily intended for identification of the critical point of water, i.e., one which provides the maximum storage time at a given temperature as well as the effects of different drying methods, considering different drying temperatures and rates, on the preservation of physical and physiological seed quality (Ballesteros and Walters, 2007; Buttler et al., 2009; Zhang et al., 2010). However, there are few studies that examine the dynamic equilibrium of the air inside the airtight flasks during drying and the time required for the seed to achieve hygroscopic equilibrium. It is also necessary to know whether this equilibrium depends on the species, the seed water content and the amount of saturated salt solutions in sealed flasks.

Brazilwood seeds have an orthodox behavior and can maintain viability for up to 18 months when stored at 7 °C (Barbedo et al., 2002) and up to five years at -18 °C (Mello et al., 2013), at water content of about 10% in both cases. However, studies are inconclusive with regard to the storage of these seeds at room temperature, hindering the use of

this low-cost way of storage (Lamarca and Barbedo, 2012). Therefore, knowledge of the dynamics of water between *C. echinata* seeds and their storage environment at different controlled atmospheres is of critical importance for the development of appropriate conditions for storage and thus for the preservation of the species by *ex situ* conservation. This study aimed to assess the time required for the equilibrium of RH in airtight flasks after the introduction of both saturated salt solutions and *Caesalpinia echinata* seeds (Brazilwood), as well as to assess the ability to maintain the viability of these seeds in environments with different hygroscopic equilibriums.

Material and Methods

Obtaining plant material: the seeds were obtained after natural shedding (less than 24 hours of dispersion and without rain in the 24 hours prior to harvesting), from approximately 25 mother trees, at random, in a wood of about 100 trees at Reserva Biológica e Estação Experimental de Moji-Guaçú (Biological Reserve and Research Station), in the city of Mogi-Guaçú (47°09' W, 22°15' S, altitude 610 m) and the Santa Carolina farm, in the city of Jaú (48°30' W, 22°11' S, altitude 646 m), both in the state of São Paulo, Brazil. After collection, the seeds were transported to Laboratório de Sementes do Instituto de Botânica (Seed Laboratory of the Botanical Institute) (23°37' S, 46°32' W, altitude 798 m), in the city of São Paulo, where they were processed by eliminating cracked seeds or the ones infested by insects, but maintaining the control of the origin. After this selection, the seeds were dried to approximately 10% water content - WC (wet basis), in an oven at 40 °C (Barbedo et al., 2002). They were then stored in a freezer at -18 °C (Hellmann et al., 2006; Mello et al., 2013) until the beginning of the experiments, not exceeding seven days of the date of collection.

Physical and physiological determinations: WC was gravimetrically determined by the oven method at 103 ± 2 °C for 17 hours (Brasil, 2009) and the results are shown as a percentage, on a wet basis. The water potential (WP) of the embryos and seed coats were obtained by the potentiometer WP4 (Decagon, Pullmann), that uses the chilled mirror dewpoint method (Decagon Devices, 2000).

Germination tests were carried out in germination chambers (25 ± 1 °C and $90 \pm 5\%$) in paper rolls, with assessments every three days up to 15 days (Mello and Barbedo, 2007). Seeds that issued primary root with at least 1 cm were used to calculate the percentage of germinated seeds; those resulting in the development of normal seedlings (Barbedo et al., 2002) were used to calculate the percentage of germination.

The physiological quality of the seeds was also analyzed

by the tetrazolium test, according to the methodology described by Lamarca et al. (2009). The seeds were pre-conditioned for 2 hours in water at 25 °C, then the coats were removed and the embryos were incubated in tetrazolium solution at 0.05% for two hours at 35 °C in the absence of light. Subsequently, the embryos were washed in running water, assessed and classified according to categories soft pink color (healthy tissue), deep red (deteriorating) and natural color of the tissue (dead tissue). In addition, the estimates of frequency distribution of viable, damaged and dead tissues of the seeds from each treatment were calculated (Lamarca and Barbedo, 2012).

Obtaining isotherms of water sorption in saturated salt solutions and period for equilibrium: the salts used were potassium chloride (KCl), sodium chloride (NaCl), sodium nitrite (NaNO$_2$) and calcium bromide (CaBr$_2$). To obtain the solutions, the salts in flasks were added containing deionized water until formation of the precipitate, according to the methodology described by Vertucci and Roos (1990) and Sun (2002), which were placed in an airtight environment at 25 ± 1 °C. The RH values of the flasks were obtained using a Data Logger hygrometer with Weather Station model 30.3015 (Incoterm, Porto Alegre), programed to record data every five minutes. This equipment is based on the measurement of RH and temperature, using different peripheral bases remaining in the flasks and communicating with the central base by radio waves, enabling such measurement without opening the flask (Incoterm, 2004).

For each salt, four target ratios between salt solution volume and air flask volume (hereinafter named SVAV) were established: 1:128 (or 7.8 . 10^{-3} mL of solution . mL of air^1, hereinafter mL.mL^{-1}), 1:64 (or 1.6 . 10^{-2} mL . mL^{-1}), 1:32 (3.1 . 10^{-2} mL . mL^{-1}) and 1:16 (6.2 . 10^{-2} mL . mL^{-1}). The air flask volume (1,280 mL) was calculated from the total volume of the airtight flask, by subtracting the volume occupied by the saturated salt solution and by the peripheral base of the hygrometer. To obtain those relationships were used, respectively, 9.90; 19.66; 38.73 and 75.18 mL of solution.

The RH values obtained by the hygrometer were converted to WP from the formula $\Psi = -[(\ln RH/100).R.T.\overline{V}_a^{-1}]$ (Taiz and Zeiger, 2004), where Ψ is the air WP (in MPa), R is the gas constant, T is the temperature (in Kelvin) and $\overline{V}_a$ is the molar volume of water.

Before the beginning of the records of RH, the airtight flasks remained for a week with 100 g of silica gel in beads in order to reduce the RH to values of approximately 5%. After this period, the silica gel was removed and immediately replaced by the saturated salt solutions. The solutions remained in the airtight flask until the hygroscopic equilibrium was reached, i.e., when variations larger than 5% in RH of the flask were not registered for four hours.

Periods for hygroscopic equilibrium of the air of the flasks with saturated salt solutions in the presence of seeds: in order to verify changes in the periods necessary to achieve the hygroscopic equilibrium within the flasks when adding seeds, unviable (i.e., stored for more than two years in a natural environment and with zero germination) seeds of C. echinata of Mogi-Guaçu were used to prevent any metabolic activity that might interfere with the stability of RH, which would hinder its hygroscopic equilibrium with the salt solutions. Saturated salt solutions in the ratios described above were inserted into airtight flasks and remained for 48 hours, and then 50 seeds were placed (totaling 11.0 ± 0.8 g). In order to standardize and homogenize the water content, the seeds were stored for seven days in an airtight flask with deionized water in the bottom without contact with the seeds, which guaranteed RH of 95 ± 5%. The seeds were then placed in the sealed flasks and remained in these until the new equilibrium of moisture content was achieved, considered when a variation greater than 5% in RH of the flask was not recorded, over a four hour interval. After this period, the seeds were taken and characterized for WC and WP, as described above.

Seed storage in flasks with saturated salt solutions: seeds of C. echinata of Jaú, initially with 10% WC, were incubated for 15 days at 25 °C and 12 hours of photoperiod in flasks with saturated salt solutions of zinc nitrate (Zn (NO$_3$)$_2$), chloride calcium (CaCl$_2$), calcium bromide (CaBr$_2$) and zinc chloride (ZnCl$_2$), besides blue silica gel in beads previously dried in an oven. Different saturated salt solutions from those used in previous hygroscopic equilibrium experiments were selected, because the objective was to achieve lower WP, needed for the storage of seeds. The assembly of the flasks and salt solutions was performed as described above. At the end of 15 days, samples of the seeds were taken for analysis of WC, WP and germination, as described above. The remaining seeds were transferred to new flasks with the same saturated salt solutions described above and after 30 and 60 days (totaling 45 and 75 days of storage, respectively) they were again assayed for WP, WC and germination.

Experimental design: the experimental design for all experiments was completely randomized. The experiments of periods for equilibrium of RH were carried out in two replications and the ones involving seeds were in four replications. The experiment with seed storage in salt solutions was in a 5 x 3 factorial arrangement (storage environment versus storage period). The data obtained for WC and WP (these being always considered in module), germinated seeds and germination were submitted to analysis of variance (F test) at 5% probability. Means were compared by Tukey test at 5% (Santana and Ranal, 2004).

Results and Discussion

Obtaining isotherms of water sorption in saturated salt solutions and period needed for equilibrium:

Changes in RH of the flasks, i.e., less than 5%, were very small for all salts after approximately 12 to 16 hours (Figure 1).

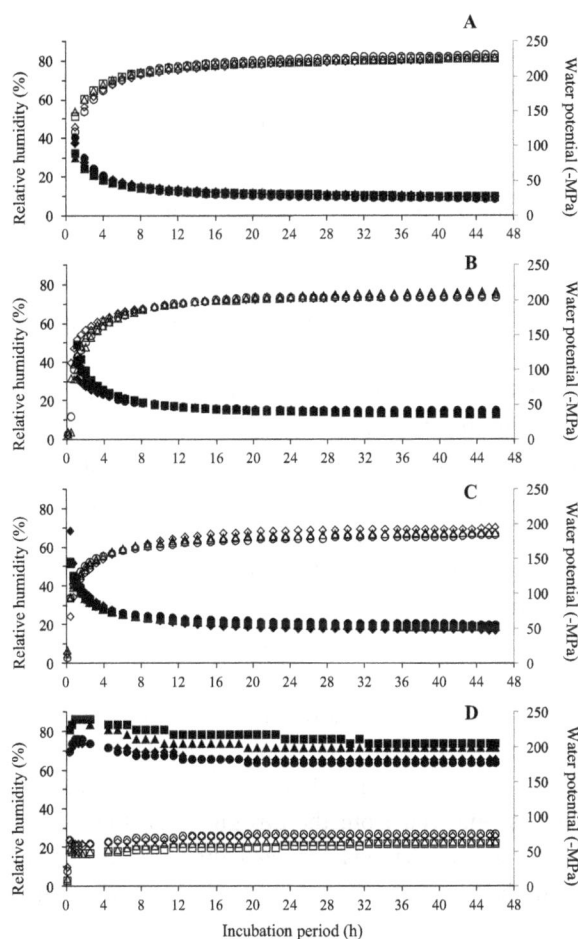

Figure 1. Relative humidity (open symbols) and water potential (filled symbols) of the sealed air chambers, empty, at 25 °C in salt solutions containing KCl (A), NaCl (B), NaNO$_2$ (C) and CaBr$_2$ (D) in the ratios of 7.8.10^{-3} (◊); 1.6.10^{-2} (□); 3.1.10^{-2} (Δ) and 6.2.10^{-2} (o), over 48 hours of incubation.

The hygroscopic equilibrium was achieved when the RH was about 82% for KCl (Figure 1A), 75% for NaCl (Figure 1B), 68% for NaNO$_2$ (Figure 1C), and 25% for CaBr$_2$ (Figure 1D). The first two values are within the range described by Sun (2002) for 25 °C, namely respectively 84 ± 2% and 75 ± 2% and the value of NaNO$_2$ was very close to that described by the author (64 ± 2%). However, the value recorded for CaBr$_2$ was above the one described by that author (16 ± 2%). WP, in turn, ranged from

approximately -200 MPa in CaBr$_2$ to -25 MPa in KCl (Figure 1). Analyzing the RH, no differences were observed in the periods to achieve the moisture content equilibrium due to the variation of SVAV. However, when WP was analyzed, there was a slight variation in this period just for the CaBr$_2$ solution (Figure 1D).

The seeds of *C. echinata*, when introduced into the flasks, contained approximately 32% WC, corresponding to approximately -15 MPa. Immediately after the introduction of these seeds there was an increase in RH values, indicating that water was easily moved from the seeds to the air of the flasks with salt solutions (Figure 2). This resulted in reduction in WC of the seeds, which was the greater as was lower the initial RH of the flasks, reaching 6.3% (< -100 MPa) in the flasks with CaBr$_2$ with a SVAV ratio of 6.2.10^{-2} mL.mL^{-1} (Table 1).

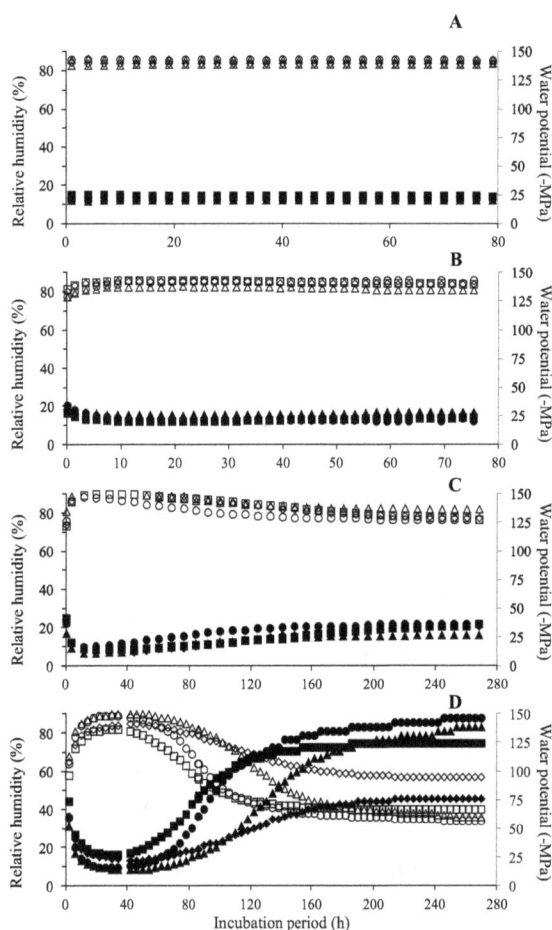

Figure 2. Relative humidity (open symbols) and water potential (filled symbols) of the airtight air chambers at 25 °C containing salt solutions of KCl (A), NaCl (B), NaNO$_2$ (C) and CaBr$_2$ (D) at the ratios of 7.8.10^{-3} (◊); 1.6.10^{-2} (□); 3.1.10^{-2} (Δ) and 6.2.10^{-2} (o), over 48 hours of incubation after addition of 50 seeds of *Caesalpinia echinata* with 32% of water (-15 MPa).

Table 1. Seed water content and water potential (-MPa) of embryo and seed coat of *Caesalpinia echinata* Lam. seeds after moisture equilibrium in an atmosphere controlled by saturated salt solutions.

Salt solution	Final water content (%)	Water potential (MPa)	
		Seed coat	Embryo
KCl	16.5 a*	-27.6 a	-34.3 a
NaCl	12.7 b	-44.5 a	-50.0 a
NaNO$_2$	10.7 c	-67.5 b	-75.0 b
CaBr$_2$ 7.8 . 10^{-3}	7.5 d	-91.5 c	-97.3 c
CaBr$_2$ 1.6 . 10^{-2}	6.5 e	-116.8 d	-131.7 d
CaBr$_2$ 3.1 . 10^{-2}	6.3 e	-128.3 d	-136.5 d
CaBr$_2$ 6.2 . 10^{-2}	6.5 e	-112.1 d	-145.8 d

*Means followed by the same letter in columns do not differ by Tukey test at 5%.

In the flasks with KCl solution, in which RH was already high (82%, Figure 1A), the introduction of the seeds practically did not change these values, rising about 4% and stabilizing at 84-85% at the end, or 270 minutes after the start of the experiment (Figure 2A), although the WC of the seeds has dropped by half (Table 1). In the flasks with NaCl solution, with RH 75% (Figure 1B), this one rose to 86% within the first 10 hours, and from this point on it showed a slight reduction, stabilizing at around 84%, i.e., not returning to the initial RH (75%, Figure 1B) even after over 70 hours of incubation (Figure 2B). The RH of the flasks with NaNO$_2$ solution, initially at 68%, increased to 92% in the first 15 hours following the addition of the seeds and then had a progressive reduction but, like in the flasks with the NaCl, it did not return to the initial values, even after over 260 hours, stabilizing at about 80% (Figure 2C).

In the flasks with CaBr$_2$ solution the greatest changes were observed. Initially with RH of 25%, after adding seeds, the RH was increased to around 80% in the first 36 hours. From this point on, the RH progressively decreased but, similar to what occurred in the flasks with the other solutions, it also did not reach the initial values, even after 270 hours (Figure 2D). However, unlike what was observed in the other flasks, the ones containing CaBr$_2$ solution had different RH stabilization values, depending on the SVAV relationship, and the less concentrated ones stabilized at higher RH. In SVAV of 7.8 . 10^{-3} mL.mL^{-1}, e.g., stabilization occurred at 57% RH, while in SVAV of 6.2 . 10^{-2} mL.mL^{-1}, RH stabilized at 34%. These differences followed the same pattern of differences observed in the final values of WC of the seeds between the lower SVAV ratio (7.8 . 10^{-3} mL.mL^{-1}) and the others, as well as in the WP of embryos and coats (Table 1).

The changes in RH with the addition of seeds were accompanied by changes in WP of the air. The smaller the difference between the WP of the air in the flasks and of the seeds added to these flasks, the less variation in both WP and RH (Figure 2). Thus, in the flasks with KCl

solution, in which the difference between these potentials was only 10 MPa (-15 MPa of the seeds, -25 MPa in the air of the flasks), there was practically no change in RH; in those ones containing NaCl, the difference was 25 MPa (-40 MPa in the flasks), which promoted an increase of 20 MPa in the air (reaching -20 MPa); in the ones containing NaNO$_2$, with a difference of 40 MPa (-55 MPa in the flasks), the increase was 40 MPa (reaching -15 MPa); and in the ones containing CaBr$_2$, the increase was about ten-fold, reaching values between -15 and -30 MPa, depending on the SVAV relationship.

Sorption isotherms have been used for the analysis of seed water relations subjected to environments with saturated salt solutions and, in this situation, moisture content equilibrium is often defined from gravimetric assessments (Vertucci and Roos, 1990; Ballesteros and Walters, 2007; Gazor, 2010). However, usually, the sample is removed from the airtight flask for weighing, allowing gas exchanges between the air present in the airtight flask and the external atmosphere. The internal RH, previously kept in equilibrium by means of saturated salt solutions, undergoes changes and goes back into the equilibrium process after closing the chamber. In this study it was clearly shown the importance of knowing the difference between the mass of the water content of seeds to be stored (or to be subjected to drying by the use of saturated salt solutions) and the initial and/or desired RH of the storage container, as well as the difference between their WP. A small amount of seeds (10 g in 1 liter of air) caused substantial changes in the moisture content equilibrium of the system.

Seed storage in flasks with saturated salt solutions

When comparing the literature information (Sun, 2002) regarding RH of equilibrium of the saturated salt solutions used with the RH values recorded in the flasks in which the seeds were stored, it appears that there was a slight increase during the 15 days of drying, probably due to the release of water from the seeds to the environment. In the flasks with $Zn(NO_3)_2$, for example, RH increased from 35% to 50%; with $CaCl_2$, from 29% to 35%; with $CaBr_2$, from 16% to 25%. Only with $ZnCl_2$ solution, RH hardly changed, remaining at around 5%. The RH of the flasks with silica gel beads approached zero, preventing a better measure by the hygrometer.

There was no significant interaction for the water content of the seeds between the different flasks and storage periods (Table 2). This content showed slight increase from 15 to 45 days, but did not increase thereafter. Among the different flasks with salt solution, as expected, the seed water content was lower for the lower RH, reaching values of less than 6% in silica and in

$ZnCl_2$. However, even in these concentrations the germination was greater than 30% (Table 3), confirming the desiccation tolerance in seeds of C. echinata (Barbedo et al., 2002). Also confirmed was the low storage capacity of the seeds of this species in a non-refrigerated environment (Mello et al., 2013) because even with very low WC there was reduction in the germination and in the values of germinated seeds; the longer the period of storage, the lower the germinability (Table 3). After 75 days of storage almost all seeds lost the ability to produce normal seedlings and start the growth of the primary root. Importantly, the sharpest losses in the germination of seeds of C. echinata occurred in the flasks with $Zn(NO_3)_2$, i.e., the wettest. Therefore, even when keeping for a short time at non-chilled temperatures, the need to keep these seeds with water content below 10% was evident.

Table 2. Water content of *Caesalpinia echinata* Lam. seeds incubated in airtight flasks with saturated salt solutions, stored for 45 and 75 days.

Period of storage	Salt solutions					Average
	$Zn (NO_3)_2$	$CaCl_2$	$CaBr_2$	Silica gel	$ZnCl_2$	
0 days (initial)	9.95	6.98	6.19	5.62	4.78	6.70 b*
45 days	11.84	7.42	6.75	5.77	5.06	7.37 a
75 days	12.40	7.26	6.50	5.51	6.08	7.55 a
Average	11.40 A	7.22 B	6.48 BC	5.63 CD	5.30 D	

*Means followed by the same letter (lowercase in columns, uppercase in rows) do not differ by Tukey test at 5%.

Table 3. Germination and germinated seeds (%) of *Caesalpinia echinata* stored for 15, 45 or 75 days at 25 °C in airtight flasks containing silica gel or saturated salt solutions.

Content in the flasks	Storage periods					
	Germinated seeds			Germination		
	15 days	45 days	75 days	15 days	45 days	75 days
(initial values)		(90)			(50)	
$Zn (NO_3)_2$	65 aA*	22 bB	0 aC	17 bA	0 aB	0 aB
$CaCl_2$	50 abA	55 aA	8 aB	35 aA	0 aB	3 aB
$CaBr_2$	30 cAB	45 abA	15 aB	8 cA	0 aA	0 aA
Silica gel	41 bcA	38 abA	5 aB	30 aA	2 aB	0 aB
$ZnCl_2$	45 abcA	50 aA	20 aB	25 aA	0 aB	0 aB

*Means followed by the same letter (lowercase comparing storage periods, uppercase letters comparing storage environment) do not differ by Tukey test at 5%.

Analyzing the results obtained from the distribution of frequencies among viable, damaged and dead tissues, identified by the tetrazolium test, it was found that the ratios do not change much, both among the drying treatments and among the storage periods (Figure 3). This fact had already been pointed out by Lamarca and Barbedo (2012), who observed that it is possible that some non-viable seeds of C. echinata still show a large amount of viable tissues. However, the tetrazolium test identified the most severely affected tissues, both during storage and during drying. It was

observed that the seeds with the highest WC (in a solution of $Zn (NO_3)_2$) showed, after the drying period, damage in the area of the root-hypocotyl axis, but still had viable tissues in the insertion areas of the cotyledons to the axis and in the proximal and distal areas of the cotyledon. After the first 45 days of storage, the seeds already exhibited the large majority of tissues in a deteriorating state (Figure 4D) and after 75 days damage throughout the root hypocotyl area, besides the area of axis insertion to the cotyledon, plumule and the proximal area of the cotyledon (Figure 4H). The seeds subjected to the most severe drying (CaCl$_2$, CaBr$_2$ and ZnCl$_2$) showed slower deterioration of the tissues and it was possible to see damage primarily in the areas of the procambium, proceeding toward the fundamental meristem, the area of insertion of the cotyledons to the axis, plumule, and proximal and distal areas (Figures 4I-L). The damage done to both the meristem and the plumule has not always prevented seed germination ; however, it is probably responsible for the production of abnormal seedlings (Figures 4 O-P). After 75 days of storage, however, the damage could be seen throughout the insertion area of the cotyledon to the axis, and, in this case, the association with the lower germination percentage was possible (Figures 4 H and L and Table 3) .

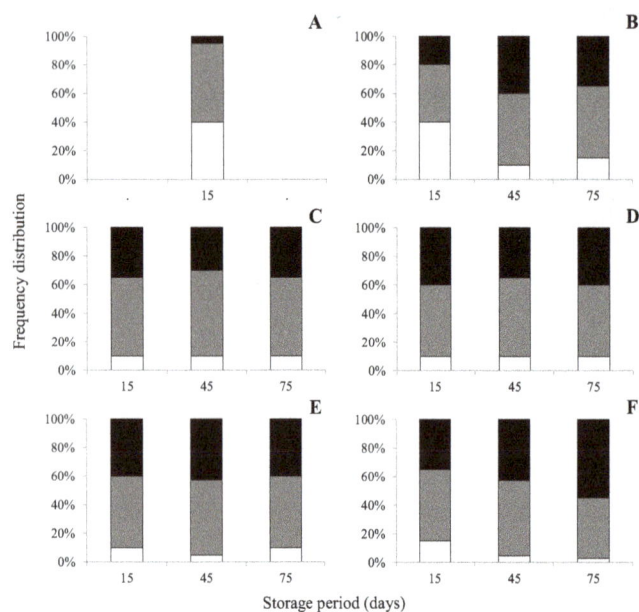

Figure 3. Frequency distribution of viable (white), deteriorated (gray) and dead (black) tissued obtained in the tetrazolium test of seeds of *Caesalpinia echinata* Lam. in control (A) or stored for 15, 45 and 75 days in a sealed environment in a controlled atmosphere by saturated salt solutions of Zn(NO$_3$)$_2$ (B), CaCl$_2$ (C), CaBr$_2$ (D), by silica gel beads (E) and ZnCl$_2$ (F).

Thus, the results regarding tolerance to desiccation of seeds of *C. echinata* corroborate the results obtained by Hellmann et al. (2006) when storing the seeds of *C. echinata* with approximately 10% moisture content (wet basis) and suggesting that the water contents of less than this, such as from 3% to 7%, typically used in gene banks, could further increase the longevity of seeds at a temperature of -18 °C. Finally, seeds of *C. echinata* still have a low period of longevity at room temperature. However, the drying does

not appear to be responsible for the deleterious effects on *C. echinata* seeds.

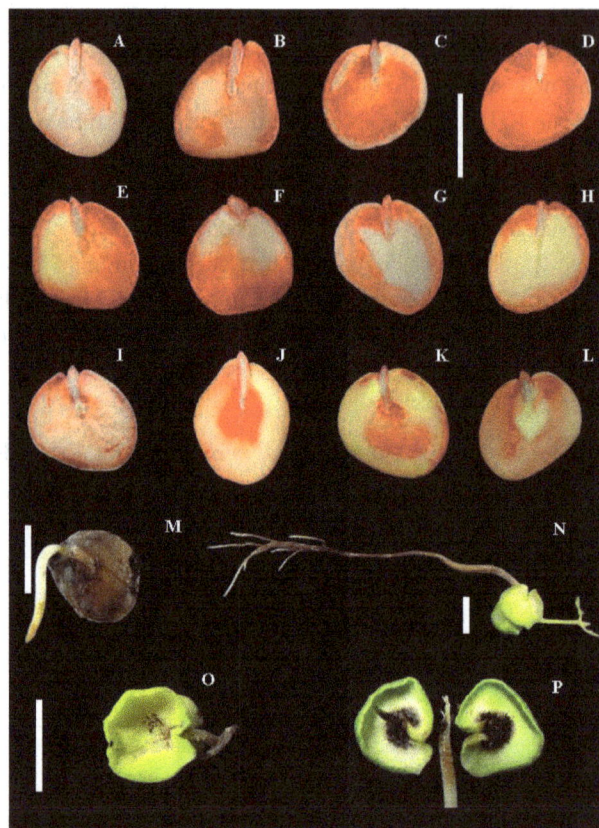

Figure 4. Classes of seeds of *C. echinata* in the tetrazolium test (A to L) and types of development in the germination test (M to P). A: viable and vigorous seed; B: seed with tissue deterioration in the distal area of the cotyledon; C: seed with deteriorating tissues throughout the proximal area of the cotyledon; D: seed with deteriorating tissues in all the cotyledon and dead tissues in the insertion area of the cotyledons to the axis; E and F: seeds with dead tissues in the proximal area of the cotyledon; G and H: seeds with dead tissues throughout the proximal area of the cotyledon; I: seed with dead tissues in the insertion area of the axis to the cotyledon; J and K: seeds with dead tissues in the distal area of the cotyledon; L: seed with dead tissues in the proximal area of the cotyledon; M: seed with primary root protrusion; N: normal seedling development; O and P: development of abnormal seedlings. Scale of 1 cm.

Conclusions

It takes 48 hours for equilibrium of the relative humidity from the inclusion of saturated salt solutions in an airtight environment without the presence of seeds, and about 12 days in the presence of seeds. The amount of salt solutions present in an airtight environment influences not only the relative humidity of the equilibrium velocity as well as the equilibrium at the end of this period.

Caesalpinia echinata seeds tolerate drying up to a water content of 5% (wet basis), but even with low water content, in the salt solutions that most seeds remained viable ($CaCl_2$ and $ZnCl_2$), these seeds maintain viability for short periods when at room temperature.

Acknowledgments

To the workers at Reserva Ecológica and Estação Experimental de Mogi Guaçu and Fazenda Santa Carolina for their assistance in the collection of Brazilwood fruits and seeds; to Instituto de Botânica and Fazenda Santa Carolina for permission to these collections; to Conselho Nacional de Desenvolvimento Científico e Tecnológico (CNPq; National Counsel of Technological and Scientific Development) for granting scholarships for master's degree (N. Martini Neto) and research productivity (C.J. Barbedo).

References

BALLESTEROS, D.; WALTERS, C. Water properties in fern spores: sorption characteristics relating to water affinity, glassy states, and storage stability. *Journal of Experimental Botany*, v.58, p.1185-1196, 2007. http://jxb.oxfordjournals.org/content/58/5/1185.full.pdf+html

BARBEDO, C.J.; BILIA, D.A.C.; FIGUEIREDO-RIBEIRO, R.C.L. Tolerância à dessecação e armazenamento de sementes de *Caesalpinia echinata* Lam. (pau-brasil), espécie da Mata Atlântica. *Revista Brasileira de Botânica*, v.25, p.431-439, 2002. http://www.scielo.br/pdf/rbb/v25n4/a07v25n4.pdf

BARBEDO, C.J.; CENTENO, D.C.; FIGUEIREDO-RIBEIRO, R.C.L. Do recalcitrant seeds really exist? *Hoehnea*, v.40, p.583-593, 2013. http://www.scielo.br/pdf/hoehnea/v40n4/01.pdf

BAZIN, J.; BATLLA, D.; DUSSERT, S.; MAAROUF-BOUTEAU, H.; BAILLY, C. Role of relative humidity, temperature, and water status in dormancy alleviation of sunflower seeds during dry after-ripening. *Journal of Experimental Botany*, v.62, p.627-640, 2011. http://jxb.oxfordjournals.org/content/62/2/627.full.pdf+html

BRASIL. Ministério da Agricultura, Pecuária e Abastecimento. *Regras para análise de sementes*. Ministério da Agricultura, Pecuária e Abastecimento. Secretaria de Defesa Agropecuária. Brasília: MAPA/ACS, 2009. 395p. http://www.agricultura.gov.br/arq_editor/file/2946_regras_analise__sementes.pdf.

BUTTLER, L.H.; HAY, F.R.; ELLIS, R.H.; SMITH, R.D. Post abscission, predispersal seeds of *Digitalis purpurea* remain in a developmental state that is not terminated by dessication *ex plant*. *Annals of Botany*, v.103, p.785-794, 2009. http://aob.oxfordjournals.org/content/early/2009/01/09/aob.mcn254.full.pdf+html

CARVALHO, N.M.; NAKAGAWA, J. *Sementes*: ciência, tecnologia e produção. 5.ed. Jaboticabal: FUNEP, 2012. 588p.

CHOUDHURY, D.; SAHU, J.K.; SHARMA, G.D. Moisture sorption isotherms, heat of sorption and properties of sorbed water of raw bamboo (*Dendrocalamus longispathus*) shoots. *Industrial Crops and Products*, v.33, p.211-216, 2011. http://www.sciencedirect.com/science/article/pii/S0926669010002591

CRUZ-SILVA, I.; NEUHOF, C.; GOZZO, A.J.; NUNES, V.A.; HIRATA, I.Y.; SAMPAIO, M.U.; FIGUEIREDO-RIBEIRO, R.C.L.; NEUHOF, H.; ARAÚJO, M.S. Using a *Caesalpinia echinata* Lam. protease inhibitor as a tool for studying the roles of neutrophil elastase, cathepsin G and proteinase 3 in pulmonary edema. *Phytochemistry*, v.96, p.235-243, 2013. http://www.sciencedirect.com/science/article/pii/S0031942213003932#

DECAGON DEVICES, I. *Dewpoint PotentiaMeter – Operator's manual*. Pullman: Decagon, 2000. 78p.

ELLIS, R.H.; HONG, T.D. Temperature sensitivity of the low-moisture-content limit to negative seed longevity-moisture content relationships in hermetic storage. *Annals of Botany*, v.97, p.785-791, 2006. http://aob.oxfordjournals.org/content/97/5/785.full.pdf+html

GAZOR, H.R. Moisture isotherms and heat of desorption of canola. *Agricultural Engineering International: the CIGR Ejournal*, v.12, manuscript 1440, 2010. http://www.cigrjournal.org/index.php/Ejounral/article/view/1440/1296

GOMES, E.C.B.S.; JIMENEZ, G.C.; SILVA, L.C.N.; SÁ, F.B.; SOUZA, K.P.C.; PAIVA, G.S.; SOUZA, I.A. Evaluation of antioxidant and atiangiogenic properties of *Caesalpinia echianata* extracts. *Journal of Cancer*, v.5, p.143-150, 2014. http://www.jcancer.org/v05p0143.pdf

HELLMANN, M.E.; MELLO, J.I.O.; FIGUEIREDO-RIBEIRO, R.C.L.; BARBEDO, C.J. Tolerância ao congelamento de sementes de pau-brasil (*Caesalpinia echinata* Lam.) influenciada pelo teor de água. *Revista Brasileira de Botânica*, v.29, p.91-99, 2006. http://www.scielo.br/pdf/rbb/v29n1/a09v29n1.pdf

INCOTERM. *Estação meteorológica controlada por sinais de rádio mod. 30.3015. Manual de instruções*. Porto Alegre: Incoterm, 2004. 24p.

LAMARCA, E.V.; BARBEDO, C.J. Short storability of *Caesalpinia echinata* Lam. seeds as a consequence of oxidative processes. *Hoehnea*, v.39, p.577-586, 2012. http://www.scielo.br/pdf/hoehnea/v39n4/06.pdf

LAMARCA, E.V.; LEDUC, S.N.M.; BARBEDO, C.J. Viabilidade e vigor de sementes de *Caesalpinia echinata* Lam. (pau-brasil – Leguminosae) pelo teste de tetrazólio. *Revista Brasileira de Botânica*, v.32, p.793-803, 2009. http://www.scielo.br/pdf/rbb/v32n4/a17v32n4.pdf

MARCOS-FILHO, J. *Fisiologia de sementes de plantas cultivadas*. Piracicaba: FEALQ, 2005. 495p.

MELLO, J.I.O.; FIGUEIREDO-RIBEIRO, R.C.L.; BARBEDO, C.J. Sub-zero temperature enables storage of seeds of *Caesalpinia echinata* Lam. *Journal of Seed Science*, v.35, p.519-523, 2013. http://www.scielo.br/pdf/jss/v35n4/14.pdf

MELLO, J.I.O.; BARBEDO, C.J. Temperatura, luz e substrato para a germinação de sementes de pau-brasil *Caesalpinia echinata* Lam., Leguminosae-Caesalpinoidae. *Revista Árvore*, v.31, p.645-655, 2007. http://www.scielo.br/pdf/rarv/v31n4/09.pdf

PILATTI, F.K.; AGUIAR, T.; SIMÕES, T.; BENSON, E.E.; VIANA, A.M. In vitro and cryogenic preservation of plant biodiversity in Brazil. *In Vitro Cellular and Developmental Biology – Plant*, v.47, p.92-98, 2011. http://link. springer.com/article/10.1007%2Fs11627-010-9302-y#page-1

SALLON, S; SOLOEW, E.; COHEN, Y.; KORCHINSKY, R.; EGLI, M.; WOODHATCH, I.; SIMCHONI, O.; KISLEV, M. Germination, genetics, and growth of an ancient date seed. *Science*, v.320, p.1464, 2008. http://www. sciencemag.org/content/320/5882/1464.full.pdf

SANTANA, D.G.; RANAL, M.A. *Análise da germinação:* um enfoque estatístico. Universidade de Brasília: Brasília, 2004. 248p.

SUN, W.Q. Methods for the study of water relations under dessication stress. In: BLACK, M.; PRITCHARD, H.W. (Eds.) *Dessication and survival in plants:* drying without dying. Wallingford/New York: CABI Publishing, 2002. 84p.

TAIZ, L.; ZEIGER, E. *Fisiologia vegetal*. 3.ed. Porto Alegre: Artmed, 2004. 613p.

VERTUCCI, C.W.; ROOS, E.E. Theoretical basis of protocols for seed storage. *Plant Physiology*, v.94, p.1019-1023, 1990. http://www.plantphysiol. org/content/94/3/1019.full.pdf+html

ZHANG, M.; ZHUO, J.; WANG, X.; WU, S.; WANG, X. Optimizing seed water content: relevance to storage stability and molecular mobility. *Journal of Integrative Plant Biology*, v.52, p.324-331, 2010. http://www.jipb.net/ Abstract.aspx?id=5666

7

Physiological potential of stylosanthes cv. Campo Grande seeds coated with different materials

Priscilla Brites Xavier[1*], Henrique Duarte Vieira[1], Cynthia Pires Guimarães[1]

ABSTRACT – The aim of this study was to assess the effect of different coatings on the physiological potential of stylosanthes cv. Campo Grande seeds. The treatments were: uncoated seeds; limestone + PVA glue; limestone + sand + PVA glue; limestone + activated carbon + PVA glue; calcium silicate + PVA glue; calcium silicate + sand + PVA glue; calcium silicate + activated carbon + PVA glue. Posteriorly, the seeds were analyzed for water content (WC), maximum diameter (MAD) and minimum diameter (MID), thousand seed weight (TSW), germination test, germination speed index (GSI), mean germination time (MGT), emergence, emergence speed index (ESI), mean emergence time (MET), shoot and root length, fresh and dry matter of shoot and root. The coating increased the TSW, MAD and MID and decreased its WC. The treatments comprising limestone + PVA glue and limestone + sand + PVA glue increased the germination time, but none of the treatments negatively affected the physiological seed quality. Treatment with calcium silicate + PVA glue was outstanding for germination speed index and fresh and dry matter of shoot and root in the stylosanthes cv. Campo Grande seeds coating.

Index terms: seed coating, seed coating machine, germination, vigor, *Stylosanthes capitata/macrocephala*.

Potencial fisiológico de sementes de estilosantes cv. Campo Grande recobertas com diferentes materiais

RESUMO – Objetivou-se com este trabalho avaliar o efeito de diferentes recobrimentos sobre o potencial fisiológico de sementes de estilosantes cv. Campo Grande. Os tratamentos foram: sementes não recobertas; calcário + cola PVA; calcário + areia + cola PVA; calcário + carvão vegetal + cola PVA; silicato de cálcio + cola PVA; silicato de cálcio + areia + cola PVA; silicato de cálcio + carvão vegetal + cola PVA. Posteriormente, as sementes foram avaliadas quanto ao teor de água (TA), diâmetro máximo (DMA) e mínimo (DMI), peso de mil sementes (PMS), teste de germinação, índice de velocidade de germinação, tempo médio de germinação, emergência, índice de velocidade de emergência, tempo médio de emergência, comprimento da parte aérea e raiz, massa fresca e seca da parte aérea e raiz. Os tratamentos constituídos por calcário + cola PVA e calcário + areia + cola PVA aumentaram o tempo de germinação. Nenhum dos tratamentos prejudicou a qualidade fisiológica das sementes. O tratamento com silicato + cola PVA se destacou para as variáveis, índice de velocidade de germinação, massa fresca e seca da parte aérea e de raiz no recobrimento de sementes de estilosantes cv. Campo Grande.

Termos para indexação: recobrimento de sementes, drageadora, germinação, vigor, *Stylosanthes*.

Introduction

The Brazilian cattle industry is characterized by having most of its flock raised on pasture; this practice is an economical way to produce and provide food for animals (Ferraz and Felício, 2010). Therefore, the role of fodder plants in this production system is extremely important, both as an income as to its sustainability.

Tropical poaceae generally have lower nutritional quality compared to the temperate ones. Thus, the introduction of tropical Fabaceae (or Leguminosae, commonly known as the legume, pea, or bean family), adapted to the pasture raising system, contributes to the solution of problems of low level of nitrogen in soils and the reduced protein quality available to the animals (Shelton et al., 2005).

Stylosanthes cv. (cultivated variety or cultivar) Campo Grande is a fodder Fabaceae, launched by EMBRAPA (Empresa Brasileira de Pesquisa Agropecuária; Brazilian Corporation of Agricultural Research) in 2007, comprising the physical mixture of improved seeds of *Stylosanthes*

[1]Universidade Estadual do Norte Fluminense Darcy Ribeiro, Setor de Tecnologia de Sementes, 28013-602 – Campos dos Goytacazes, RJ, Brasil.
*Corresponding author <pri_brites@yahoo.com.br>

capitata and *Stylosanthes macrocephala*, at the ratio of 80% e 20%, respectively (EMBRAPA Gado de Corte, 2007), and it is promising in providing good quality fodder. This legume is further characterized by having very small seeds (350-400 seeds/g) (EMBRAPA Gado de Corte, 2007), with color ranging from red to brown and black, features that hinder both its tillage as its sowing. In this context, the use of technologies such as coating the seeds, contributes to the solution of problems related to the size and shape of seeds, as it standardizes the seed size and shape, facilitating its distribution, whether manual or mechanical, providing greater precision in sowing (Baudet and Peres, 2004; Nascimento et al., 2009).

The seed coating procedure consists of depositing a dried, inert, fine-grained matter, called filler, and a binder materials, also called adhesive, to the surface of the seeds (Silva and Nakagawa, 1998a). Materials used as cementing must have affinity with the other ingredients; be readily soluble in water; act at low concentrations; become dry and not sticky when dehydrated; form a low viscous solution when rehydrated; and be non-hygroscopic, corrosive nor toxic (Nascimento et al., 2009). Moreover, the filler material should be preferably formed by spherical, uniform beads, with sizes between 0.100 and 0.200 mm, non-hygroscopic, without surface tension, non-hydrophilic, non-corrosive, non-toxic, sterile, must not be an environment for the reproduction of microorganisms, must be insoluble in water or weak acids, with a density of about 1 and must be easy to purchase, with compatible costs (Lopes and Nascimento, 2012).

Among the most studied fillers there is microcellulose (Microcrystalline cellulose (MCC)), sand, dolomite, kaolin, activated carbon, vermiculite, maize flour, wheat flour, tapioca starch, maize starch, Celite® and diatomaceous earth (Silva et al., 2002; Oliveira et al., 2003a; Mendonça et al., 2007; Nascimento et al., 2009; Pereira et al., 2011). The materials used as adhesives are generally organic polymers, starches, natural resins, sugars, animal glue and vegetable mucilages, which are dispersed in water to produce a fluid that can be sprayed (Baudet and Peres, 2004). While improving the morphological characteristics of the seeds, the main obstacle to the use of coated seed is the delayed germination and emergence. Several studies indicate the occurrence of this delay in relation to uncoated seeds, as noted in coated seeds of lettuce (Silva et al., 2002), bell pepper (Oliveira et al., 2003a), supersweet maize (Mendonça et al., 2007), maize (Conceição and Vieira, 2008) and carrot (Nascimento et al., 2009). This delay is due to the arrangement of fine particles and the filling of the pores of the pellet by the cementing agent and water supplied during the procedure, which forms a barrier to the gaseous exchange between the seed and the environment external to the pellet (Silva and Nakagawa, 1998a). However, despite this delay, the final germination rates are similar to those of the uncoated seeds (Silva et al., 2002; Lopes and Nascimento, 2012; Pereira et al., 2011).

With regard to seeds of fodder legumes such as stylosanthes, there is still a scarcity of studies on the coating of these seeds. Considering the size of stylosanthes seeds, this fodder seems promising in the study of seed coating. Thus, the aim of this study was to assess the effect of different types of coating materials on the physiological potential of stylosanthes cv. Campo Grande seeds.

Material and Methods

Commercial seeds of *Stylosanthes capitata/macrocephala* cv. Campo Grande I and II were used, previously submitted to mechanical chisel plow with sandpaper number 100.

As fillers were used: dolomitic limestone (0.25 mm), calcium silicate (< 0.20 mm), sand (0.25 mm) and activated carbon powder. And as a cementing material was used a solution of water and extra Cascorez glue based on Polyvinyl acetate (PVA).

The seed coating treatments were as follows: US – uncoated seeds; L + PVA (dolomitic limestone + PVA glue); L + S + PVA (dolomitic limestone + sand + PVA glue); L + AC + PVA (dolomitic limestone + activated carbon + PVA glue); CS + PVA (calcium silicate + PVA glue); CS + S + PVA glue (calcium silicate + sand + PVA); CS + AC + PVA (calcium silicate + activated carbon + PVA glue).

The ratios of filler and seeds used were: limestone 3:1 (w/w), calcium silicate 3:1 (w/w), sand 1:1 (w/w) and activated carbon 0.08:1 (w/w). The PVA-based glue was diluted with water, preheated to 70 °C (Mendonça et al., 2007), at the ratio of 1:1 (v/v) for use as a binder solution. For the application of the filler materials to be done into layers, the quantities of limestone, calcium silicate and sand were divided into portions of 12.5 g each, while the activated carbon was divided into portions of 2 g each.

For the coating procedure, a bench seed coating machine was used, model N10 Newpack, equipped with a stainless steel vat, nozzle for application of the cement solution powered by compressed air at a pressure of 4 bars, a hot air blower and a timer to regulate the length of time of the spray and the blower. The settings used in the procedure were as follows: vat speed of 90 rpm, cementing solution spray duration time of 1 second, air blower temperature of 40 °C and duration time of the connected blower of 1 minute.

One hundred grams of seeds, limestone or calcium silicate, depending on the type of coating used, were placed inside the seed coating machine vat together with a portion of filler. Then, the binding solution spray was powered three consecutive times and a portion of the filler material was added again on the seed mass, followed by another application of cement solution. Subsequently, the air blower was activated for 1 minute. This procedure corresponded to a coating layer, and was carried out until the amount of filler ended. For the treatments to which were added more than one filler material, in the case of sand and activated carbon, they were added in the same way, but in the intermediary layers. The layers of activated carbon were added after the third layer with limestone or calcium silicate. The sand layers were added after the seventh layer with these materials. This order was established according to the granulometry of the sand and charcoal, where finer grain size materials should be used in layers closer to the core of the pellet, whereas thicker granulated materials such as sand must be applied in outer layers (Silva and Nakagawa, 1998a).

After coating, the seeds were assessed for physical and physiological characteristics in a laboratory and in a greenhouse. In the laboratory, the physical characteristics were assessed for water content (WC), maximum diameter (MAD), minimum diameter (MID) thousand uncoated and coated seed weight (TSW), while the physiological ones were assessed by means of germination test on paper, first count of germination (FCG), germination speed index (GSI) and mean germination time (MGT). In a greenhouse, the physiological characteristics were assessed by means of the emergence test, emergence speed index (ESI), mean emergence time (MET), shoot length (SL), root length (RL) , fresh and dry mass of the shoot (FMS and DMS) and fresh and dry mass of the root (FMR and DMR). The procedures for assessment of the above characteristics were:

Water Content (WC) – determined with two repetitions, by the oven method at 105 °C for 24 hours (Brasil, 2009) and the results expressed in percentage (wet basis).

Maximum diameter (MAD) and Minimum diameter (MID) – determined with four replications of 50 seeds each for each treatment of coating, which were analyzed by SAS equipment (Seed Analysis System), for the supply of the larger and smaller diameter (MAD and MID, respectively), and the results expressed in millimeters (mm).

Thousand uncoated and coated seed weight (TSW) – eight repetitions of 100 seeds each were used, for each coating treatment, according to the methodology (Brasil, 2009).

Germination test – conducted under the Rules for Seed Analysis (Brasil, 2009) with four replications of 50 seeds each, for each coating. The seeds were sown in germination boxes containing two sheets of paper for germination, moistened with distilled water in a volume of 2.5 times the weight of the paper and kept for 10 days in a B.O.D. (Biochemical Oxygen Demand)-type germination chamber with a photoperiod of 16/8 h (dark/light) and alternating temperature of 20-35 °C. The assessments were performed at 4 and 10 days after the start of the test (first count (FCG) and final germination (G), respectively), computing the number of normal seedlings, abnormal seedlings (AS) and non-germinated seeds (NGS), according to the criteria established by Brasil (2009), and the results expressed as a percentage. Throughout the test, daily counts were performed to determine the germination speed index (GSI) by the formula proposed by Maguire (1962) and the mean germination time (MGT), according to Edmond and Drapala (1958).

Emergence test – conducted in a greenhouse where the seeds were sown in plastic trays with a capacity of 2.2 liters of substrate, containing a mixture of sand and soil (2:1) (v/v). Daily counts of the number of emerged plants were held for a period of 30 days. At the end, were determined the percentage of emergence (E), the emergence speed index (ESI), according to a formula by Maguire (1962), and the mean emergence time (MET), according to an adaptation by Edmond and Drapala (1958). At the end of 30 days, the plants were carefully removed from the trays and had their roots properly washed for measuring shoot length (SL) and root length (RL). Subsequently, the shoot was separated from the root and both were stored in paper bags and weighed on precision scales to determine the fresh mass of the shoot (FMS) and the fresh mass of the root (FMR). Then the bags were kept in an air forced circulation stove at 65 °C for 72 hours to determine the dry mass of the shoots (DMS) and the dry mass of the roots (DMR) (Silva and Queiroz, 2006).

Statistical analysis was performed using a completely randomized design model for the variables: water content, maximum and minimum diameters, first count of germination, germination percentage, abnormal seedlings and non-germinated seeds, germination speed index, mean germination time. For the variables percentage of emergence, emergence speed index, mean germination time, shoot and root length and fresh and dry mass of shoot and root, a randomized block design model was employed, with four replications of 50 seeds each. For determining the weight of a thousand seeds, a descriptive analysis of the data was performed. The data for the MET were transformed to $\sqrt{x}$ due to not following a normal distribution.

The data were submitted to the analysis of variance and the means were compared by the Duncan test at 5% probability with the help of the free software Assistência Estatística (ASSISTAT) 7.6 beta (Silva, 2013).

Results and Discussion

The coating of seeds with the proposed materials promoted increases in TSW of 1.6 to 2.3 times in the coated seeds compared to the uncoated ones (Table 1). Medeiros et al. (2004) found increases of 2.5 and 4.0 times in carrot TSW, due to the ratio between the filler material (vermiculite) and seeds. The great

advantage in increasing the seed weight and consequently the seed size is in the ease of sowing these seeds, whether manually or mechanically (Medeiros et al., 2004; Nascimento et al., 2009; Gadotti and Puchala, 2010). Therefore, the different coating treatments assessed in this work would facilitate the sowing of stylosanthes seeds, not only by the increase in the TSW, but also for changing and standardizing their color and shape.

Table 1. Thousand seed weight (TSW) (g), maximum diameter (MAD) (mm), minimum diameter (MID) (mm) and water content (WC) (%) of stylosanthes cv. Campo Grande seeds coated with the following treatments: US – uncoated seeds; L + PVA (dolomitic limestone + PVA); L + S + PVA (dolomitic limestone + sand + PVA); L + AC + PVA (dolomitic limestone + activated carbon+ PVA); CS + PVA (calcium silicate + PVA); CS + S + PVA (calcium silicate + sand + PVA); CS + AC + PVA (calcium silicate + activated carbon + PVA).

Treatment	TSW (g)	MAD (mm)	MID (mm)	WC (%)
US	2.42	2.48 e	1.53 e	10.5 a
L + PVA	4.15	2.67 cd	1.80 c	6.4 bc
L + S + PVA	5.59	2.75 b	1.95 a	4.9 e
L + AC + PVA	4.64	2.64 d	1.79 c	6.3 cd
CS + PVA	3.78	2.75 b	1.80 c	6.7 b
CS + S + PVA	4.53	2.87 a	1.86 b	5.9 d
CS + AC + PVA	3.76	2.72 bc	1.74 d	6.6 bc
Average	-	2.70	1.78	6.76
CV (%)	-	1.31	1.60	2.57

The coating treatments in which there was the addition of limestone and/or sand (L + PVA; L + S + PVA; L + AC + PVA and CS + S + PVA) showed TSW above 4 g, while those to which were added calcium silicate and active charcoal showed TSW less than 4 g. Probably, this is due to the density of the materials used in the coating, since the limestone (2.86 g/cm^3) and sand (1.53 g/cm^3) have a density that is greater than the one for calcium silicate (2.10 g/cm^3) and charcoal (0.25 g/cm^3). Therefore, seeds coated with limestone and sand resulted in more dense layers and hence higher TSW than those coated with calcium silicate and charcoal, which provided lower TSW.

Regarding the results found for MAD and MID, it was possible to see the relation of the sand on the thickness of the coating (Table 1). The highest MAD value was observed for coating comprising CS + S + PVA. The coating based on L + S + PVA provided the greatest MID, followed by the coating with CS + S + PVA. The shape of the sand granules may have contributed to the formation of pellets with higher diameters, since limestone and sand of 0.25 mm were used.

Coated seeds showed lower WC values when compared to uncoated seeds (US) (Table 1). This indicates that the materials used in the coating did not retain moisture and that the temperature of 40 °C was sufficient for drying the water applied during the coating procedure. It is noteworthy that this

reduction was more pronounced in treatments where sand was used as a filler (L + S + PVA and CS + S + PVA); however, these did not differ from the treatment with L + AC + PVA. The addition of this material resulted in higher exposure to hot air for drying the pellet, due to the application of a larger number of layers to them. Probably these seeds showed lower WC values due to their exposure to heat for more times compared to the seeds coated with the other materials.

Conceição et al. (2009) and Lagôa et al. (2012) also found WC values significantly lower than those found with uncoated seeds. Conceição et al. (2009) have attributed the low water content values to the coating and not to the seeds, since the seeds subjected to removal of the coating showed water content similar to the one for uncoated seeds.

As for the physiological characteristics, it was observed that in FCG there was no significant difference among treatments (Table 2). The same was observed by Nascimento et al. (2009) for carrot seeds coated with sand and microcellulose, by Peske and Novembre (2011) for millet seeds coated with different fillers (microcellulose, magnesium thermophosphate, reactive phosphate, phytic acid, dicalcium phosphate, gypsum and vermiculite) and PVA and by Tunes et al. (2014) for irrigated rice seeds coated with kaolin and carbonized rice husk.

Table 2. First count of germination (FCG) (%), germination speed index (GSI), mean germination time (MGT), germination (%), abnormal seedlings (AS) (%) and non-germinated seeds (NGS) (%) of stylosanthes cv. Campo Grande seeds coated with the following treatments: US – uncoated seeds; L + PVA (dolomitic limestone + PVA); L + S + PVA (dolomitic limestone + sand + PVA); L + AC + PVA (dolomitic limestone + activated carbon+ PVA); CS + PVA (calcium silicate + PVA); CS + S + PVA (calcium silicate + sand + PVA); CS + AC + PVA (calcium silicate + activated carbon + PVA).

Treatment	FCG (%)	GSI	MGT	G (%)	AS (%)	NGS (%)
US	63 a	16.77 ab	2.23 a	66 a	2 a	14 a
L + PVA	51 a	13.33 b	2.66 bc	53 b	8 b	11 a
L + S + PVA	54 a	13.18 b	2.92 c	61 ab	7 ab	10 a
L + AC + PVA	59 a	16.42 ab	2.49 ab	60 ab	9 b	7 a
CS + PVA	66 a	18.97 a	2.23 a	68 a	7 ab	12 a
CS + S + PVA	54 a	14.80 b	2.40 ab	56 ab	10 b	14 a
CS + AC + PVA	55 a	15.02 b	2.47 ab	59 ab	6 ab	11 a
Average	57.7	15.50	2.49	60.3	7.3	11.07
CV (%)	14.57	15.74	9.41	13.00	56.95	41.76

Means followed by the same letter are not statistically different from each other by Duncan test ($p < 0.05$).

In general, seed coating causes a delay in germination rate, as reported by various authors (Silva et al., 2002; Oliveira et al., 2003a; Oliveira et al., 2003b; Mendonça et al., 2007; Conceição and Vieira, 2008). This behavior, however, was not observed for the treatments assessed in this study, since none of the treatments differed from the US. It is worth noting that treatment with CS + PVA stood out among the coatings, but did not differ from L + AC + PVA.

Delay in germination can occur due to the material used in the coating procedure, which requires a physical barrier that must be overcome by the seed. However, some materials allow better diffusion of gases and water between the seed and the external environment (Nascimento et al., 2009). During the experiment it was observed that the coating with calcium silicate would cut up easily when in contact with water. Thus, the barrier imposed by coating with this material was quickly undone and, therefore, gas exchange and the water uptake by the seeds were more facilitated compared to seeds coated with other materials. This may have led to faster water absorption by these seeds.

On the other hand, the fact that the treatment with L + AC + PVA did not differ statistically from the silicate may be related to charcoal characteristics. Activated carbon is obtained by pyrolysis (chemical decomposition by action of heat) of carbonaceous materials of vegetable origin, followed by chemical activation. At the end of the procedure an adsorbent material is obtained, exhibiting a highly porous structure (Monocha, 2003). Thus, the use of charcoal in such treatment may have probably granted gas exchanges, improving oxygen supply, which is essential to germination, because the limestone promotes a waterproofing effect of the coating, as has been observed in tomato seeds (Oliveira et al., 2003b).

Best MGT is related to the lowest values found for this variable, as verified for US and CS + PVA; however, these treatments did not differ from L + AC + PVA, CS + S + PVA and CS + AC + PVA (Table 2). The use of calcium silicate as the filler did not cause major impediment to water absorption and gas exchanges, which gives good characteristics as a filler. Conversely, the greatest MGT was observed for seeds coated with L + S + PVA, and the treatment with L + PVA did not differ from this one. Materials such as limestone may result in sealing the gas exchanges, associated to a high moisture retention rate, imposed by the thickness of the material layer (Silva and Nakagawa, 1998a). This limestone waterproofing effect was also observed in the coating of tomato seeds (Oliveira et al., 2003b).

The main obstacle to the use of coated seeds is in the delay generated in seed germination (Silva et al., 2002). However, despite these delays, the final germination rates are similar to those of uncoated seeds (Silva et al., 2002, Tavares et al., 2012). Although the different filling materials assessed in this study have entailed delays in germination time, only the limestone + PVA provided the small percentage of germination, 53% (Table 2). However, the treatments plus sand and charcoal (L + S + PVA, L + AC + PVA, CS + S + PVA and CS + AC + PVA) showed no differences compared to limestone + PVA.

Likewise, the use of limestone in lettuce seed coating promoted reduction in seed germination (only 47%), regardless of granulometry and pellet size formed (Silva and Nakagawa, 1998a). On the other hand, Tavares et al. (2012) have found no negative effect of limestone or aluminum silicate at a percentage of seed germination of two rice cultivars.

This divergence noticed among different studies on seed coating occurs depending on the type of material used in the procedure and the layer thickness of coating deposited on the seeds (Pereira et al., 2011). Normally, after the seeds overcome the barrier imposed by coating, the seedlings are equal in growth velocity, forming uniform seedlings (Silva and Nakagawa, 1998b).

The seeds that have not received any coating (US) showed lower percentages of abnormal seedlings (2%) compared to treatment with L + PVA, L + AC + PVA and CS + S + PVA (Table 2). Probably the arrangement formed by the particles of the fillers and the adhesive solution have promoted difficulty in gas exchanges between the seeds and the external environment, besides the need to break the barrier imposed by coating. These impediments may have boosted the malformation of seedlings,

resulting in abnormal seedlings. On the other hand, the percentage of non-germinated seeds (NGS) was not affected by the coating treatments (Table 2) and therefore this result is probably due to an intrinsic characteristic of the species itself or seed lot.

With regard to assessments carried out in a greenhouse, there was a significant difference only for mean emergence time (MET) where none of the treatments differed from the US, except treatment with L + PVA; however, this delay did not affect the final quality of plants derived (Table 3). From these results it is possible to see the negative effect of coating only with limestone as a filler in seedling emergence time. When there is the addition of sand and charcoal to the coating with limestone, this effect is not observed, which may be related to the higher density of coating comprising only limestone, providing a major impediment to the emergence.

Table 3.   Emergence (%), emergence speed index (ESI), mean emergence time (MET), shoot length (SL) (cm), root length (RL) (cm), fresh mass of shoot (FMS) (mg/plant), dry mass of shoot (DMS) (mg/plant), fresh mass of root (FMR) (mg/plant) and dry mass of root (DMR) (mg/plant) of stylosanthes cv. Campo Grande plants derived from seeds coated with the following treatments: US – uncoated seeds; L + PVA (dolomitic limestone + PVA); L + S + PVA (dolomitic limestone + sand + PVA); L + AC + PVA (dolomitic limestone + activated carbon+ PVA); CS + PVA (calcium silicate + PVA); CS + S + PVA (calcium silicate + sand + PVA); CS + AC + PVA (calcium silicate + activated carbon + PVA).

Treatment	E (%)	ESI	MET	SL	RL	FMS	DMS	FMR	DMR
US	64 a	8.12 a	4.91 a	1.70 a	11.07 a	46.31 a	5.66 a	17.88 a	4.90 a
L + PVA	60 a	6.72 a	6.24 b	1.87 a	10.55 a	48.65 a	4.87 a	19.44 a	4.89 a
L + S + PVA	52 a	5.92 a	5.74 ab	1.98 a	10.22 a	44.21 a	2.41 a	18.96 a	5.10 a
L + AC + PVA	57 a	6.69 a	5.78 ab	1.88 a	11.04 a	45.01 a	3.90 a	19.30 a	5.28 a
CS + PVA	61 a	7.70 a	5.25 ab	1.79 a	11.01 a	59.72 a	7.92 a	23.67 a	5.74 a
CS + S + PVA	56 a	6.99 a	5.34 ab	1.94 a	10.09 a	47.59 a	4.05 a	17.78 a	4.68 a
CS + AC + PVA	51 a	7.03 a	4.86 a	1.85 a	11.11 a	46.38 a	5.15 a	18.31 a	4.77 a
Average	57.14	7.02	5.45	1.86	10.73	48.27	4.85	19.33	5.05
CV (%)	13.84	19.43	6.09	11.30	12.20	20.48	72.81	24.86	15.67

Means followed by the same letter are not statistically different from each other by Duncan test (p < 0.05).

As mentioned, none of the assessed treatments has damaged the seedling emergence in this work; however, some authors have observed a negative effect of coating for this variable. The use of limestone in lettuce seed coating has promoted a decrease in the percentage of emerged seedlings in relation to the control and coating with sand or sand + limestone (Silva and Nakagawa, 1998b) and such results differ from that observed in this study. Likewise, Santos et al. (2010) have also found an adverse effect of limestone in the emergence of brachiaria cv. Marandu plants. According to the authors, the coatings that resulted in worst emergence percentages were limestone + PVA; limestone + bentonite + PVA and sand + bentonite + polymer (19, 16 and 18%, respectively), and the latter also showed the lowest values of ESI (0.99).

At 30 days after sowing, the plants did not differ in length (SL and RL), fresh mass (FMS and FMR) and dry mass (DMS and DMR) of shoot and root compared to plants obtained from US. The results have permitted to infer that none of the treatments has damaged plant emergence and early seedling development; therefore, they are promising in the coating of the seeds of this Fabaceae. Therefore, advances in the use of these coatings are on the incorporation of micronutrients, insecticides, fungicides and growth regulators among others, to improve the efficiency of application thereof, as well as the yield of the crop. Silva et al. (2002) have also found no differences in dry mass of shoot and root of seedlings derived from lettuce seeds coated with bentonite and PVA.

With regard to ESI and MET, it is expected that their responses be equivalent, since these variables are about speed and time, which are closely related. However, it was possible to find a different behavior for the treatment with L + PVA for MET, which differed from the control, which was not observed for ESI (Table 3). This happened probably due to the large variability observed for ESI (AC = 19.43%) in relation to MET (AC = 6.09%) (Table 3). Which justifies why relatively large differences in the ESI values were not statistically significant, indicating that the coating treatments promoted a uniformity in the initial seedling growth.

The use of glue based on polyvinyl acetate (PVA) in this study has not proved to be a problem for seed germination and plant emergence. On the other hand, Silva et al. (2002) have observed that the use of 75% of bentonite + 25% of PVA has resulted in lower values of GSI (33), while the use of 100% of PVA has not differed from control for the coated seeds of lettuce.

From the data observed in Tables 2 and 3 it can be seen that the behavior of the coated seeds has not followed the same pattern for testing in germination chamber and greenhouse. Probably this was due to the fact that in a germination chamber the conditions are controlled throughout the test, unlike the conditions in a greenhouse, where the seedlings are exposed to temperature and humidity variations. In addition, the moisture provided by the substrate in a greenhouse around the seeds, and a larger contact surface between seeds and substrate, may have contributed for the coating to fall apart more easily, contrary to what is expected from the paper substrate in a gerbox.

Conclusions

The coating reduces the water content.

The coating with limestone + PVA glue and limestone + sand + PVA glue increases the germination time; however, no treatment affects the physiological quality of seeds.

The coating comprising calcium silicate and PVA glue stood out for the variables germination speed index and fresh and dry mass of shoot and root in the coating of stylosanthes cv. Campo Grande seeds.

Referências

BAUDET, L.; PERES, W. Recobrimento de sementes. *Seed news*, v.8, p.20-23. 2004.

BRASIL. Ministério da Agricultura Pecuária e Abastecimento. *Regras para análise de sementes*. Ministério da Agricultura Pecuária e Abastecimento. Secretaria de Defesa Agropecuária. Brasília: MAPA/ACS, 2009. 395p. http://www.agricultura.gov.br/arq_editor/   file/laborat%c3%b3rio/ sementes/regras%20para%20analise%20de%20sementes.pdf

CONCEIÇÃO, P.M.; VIEIRA, H.D. Qualidade fisiológica e resistência do recobrimento de sementes de milho. *Revista Brasileira de Sementes*, v.30, p.48-53, 2008. http://www.scielo.br/pdf/rbs/v30n3/07.pdf

CONCEIÇÃO, P.M., VIEIRA, H.D., SILVA, R.F., CAMPOS, S.C. Germinação e vigor de sementes de milho recobertas e viabilidade do inóculo durante o armazenamento. *Ciência e Agrotecnologia*, v.33, p.765-772, 2009. http://www.scielo.br/pdf/cagro/v33n3/a15v33n3.pdf

EDMOND, J.B, DRAPALA, W.J. The effects of temperature, sand and soil, and acetone on germination of okra seed. *Proceedings of the American Society for Horticultural Science*, v.71, p.428-434, 1958.

EMBRAPA GADO DE CORTE. *Cultivo e uso do estilosantes*. EMBRAPA Gado de Corte – Campo Grande, 2007. 11p. (Comunicado Técnico, 105). http://www.cnpgc.embrapa.br/publicacoes/cot/pdf/Cot105.pdf

FERRAZ, J.B.S.; FELÍCIO, P.E. Production systems – An example from Brazil. *Meat Science*, v.84, p.238-243, 2010.http://www.usp.br/gmab/ publica/msjbsf2010.pdf

GADOTTI, C.; PUCHALA, B. Revestimento de sementes. *Informativo Abrates*, v.20, p.70-71, 2010. http://www.abrates.org.br/portal/images/ stories/informativos/v20n3/minicurso03.pdf

LAGÔA, A.O.; FERREIRA, A.C; VIEIRA, R.D. Plantability and moisture content of naked and pelleted seeds of supersweet (Sh_2) corn during cold storage conditions. *Revista Brasileira de Sementes*, v.34, n.1, p.39 – 46, 2012. http://www.scielo.br/pdf/rbs/v34n1/a05v34n1.pdf

LOPES, A.C.A.; NASCIMENTO, W.M. *Peletização em sementes de hortaliças*. EMBRAPA Hortaliças – Brasília, 2012. 28p. (Documentos, 137). http://www.cnph.embrapa.br/paginas/serie_documentos/publicacoes2012/ doc_137.pdf

MAGUIRE, J.D. Speeds of germination-aid selection and evaluation for seedling emergence and vigor. *Crop Science*, v.2, p.176-177, 1962.

MEDEIROS, E.M.; BAUDET, L.; PERES, W.B.; EICHOLZ, E.D. Modificações na condição física das sementes de cenoura em equipamento de recobrimento. *Revista Brasileira de Sementes*, v.26, n.2, p.70-75, 2004. http://www.scielo.br/pdf/rbs/v26n2/24491.pdf

MENDONÇA, E.A.F.; CARVALHO, N.M.; RAMOS, N.P. Revestimento de sementes de milho superdoce (Sh_2). *Revista Brasileira de Sementes*, v.29, n.2, p.68-79, 2007. http://www.scielo.br/pdf/rbs/v29n2/v29n2a10.pdf

MONOCHA, S.M. Porous carbons. *Sadhana*, v.28, p.335-348, 2003. http://www.ias.ac.in/sadhana/Pdf2003Apr/Pe1070.pdf

NASCIMENTO, W.M.; SILVA, J.B.C.; SANTOS, P.E.C., CARMONA, R. Germinação de sementes de cenoura osmoticamente condicionadas e peletizadas com diversos ingredientes. *Horticultura Brasileira*, v.27, p.12-16, 2009. http://www.scielo.br/pdf/hb/v27n1/03.pdf

OLIVEIRA, J.A.; PEREIRA, C.E.; GUIMARÃES, R.M.; VIEIRA, A.R.; SILVA, J.B.C. Desempenho de sementes de pimentão revestidas com diferentes materiais. *Revista Brasileira de Sementes*, v.25, n.2, p.36-47, 2003a. http://www.scielo.br/pdf/rbs/v25n2/19647.pdf

OLIVEIRA, J.A.; PEREIRA, C.E; GUIMARÃES, R.M; VIEIRA, A.R.; SILVA, J.B.C. Efeito de diferentes materiais de peletização na deterioração de sementes de tomate durante o armazenamento. *Revista Brasileira de Sementes*, v.25, n.2, p.20-27, 2003b. http://www.scielo.br/ pdf/rbs/v25n2/19645.pdf

PEREIRA, C.E.; OLIVEIRA, J.A.; GUIMARÃES, R.M.; VIEIRA, A.R.; EVANGELISTA, J.R.E.; OLIVEIRA, G.E. Tratamento fungicida e peliculização de sementes de soja submetidas ao armazenamento. *Ciência e Agrotecnologia*, v.35, p.158-164, 2011. http://www.scielo.br/pdf/cagro/v35n1/a20v35n1.pdf

PESKE, F.B.; NOVEMBRE, A.D.L.C. Pearl millet seed pelleting. *Revista Brasileira de Sementes*, v.33, n.2, p.352-362, 2011. http://www.scielo.br/pdf/rbs/v33n2/18.pdf

SANTOS, F.C.; OLIVEIRA, J.A.; VON PINHO, E.V.R..; GUIMARÃES, R.M.; VIEIRA, A.R. Tratamento químico, revestimento e armazenamento de sementes de *Brachiaria brizantha* cv. Marandu. *Revista Brasileira de Sementes*, v.32, n.3, p.69-78, 2010. http://www.scielo.br/pdf/rbs/v32n3/v32n3a08.pdf

SHELTON, H.M.; FRANZEL, S.; PETERS, M. Adoption of tropical legume technology around the world: analysis of success. *Tropical Grasslands*, v.39, p.198-209, 2005. http://www.tropicalgrasslands.asn.au/Tropical%20Grasslands%20Journal%20archive/PDFs/Vol_39_2005/Vol_39_04_2005_pp198_209.pdf

SILVA, F.A.S. *ASSISTAT* – Assistência Estatística, versão 7.6. Universidade Federal de Campina Grande – PB. 2013.

SILVA, J.B.C.; NAKAGAWA, J. Metodologia para avaliação de materiais cimentantes para peletização de sementes. *Horticultura Brasileira*, v.16, p.31-37, 1998a. http://www.abhorticultura.com.br/biblioteca/arquivos/Download/biblioteca/hb_16_1.pdf

SILVA, J.B.C.; NAKAGAWA, J. Confecção e avaliação de péletes de sementes de alface. *Horticultura Brasileira*, v.16, p.151-158, 1998b. http://www.abhorticultura.com.br/biblioteca/arquivos/Download/biblioteca/hb_16_2.pdf

SILVA, D. J., QUEIROZ, A.C. *Análise de alimentos*: métodos químicos e biológicos. Viçosa, MG: UFV, 2006. 235p.

SILVA, J.B.C.; SANTOS, E.C.; NASCIMENTO, W.M. Desempenho de sementes peletizadas de alface em função do material cimentante e da temperatura de secagem dos péletes. *Horticultura Brasileira*, v.20, p.67-70, 2002. http://www.scielo.br/pdf/hb/v20n1/14420

TAVARES, L.C.; RUFINO, C.A.; DÖRR, C.S.; BARROS, A.C.S.A.; PESKE, S.T. Performance of lowland rice seeds coated with dolomitic limestone and aluminum silicate. *Revista Brasileira de Sementes*, v.34, n.2, p.202-211, 2012. http://www.scielo.br/pdf/rbs/v34n2/03.pdf

TUNES, L.V.M.; FONSECA, D.A.R.; MENEGHELLO, G.E.; REIS, B.B.; BRASIL, V.D.; RUFINO, C.A.; VILLELA, F.A. Qualidade fisiológica, sanitária e enzimática de sementes de arroz irrigado recobertas com silício. *Revista Ceres*, v.61, p.675-685, 2014. http://www.ceres.ufv.br/ceres/revistas/V61N005P01013.pdf

8

The effect of drying temperatures and storage of seeds on the growth of soybean seedlings

Cesar Pedro Hartmann Filho[1*], André Luís Duarte Goneli[1], Tathiana Elisa Masetto[1], Elton Aparecido Siqueira Martins[1], Guilherme Cardoso Oba[1]

ABSTRACT – Drying of seeds reduces their moisture content to levels appropriate for storage. However, care in the temperatures applied in the process is necessary to avoid damage to the seeds that are dried. The aim of this study was to evaluate the effect of different drying temperatures and storage on the growth of soybean seedlings. Harvested with a moisture content of approximately 23% (w.b.), the seeds were dried at different temperatures (40, 50, 60, 70, and 80 °C) until reaching moisture content of $12.5 \pm 0.7\%$ (w.b.), and they were subsequently stored for 180 days in an environment without climate control. A germination test and evaluations of seedling performance were carried out every 45 days, determining the full length of the seedling and hypocotyl and root lengths, along with their respective dry matter weights. The results showed that: a) the increase in the temperature of drying air affects the physiological quality of soybean seeds, and this effect is accentuated over time, especially on root length; and b) the air temperature of 40 °C can be recommended for drying of soybean seeds in association with the storage time of 180 days under storage conditions without climate control.

Index terms: *Glycine max* L., germination, seed vigor, post-harvest.

Temperaturas de secagem e armazenamento de sementes no crescimento de plântulas de soja

RESUMO - A secagem de sementes permite reduzir o teor de água a níveis adequados para o armazenamento. Porém, cuidados com as temperaturas aplicadas no processo são necessários para evitar danos às sementes submetidas à secagem. O trabalho foi realizado com o objetivo de avaliar o efeito de diferentes temperaturas de secagem e do armazenamento sobre o crescimento de plântulas de soja. As sementes foram colhidas com aproximadamente 23% de teor de água, e submetidas a diferentes temperaturas de secagem (40, 50, 60, 70 e 80 °C) até atingirem $12,5 \pm 0,7\%$ (b.u.), sendo posteriormente armazenadas, em ambiente não controlado, durante 180 dias. A cada 45 dias foi realizado o teste de germinação e as avaliações referentes ao desempenho de plântulas, determinando-se os comprimentos total, do hipocótilo e da raiz, e ainda verificando suas respectivas massas de matéria seca. Com base nos resultados pôde-se concluir que: a) o aumento da temperatura do ar de secagem afeta a qualidade fisiológica das sementes, sendo esse efeito potencializado com o tempo de armazenamento, principalmente sobre o comprimento da raiz; b) a temperatura do ar de 40 °C pode ser recomendada para a secagem de sementes de soja, sendo associada ao período de armazenamento de 180 dias em condições não controladas.

Termos para indexação: *Glycine max* L., germinação, vigor de sementes, pós-colheita.

Introduction

Obtaining quality soybean seeds has always been one of the main challenges within the production system (Marcos-Filho, 2013). Thus, due to this importance, there has been an incessant search for effective methods of characterizing the quality of a determined seed lot within the crop production process (Vieira et al., 2013b).

Due to the need for increasingly practical methods in regard to determination of qualitative parameters, generally because of little time available to laboratories, more dynamic tests are chosen, and thus, more elaborate evaluations, for example, an evaluation relative to initial growth of seedlings, are often not chosen. As a result, some problems observed during the seedling phase end up being overlooked, and, depending on the levels of severity, crop performance in the

[1]Universidade Federal da Grande Dourados, Faculdade de Ciências Agrárias, Caixa Postal 533, 79804-970 - Dourados, MS, Brasil.
*Corresponding author <cphartmann21@hotmail.com>

field and, consequently, crop yield may be affected.

The evaluation of seedling performance and structures, more specifically, the hypocotyl and the root, can provide useful results regarding seed quality, considering their ability to manifest damage in some points of the production chain, such as that generated at harvest and during the processing phase (Barbosa et al., 2014; Faria et al., 2014).

After harvest, it is common for seed lots to have moisture contents considered inadequate for safe and effective storage, such as contents higher than 12%. Given this situation, there is clearly a need for reduction of this characteristic to preserve the physiological quality of seeds for at least eight months, impeding possible chemical and physical changes that may come about during storage up to sale of the seeds (Peske et al., 2013; Barrozo et al., 2014; Carvalho et al., 2016).

For efficient storage, not only good initial quality is necessary, but seeds need to have gone through drying conditions that avoid loss of physiological quality and only then be stored under what are considered to be ideal conditions, such as temperatures below 20 °C and relative humidity below 60% (França-Neto et al., 2010; Mbofung et al., 2013; Smaniotto et al., 2014). Nevertheless, when carried out without due care, both drying and storage can have a negative impact on seed quality, immediately affecting, for example, their germination and vigor through improper drying or, moreover, worsening some such detrimental effect in accordance with storage time and conditions (Deliberali et al., 2010; Ullmann et al., 2010; Schuh et al., 2011; Resende et al., 2012; Faria et al., 2014; Rathinavel, 2014; Mahjabin and Abidi, 2015; Paraginski et al., 2015).

Drying soybean seeds at temperatures of 45 and 55 °C directly affect viability, germination, and vigor of soybean seeds, with a further negative impact on the storage potential of the material. The prejudicial effect worsens the immediate problems caused by drying, especially under temperatures greater than 20 °C and relative humidities that are not controlled, which reduce the potential for nine months of seed storage to approximately four months (Surki et al., 2012).

In this context, evaluation of seedling performance is an extremely important tool because determination of parameters for growth and initial development, such as assimilation of cotyledonary reserves on the part of the embryonic axis, can portray not only the immediate effects caused by drying, but also their implications for storage, increasing reliability regarding the physiological potential of seeds at different periods. Based on the literature, the seed lots with high germination and vigor have larger seedlings that are able to accumulate more dry matter (Henning et al, 2010; Pereira et al., 2013; Pereira et al., 2015). These discoveries may be related to the fact that high vigor soybean seeds have not only greater physical integrity (as long as the drying operations were well performed (Afrakhteh et al., 2013) but also higher contents of starch, proteins, and soluble sugars and greater ability to mobilize reserves during the germination period (Henning et al., 2010).

Thus, the aim of this study was to evaluate the effects of different drying temperatures, associated with storage of seeds, on the initial growth of soybean seedlings.

Materials and Methods

Seeds of the soybean cultivar SYN 1059 RR (V-TOP) were produced under the no-till system and center pivot irrigation from January to May 2014 on the São Lourenço Farm at 22°11'58.06" S, 54°53'24.32" W, and 452 meters altitude, in Dourados, Mato Grosso do Sul state, Brazil. They were analyzed at the Preprocessing and Storage of Agricultural Products Laboratory belonging to the School of Agrarian Sciences of the Federal University of Grande Dourados (UFGD) in the municipality of Dourados, MS, Brazil.

Preparation of the area involved only desiccation (Paraquat + Diuron - 400 g + 200 g.ha^{-1}; 2,4-Dicholorophenoxy - 806 g.ha^{-1}), and the fertilization process, for its part, was carried out at the same time as sowing, using 260 kg.ha^{-1} of the fertilizer formulation 02-20-20 (N-P-K), also containing 8% Ca, 4% S, 0.2% Zn, and 0.1% Bo. The seeds were treated with insecticides (Fipronil – 12.5 g.50 kg of seeds^{-1}; Tiamethoxan – 17.5 g.50 kg of seeds^{-1}) and fungicides (Metalaxil-M + Fludioxonil – 0.5 g + 1.25 g.50 kg of seeds^{-1}), and were then sown mechanically at a spacing of 0.45 m between the plant rows and a density of 18 seeds.m^{-1}, obtaining a final population of 355,555 plants.ha^{-1}.

The seeds produced were harvested manually close to the stage of physiological maturity (R7 + 15 days), according to the determination of Fehr and Caviness (1977), with a moisture content of approximately 23 ± 0.5% (w.b.) (0.30 ± 0.005 decimal d.b.).

After that, drying occurred in an experimental fixed bed dryer, with a drying chamber of 0.080 m diameter by 1.0 m height. The experimental dryer used a set of electrical resistors as a heating source, which had a total of 12 kW of power and, associated with them, a centrifuge type fan of 0.75 kW of power, Ibram VSI-160. Temperature was controlled by a universal process controller, Novus N1200, working with a Proportional-Integral-Derivative (PID) controller. The air flow used was 0.2 m³.s^{-1}.m^{-2}, selected by a frequency inverter connected to the fan motor.

The drying temperatures used were 40, 50, 60, 70, and

80 °C to obtain seeds with moisture content of 12.5 ± 0.7% (w.b.) (0.14 ± 0.007 decimal d.b.). Drying was monitored by the gravimetric method, and loss in weight of the product was monitored in a stepwise and prudent manner, considering the final pre-established moisture content. In this step, the times required so that each drying temperature reached such a final moisture content were 230, 160, 112, 75, and 57 minutes, respectively. For determination of moisture contents, the laboratory oven method was used at 105 ± 3 °C for 24 hours, with two replications of 200 seeds (approximately 25 g) (Brasil, 2009). With monitoring of drying, the water reduction rate (WRR) of the soybean seeds was determined for each drying temperature used, being established by Equation 1, according to Corrêa et al. (2001), who conceive of this calculation as the amount of water that a determined product loses per unit of dry matter per unit of time.

$$WRR = \frac{W_0 - W_i}{DM(t_i - t_0)} \tag{1}$$

in which,

WRR: water reduction rate, in kg.kg^{-1}.h^{-1};

W_0: total previous water weight, in kg;

W_i: total current water weight, in kg;

DM: dry matter weight, in kg;

t_0: total previous drying time, in h; and

t_i: total current drying time, in h.

After drying, the soybean seeds were placed according to the drying treatment in non-hermetic metallic containers. These closed containers were kept in an environment with temperature and relative humidity not controlled for a period of 180 days; germination and seedling performance, along with seed moisture content, was evaluated immediately after drying and every subsequent 45 days.

Two thermo hygrometers were installed near the containers to monitor and register variations in heat and humidity in the environment during the storage period (Figure 1). Mean temperature throughout the experiment was 21.4 °C, with a maximum of 28.4 °C and minimum of 14.5 °C. The mean value of relative humidity was 57.9%, with a maximum of 83.4% and minimum of 47.3%.

The germination test was carried out in accordance with the criteria established in the Rules for Seed Testing (Brasil, 2009). Four 50-seed subsamples for each treatment were distributed in rolls of paper toweling moistened with distilled water at two and a half times the weight of the dry paper and kept at 25 °C in a Mangelsdorf germinator. The percentage of normal seedlings was calculated eight days after the test was set up.

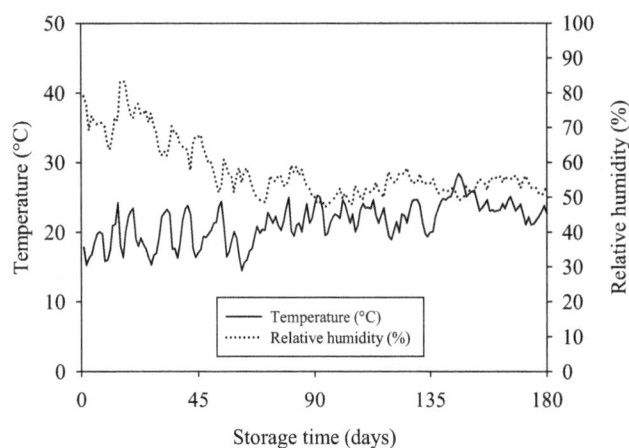

Figure 1. Mean daily values of temperature and relative humidity over 180 days of storage of soybean seeds in an environment without climate control.

Seedling performance was evaluated through determination of the length of seedlings and their structures, and through total dry matter weight and weight of the hypocotyl and roots (Nakagawa, 1999, Benincasa, 2003; Ferreira and Borghetti, 2004; Marcos-Filho, 2015).

Seedling length was determined using four 15-seed subsamples that were arranged in two rows lengthwise on the sheets of paper toweling, which were moistened with distilled water at two and a half times the weight of the dry paper. The seeds were placed with the point of the radicle turned downward to direct seedling growth in as straight a line as possible to favor measurement of length. After this procedure, the paper rolls were arranged similar to the germination test and were placed in a Mangelsdorf germinator regulated to 25 ± 1 °C for seven days.

Total length of the seedlings was measured from the tip of the radicle up to the point of connection of the apical bud to the stem. Length of the hypocotyl was measured from point of connection of the hypocotyl with the root up to its point of connection with the cotyledons. The length of the primary root was measured from the tip of the root up to its point of connection with the hypocotyl. All lengths were measured with a digital caliper with 0.01 mm resolution, and results were obtained by adding up the measurements taken of each normal seedling in each replication and then dividing by the number of normal seedlings measured; results were expressed in mm.seedling^{-1}.

The seedlings measured in the length test were used for determination of dry matter weight (Benincasa, 2003). A scalpel was used to remove the cotyledons and to divide the seedling into hypocotyl and root. Soon after this procedure, the hypocotyls and the roots for each replication were placed separately in previously weighed aluminum containers, which, in turn, were placed in a previously regulated laboratory oven with circulation and renewal of air and kept at 80 °C for 24 hours (Benincasa, 2003). After this period, the samples were removed from the laboratory oven and allowed to cool in a desiccator for 20 minutes, followed by weighing on an analytic balance with 0.001 g resolution, thus determining the dry matter weights of the hypocotyls and roots, and the total dry matter weight. The weight obtained was divided by the number of normal seedlings, resulting in dry matter weight per seedling, expressed in mg.seedling⁻¹.

The experiment was conducted in a 5 x 5 split plot arrangement, with five temperatures of drying air in the plots and five different storage periods in the split plots in a completely randomized design. To evaluate the latent effect of the drying temperatures, the data were subjected to polynomial regression analysis. The models were selected considering the magnitude of the coefficient of determination (R^2), the significance of regression by the F test, and the phenomenon under study using the Sisvar program (Ferreira, 2011).

Results and Discussion

Through curves of the water reduction rate (WRR) obtained during the drying of soybean seeds, it was observed that in accordance with higher temperatures, such as 60, 70 and 80 °C, the WRR was clearly more accentuated already at the beginning of the process (Figure 2). However, over time, this rate became more homogeneous among the drying treatments since the water present at the surface of the product was gradually substituted by an evaporation front that moved inward in the product. In addition, due to the involvement of more complex mechanisms in movement of water from the inside to the outside of the seed, such as liquid diffusion and capillary action, the speed of the process declined, thus manifesting similarity of the WRR among the thermal treatments applied as of approximately 50 minutes from the start of drying.

Due to hygroscopicity of the seeds, oscillations of moisture contents were observed during storage, above all because lack of control of temperature and relative humidity conditions lead to the occurrence of phenomena such as sorption and desorption (Table 1) (Tiecker Junior et al., 2014; Bessa et al., 2015).

There was a reduction in moisture content in all the seed lots evaluated (Table 1), probably due to reduction in the values of relative humidity in the air of the environment over time (Figure 1) and possibly also due to the centesimal composition of the seeds since they are aleuro-oleaginous and tend to retain less water internally because their oil content is less hydrophilic (Dios, 1984).

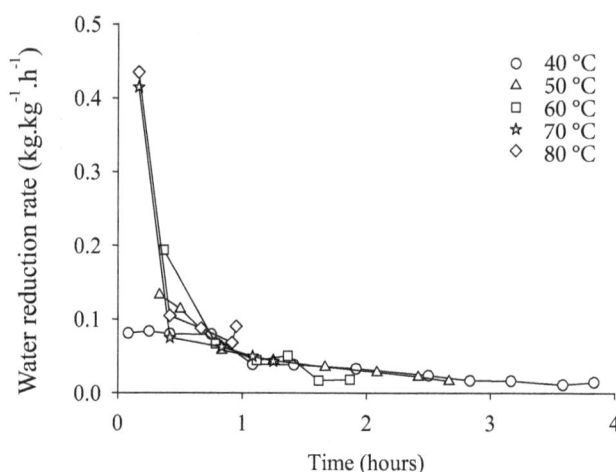

Figure 2. Water reduction rate for soybean seeds (cv. SYN 1059 RR) during the drying process at different temperatures.

Table 1. Mean values of moisture content (% w.b.) of soybean seeds as a function of drying temperature and of storage time.

Storage time (days)	Drying temperature (°C)				
	40	50	60	70	80
0	13.2	12.7	12.4	12.5	12.0
45	13.4	12.8	12.4	12.6	12.3
90	12.6	12.1	12.0	11.8	11.4
135	11.1	10.6	10.6	10.5	10.2
180	11.3	11.0	10.8	10.8	10.4

For seed germination, it was observed that due to the increase in temperature of drying air, the percentage of normal seedlings already decreased immediately after drying, and this deleterious effect intensified in a linear manner over the storage time (Figure 3).

Immediately after drying, the germination observed for the temperatures of 40, 50, 60, 70, and 80 °C were 100, 97, 88, 28, and 1 %, respectively (Figure 3). However, due to the occurrence of this latent response, in the evaluation after 45 days, the dissimilarity between the two treatments was evident, with daily reduction in the percentage of normal seedlings caused by the temperature of 50 °C being twice that brought about by the temperature of 40 °C; greater daily reductions were also observed for the temperatures of 60, 70, and 80 °C.

Consequently, at the end of storage, the values of the germination observed for the temperatures of 40, 50, 60, 70, and 80 °C were 86, 64, 55, 3, and 1%, respectively (Figure 3). Thus, if the parameters established by regulatory standard number 45 instituted by Brasil (2013) were adopted, which establishes 80% germination as the minimum limit for sale of soybean seeds, only the lot of seeds dried at 40 °C could be fit to the standard commercialization, given that the lots dried at 50 and 60 °C lose their capacity at approximately 150 and 101 days after drying, and the others already immediately after the drying operation.

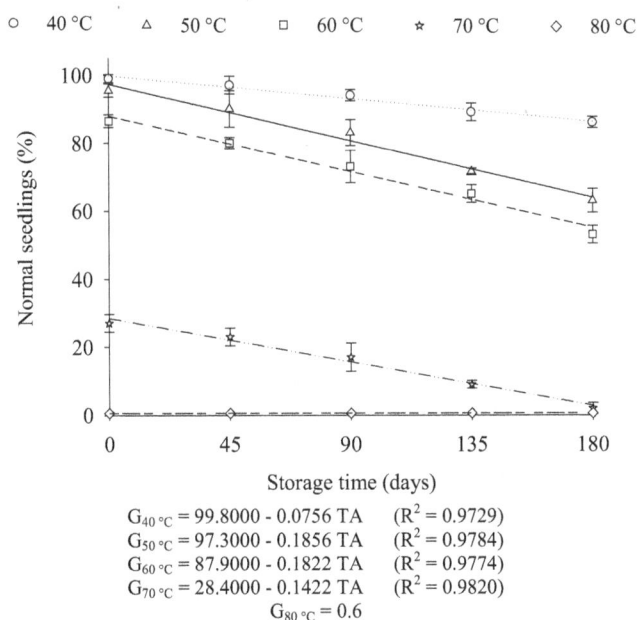

$$G_{40\,°C} = 99.8000 - 0.0756\ TA \quad (R^2 = 0.9729)$$
$$G_{50\,°C} = 97.3000 - 0.1856\ TA \quad (R^2 = 0.9784)$$
$$G_{60\,°C} = 87.9000 - 0.1822\ TA \quad (R^2 = 0.9774)$$
$$G_{70\,°C} = 28.4000 - 0.1422\ TA \quad (R^2 = 0.9820)$$
$$G_{80\,°C} = 0.6$$

Figure 3. Results of percentage of normal seedlings for the germination test of soybean seeds as a function of drying temperature and storage time.

The effect of different drying temperatures on the physiological quality of seeds was also observed by Ullmann et al. (2015) in evaluation of sweet sorghum seeds, in which a reduction in the germination was observed in accordance with the temperature factor, especially at temperatures above 40 °C; each increase in degree of temperature led to reduction of half a percentage point in germination. As suggested by Afrakhteh et al. (2013) and also observed by Menezes et al. (2012a,b), and furthermore in this study, this is generally a result of the increase in the rate of water removal (WRR) from the product, which is stimulated by the increase in temperature of the drying air. This increase in WRR, upon creating a stark difference between the peripheral and inner part of the product, promotes the formation of seed coat cracks and microfissures in the cotyledons, affecting the quality of the seed. In addition, situations such as these can increase the susceptibility of the material to latent damage, or even worsen deterioration, thus reducing the storage potential of the product and its physiological quality (Mbofung et al., 2013).

The deleterious effect portraying the damage caused by drying in soybean seeds was also observed by Silva et al. (2007) upon finding that the use of drying temperatures higher than 40 °C led to cell damage, such as membrane disarray and leaching of solutes, mainly in the region of the embryonic axis, harmful to the development of seedlings and thus reducing the two main physiological properties, germination and vigor. In addition, as found by Vieira et al. (2013a), the initial quality of a soybean seed lot and the conditions during storage determine maintenance of seed physiological potential since undamaged seeds undergo less reduction in seed coat thickness when subjected to either storage under cold conditions (10 °C) or conditions already considered harmful to soybean physiological potential (25 °C).

Evaluations of seedling performance also manifested the negative effects of the increase in drying temperature and of storage on the physiological quality of soybean seeds. It was observed that the total length of seedlings (Figure 4a) and the length of the hypocotyl (Figure 4b) and of the root (Figure 4c) showed decreasing results in accordance with an increase in temperature of drying air, and linear degradation of seed response over the storage time .

Evaluation of the length of seedlings and their structures compared to germination percentage exhibited greater sensitivity in differentiating the lots in regard to physiological quality immediately after drying. Although it was found that both analyses were negatively affected by the drying temperatures, associated with prolongation of the storage period, for germination, the immediate difference found was evident after the application of the temperature of 60 °C (Figure 3), whereas by evaluation of the seedling characteristics, the differences between the thermal treatments applied to the seeds were immediately observable from the temperatures of 40 and 50 °C (Figure 4). According to the review of Garcia et al. (2004), thermal damages might not manifest immediate effects on germination; however, after a period of storage, seed vigor may undergo considerable reductions.

It should be noted that analysis of seedling growth and other complementary tests compose a vigor index for evaluation of soybean seed quality (Vanzolini et al., 2007), and, from the results obtained, it may be inferred that the deleterious effects of high temperatures during drying were more expressive on the growth of the shoots and root, which may have exhibited greater sensitivity to the treatments.

Increasing the temperature factor in soybean seeds, especially at temperatures above 40 °C, compromises seedling growth since an increase in drying rate leads to damages to meristematic tissues and, consequently, injured development of the embryonic axis (Afrakhteh et al., 2013). In this context, Costa et al. (1999) highlight that the greater the length of the hypocotyl, the greater the seedling capacity to emerge and overcome the resistance that may be posed by depth of sowing and/or hardened soil.

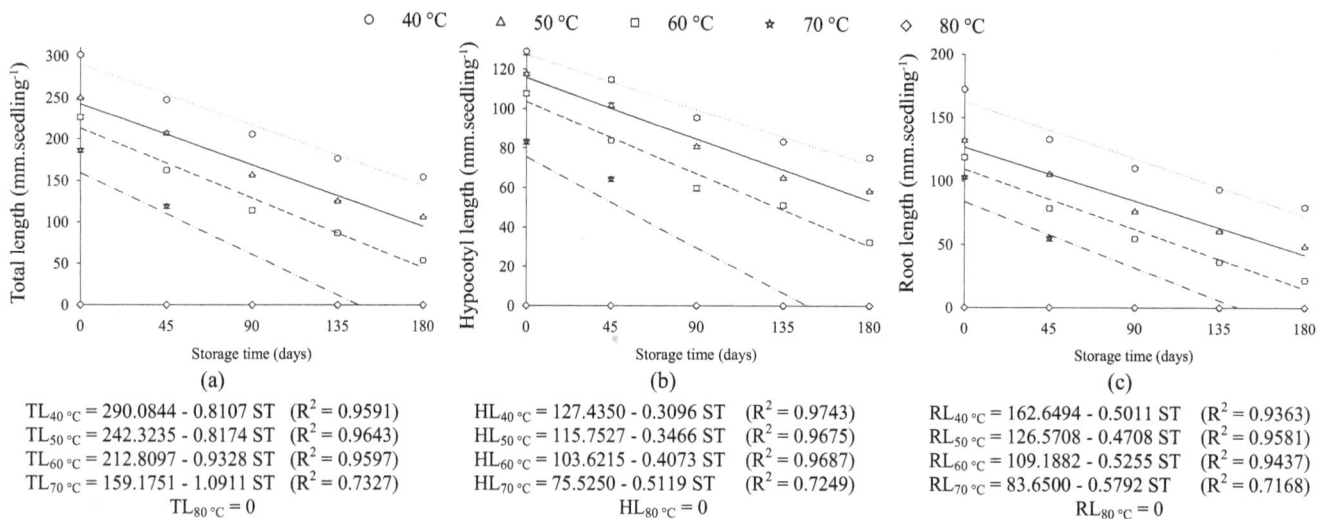

$TL_{40\,°C} = 290.0844 - 0.8107\ ST\quad (R^2 = 0.9591)$
$TL_{50\,°C} = 242.3235 - 0.8174\ ST\quad (R^2 = 0.9643)$
$TL_{60\,°C} = 212.8097 - 0.9328\ ST\quad (R^2 = 0.9597)$
$TL_{70\,°C} = 159.1751 - 1.0911\ ST\quad (R^2 = 0.7327)$
$TL_{80\,°C} = 0$

$HL_{40\,°C} = 127.4350 - 0.3096\ ST\quad (R^2 = 0.9743)$
$HL_{50\,°C} = 115.7527 - 0.3466\ ST\quad (R^2 = 0.9675)$
$HL_{60\,°C} = 103.6215 - 0.4073\ ST\quad (R^2 = 0.9687)$
$HL_{70\,°C} = 75.5250 - 0.5119\ ST\quad (R^2 = 0.7249)$
$HL_{80\,°C} = 0$

$RL_{40\,°C} = 162.6494 - 0.5011\ ST\quad (R^2 = 0.9363)$
$RL_{50\,°C} = 126.5708 - 0.4708\ ST\quad (R^2 = 0.9581)$
$RL_{60\,°C} = 109.1882 - 0.5255\ ST\quad (R^2 = 0.9437)$
$RL_{70\,°C} = 83.6500 - 0.5792\ ST\quad (R^2 = 0.7168)$
$RL_{80\,°C} = 0$

Figure 4. Results of the total length of the seedling (a) and the length of its structures, the hypocotyl (b) and root (c), of soybean seeds subjected to different drying temperatures as a function of storage time.

In regard to drying temperatures of 70 and 80 °C, not only was there a response of smaller seedlings but also decreased quality of the seed lots (Figure 4). Seeds dried at 70 °C, for example, led to formation of smaller seedlings and shorter total length (159.2 mm.seedling⁻¹) and length of the shoots (75.5 mm.seedling⁻¹) and of the root (83.6 mm.seedling⁻¹), but they also exhibited the highest results of latent damage, which possibly impeded formation of normal seedlings that could be measured at 150 days after the storage beginning . Drying at the temperature of 70 °C also damage the physiological quality of crambe seeds, which exhibited results analogous to those found here and the occurrence of smaller seedlings, given the higher rate of injuries to the seeds (Faria et al., 2014). The soybean seeds dried at 80 °C, both immediately after drying and during storage, were not able to generate normal seedlings, indicating the harmful effects of high temperatures on the quality of soybean seeds.

Although the seedling structures exhibit similar responses in accordance with an increase in drying temperature and storage time, the susceptibility of the primary root immediately after drying and over the storage time should be noted (Figure 4c). Possibly, since the radicle is located in a region slightly more external in the seed compared to the hypocotyl, its exposure to drying and to the deterioration processes was more evident, indicating that the performance of the root constitutes a good indication in distinguishing among seed lots, as has already been proposed by Vanzolini et al. (2007) and recommended by Chauhan (1985) precisely for soybean seeds, and as observed for seeds of wheat (Maia et al., 2007), maize (Venancio et al., 2012), and papaya (Mengarda et al., 2015).

The harmful effects of drying temperatures above 40 °C, along with seed storage, were also observed in the dry matter weight of seedlings and their structures since the dry matter weights of the seedling, the hypocotyl, and the root coming from seeds dried at 40 °C were greater by 5.6, 3.3, and 2.3 mg.seedling⁻¹ in relation to the results observed for the temperature of 50 °C (Figure 5).

By the multiplicative coefficients of the equations fitted to the data observed for total dry matter weight of the seedling, it was found that the immediate effect arising from increasing the temperature of drying air was the determining factor in maintaining the quality of seeds during storage because, in accordance with the increase in temperature, there was greater latent loss of total accumulated dry matter (Figure 5). Thus, evaluation allowed confirmation of the ability of the test in showing both the immediate effects and latent effects caused by an increase in the temperature of drying air, as already described by Surki et al. (2012), also for soybean seeds, upon manifesting the interference of both factors on the dynamic of the reserves during seedling development.

The daily loss brought about by the temperature of 70 °C, for example, was twice that observed for seeds dried at 40 °C, indicating the prejudicial effects on the capacity for transferring reserves for growth of the embryonic axis in seeds dried at high temperatures and then stored. Possibly, reduction in the initial quality of the lots attributed to the increase in temperature may have increased the susceptibility of the seeds to deterioration during storage, negatively affecting accumulation of dry matter during the seedling phase. In this respect, with the increase in drying temperature and in storage time, the mechanisms directly connected with the processes of translocation and transformation of cotyledonary reserves into substances that can be assimilated by the embryonic axis were affected, restricting the accumulation of dry matter coming from the cotyledons.

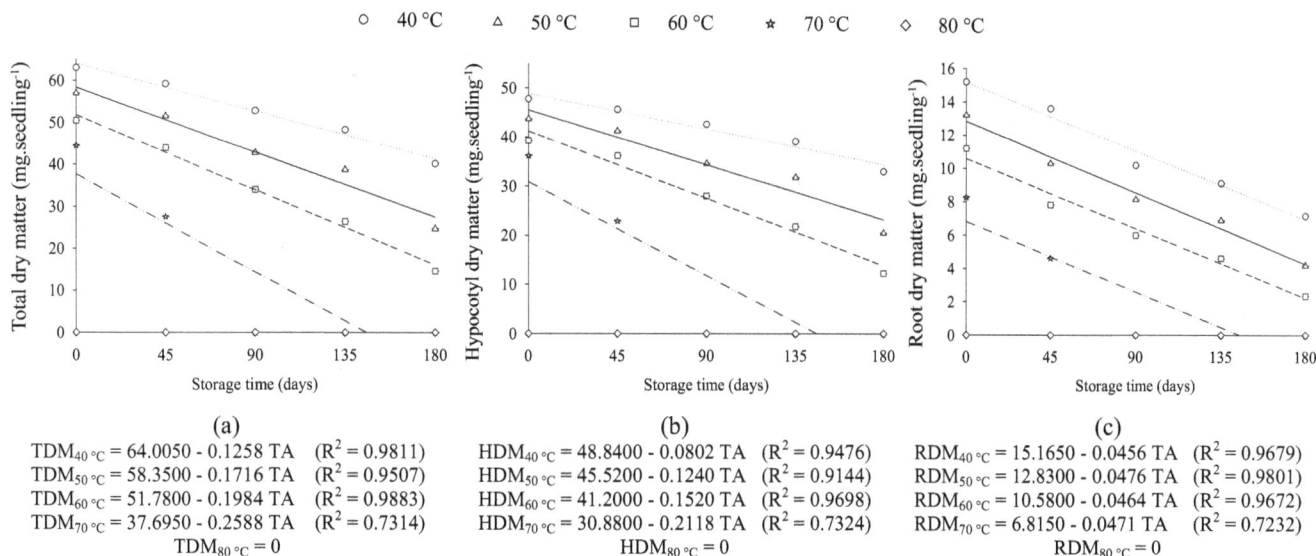

Figure 5. Results of the total dry matter of the seedling (a) and of its structures, the hypocotyl (b) and root (c), of soybean seeds subjected to different drying temperatures as a function of storage time.

Conclusions

The increase in the temperature of drying air above 40 °C affects seedling performance, and this effect is accentuated over the time of storage, above all on root growth.

The air temperature of 40 °C can be recommended for drying soybean seeds along with a storage period of 180 days under storage conditions that are not climate controlled.

Acknowledgements

To the Coordenação de Aperfeiçoamento de Pessoal de Nível Superior (Capes) and the Fundação de Apoio ao Desenvolvimento do Ensino, Ciência e Tecnologia do Estado de Mato Grosso do Sul (Fundect), due the financial support for research.

References

AFRAKHTEH, S.; FRAHMANDFAR, E.; HAMIDI, A., RAMANDI, H.D. Evaluation of growth characteristics and seedling vigor in two cultivars of soybean dried under different temperature and fluidized bed dryer. *International Journal of Agriculture and Crop Sciences*, v.5, n.21, p.2537-2544, 2013. http://ijagcs.com/wp-content/uploads/2013/09/2537-2544.pdf

BARBOSA, R.M.; VIEIRA, B.G.T.L.; MARTINS, C.C.; VIEIRA, R.D. Qualidade fisiológica e sanitária de sementes de amendoim durante o processo de produção. *Pesquisa Agropecuária Brasileira*, v.49, n.12, p.977-985, 2014. http://seer.sct.embrapa.br/index.php/pab/article/view/19217/12839

BARROZO, M.A.S.; MUJUMDAR, A.; FREIRE, J.T. Air-Drying of seeds: A review. Drying Technology: *An International Journal*, v.32, n.10, p.1127-1141, 2014. http://www.tandfonline.com/doi/pdf/10.1080/07373937.2014.915220

BENINCASA, M.M.P. Análise de crescimento de plantas: noções básicas. 2.ed. Jaboticabal: FUNEP, 2003. 41p.

BESSA, J.F.V.; DONADON, J.R.; RESENDE, O.; ALVES, R.M.V.; SALES, J.F.; COSTA, L.M. Armazenamento do crambe em diferentes embalagens e ambientes: parte I - qualidade fisiológica. *Revista Brasileira de Engenharia Agrícola e Ambiental*, v.19, n.3, p.224-230, 2015. http://www.agriambi.com.br/revista/v19n03/v19n03a05.pdf

BRASIL. MAPA, *Instrução Normativa nº45, de 17 de setembro de 2013.* Padrões de Identidade e Qualidade para a produção e a comercialização de sementes. Seção 1. Ministério da Agricultura, Pecuária e Abastecimento. *Diário Oficial da União* de 20/09/2013. Brasília. http://www.abrasem.com.br/wp-content/uploads/2012/10/Instrução-Normativa-nº-45-de-17-de-Setembro-de-2013-Padrões-de-Identidade-e-Qualiidade-Prod-e-Comerc-de-Sementes-Grandes-Culturas-Republicação-DOU-20.09.13.pdf

BRASIL. Ministério da Agricultura, Pecuária e Abastecimento. *Regras para análise de sementes*. Ministério da Agricultura, Pecuária e Abastecimento. Secretaria de Defesa Agropecuária. Brasília: MAPA/ACS, 2009. 395p. http://www.agricultura.gov.br/arq_editor/file/2946_regras_analise__sementes.pdf

CHAUHAN, K.P.S. The incidence of deterioration and its localization in aged seeds of soybean and barley. *Seed Science and Technology*, v.13, n.3, p.769-773, 1985.

CARVALHO, E.R.; OLIVEIRA, J.A.; MAVAIEIE, D.P.R.; SILVA, H.W.; LOPES, C.G.M. Pre-packing cooling and types of packages in maintaining physiological quality of soybean seeds during storage. *Journal of Seed Science*, v.38, n.2, p.129-139, 2016. http://dx.doi.org/10.1590/2317-1545v38n2158956

CORRÊA, P.C.; MACHADO, P.F.; ANDRADE, E.T. Cinética de secagem e qualidade de grãos de milho-pipoca. *Ciência e Agrotecnologia*, v.25, n.1, p.134-142, 2001. http://www.editora.ufla.br/index.php/revistas/ciencia-e-agrotecnologia/artigos-publicados/10-volumes-revista/53-vol25numero1

COSTA, J.A.; PIRES, J.L.F.; THOMAS, A.L.; ALBERTON, M. Comprimento e índice de expansão radial do hipocótilo de cultivares de soja. *Ciência Rural*, v.29, n.4, p.609-612, 1999.http://www.scielo.br/pdf/cr/v29n4/a06v29n4.pdf

DELIBERALI, J.; OLIVEIRA, M.; DURIGON, A.; DIAS, A.R.G.; GUTKOSKI, L.C.; ELIAS, M.C. Efeitos de processo de secagem e tempo de armazenamento na qualidade tecnológica de trigo. *Ciência e Agrotecnologia*, v.34, n.5, p.1285-1292, 2010.http://www.scielo.br/pdf/cagro/v34n5/29.pdf

DIOS, C.A.D. Recomendaciones sobre el manejo y poscosecha del girasol. Pergamino: Estación Experimental Agropecuária de Pergamino, INTA, p.251-261, 1984.

FARIA, R.Q.; TEIXEIRA, I.R.; CUNHA, D.A.; HONORATO, J.M.; DEVILLA, A. Qualidade fisiológica de sementes de crambe submetidas à secagem. *Revista Ciência Agronômica*, v.45, n.3, p.453-460, 2014.http://www.ccarevista.ufc.br/seer/index.php/ccarevista/article/view/2306/971

FEHR, W.R.; CAVINESS, C.E. Stages of soybean development. Special Report, n.80, Ames: Iowa State University of Science and Technology, 1977. 11p.

FERREIRA, A.G.; BORGHETTI, F. (Orgs). Germinação: do básico ao aplicado. Porto Alegre: Artmed, 2004. 324p.

FERREIRA, D.F. Sisvar: um sistema computacional de análise estatística. *Ciência e Agrotecnologia*, v.35, n.6, p.1039-1042, 2011. http://www.scielo.br/pdf/cagro/v35n6/a01v35n6.pdf

FRANÇA-NETO, J.B.; KRZYZANOWSKI, F.C.; HENNING, A.A.; PÁDUA, G.P. Tecnologia da produção de semente de soja de alta qualidade. *Informativo ABRATES*, v.20, n.3, p.26-32, 2010.

GARCIA, D.C.; BARROS, A.C.S.A.; PESKE, S.T.; MENEZES, N.L. A secagem de sementes. *Ciência Rural*, v.34, n.2, p.603-608, 2004.http://www.scielo.br/pdf/cr/v34n2/a45v34n2.pdf

HENNING, F.A.; MERTZ, L.M.; JACOB JUNIOR, E.A.; MACHADO, R.D.; FISS, G.; ZIMMER, P.D. Composição química e mobilização de reservas em sementes de soja de alto e baixo vigor. *Bragantia*, v.69, n.3, p.727-734, 2010. http://www.scielo.br/pdf/brag/v69n3/26.pdf

MAHJABIN, S.B.; ABIDI, A.B. Physiological and biochemical changes during seed deterioration: a review. *International Journal of Recent Scientific Research*, v.6, n.4, p.3416-3422, 2015. http://www.recentscientific.com/sites/default/files/2185.pdf

MAIA, A.R.; LOPES, J.C.; TEIXEIRA, C.O. Efeito do envelhecimento acelerado na avaliação da qualidade fisiológica de sementes de trigo. *Ciência e Agrotecnologia*, v.31, n.3, p.678-684, 2007. http://www.scielo.br/pdf/cagro/v31n3/a12v31n3.pdf

MARCOS-FILHO, J. *Fisiologia de sementes de plantas cultivadas*. 2ed. Londrina, PR: ABRATES, 2015. 660p.

MARCOS-FILHO, J. Importância do potencial fisiológico da semente de soja. *Informativo ABRATES*, v.23, n.1, p.21-24, 2013.http://www.abrates.org.br/images/Informativo/v23_n1/02._Julio_Importancia_Fisiologia.pdf

MBOFUNG, G.C.Y.; GOGGI, A.S.; LEANDRO, L.F.S.; MULLEN, R.E. Effects of storage temperature and relative humidity on viability and vigor of treated soybean seeds. *Crop Science*, v.53, n.3, p.1086-1095, 2013.https://dl.sciencesocieties.org/publications/cs/articles/53/3/1086

MENEZES, N.L.; CICERO, S.M.; VILLELA, F.A.; BORTOLOTTO, R.P. Using x-rays to evaluate fissures in rice seeds dried artificially. *Revista Brasileira de Sementes*, v.34, n.1, p.70-77, 2012a. http://www.scielo.br/pdf/rbs/v34n1/a09v34n1.pdf

MENEZES, N.L.; PASQUALLI, L.L.; BARBIERI, A.P.P.; VIDAL, M.D.; CONCEIÇÃO, G.M. Temperaturas de secagem na integridade física, qualidade fisiológica e composição química de sementes de arroz. *Pesquisa Agropecuária Tropical*, v.42, n.4, p.430-436, 2012b.https://www.revistas.ufg.br/index.php?-journal=pat&page=article&op=view&path%5B%5D=18457&path%5B%5D=12715

MENGARDA, L.H.G.; LOPES, J.C.; ZANOTTI, R.F.; ALEXANDRE, R.S. Desempenho de genótipos de mamoeiro quanto à qualidade física e fisiológica de sementes e análises de diversidade. *Bioscience Journal*, v.31, n.3, p.719-729, 2015.http://www.seer.ufu.br/index.php/biosciencejournal/article/view/23852/16457

NAKAGAWA, J. Teste de vigor baseados no desempenho de plântulas. In: KRZYZANOWSKI, F.C.; VIEIRA, R.D.; FRANÇA-NETO, J.B. (Eds.). *Vigor de sementes*: conceitos e testes. Londrina: ABRATES, 1999. p.2-1 a 2-24.

PARAGINSKI, R.T.; ROCKENBACH, B.A.; SANTOS, R.F.; ELIAS, M.C.; OLIVEIRA, M. Qualidade de grãos de milho armazenados em diferentes temperaturas. *Revista Brasileira de Engenharia Agrícola e Ambiental*, v.19, n.4, p.358-363, 2015.http://www.agriambi.com.br/revista/v19n04/v19n04a09.pdf

PEREIRA, W.A.; PEREIRA, S.M.A.; DIAS, D.C.F.S. Dynamics of reserves of soybean seeds during the development of seedlings of different commercial cultivars. *Journal of Seed Science*, v.37, n.1, p.63-69, 2015. http://dx.doi.org/10.1590/2317-1545v37n1142202

PEREIRA, W.A.; PEREIRA, S.M.A.; DIAS, D.C.F.S. Influence of seed size and water restriction on germination of soybean seeds and on early development of seedlings. *Journal of Seed Science*, v.35, n.3, p.316-322, 2013. http://www.scielo.br/pdf/jss/v35n3/07.pdf

PESKE, S.T.; BAUDET, L.M.; VILLELA, F.A. Tecnologia de pós-colheita para sementes. In: SEDIYAMA, T. (Ed.) *Tecnologias de produção de sementes de soja*. Londrina. 2013. p.327-344.

RATHINAVEL, K. Influence of storage temperature and seed treatments on viability of cotton seed (*Gossypium hirsutum* L.). *Cotton Research Journal*, v.6, n.1, p.1-6, 2014. http://www.cicr.org.in/isci/6-1/Paper_1.pdf

RESENDE, O.; ALMEIDA, D.P.; COSTA, L.M.; MENDES, U.C.; SALES, J.F. Adzuki beans (*Vigna angularis*) seed quality under several drying conditions. *Ciência e Tecnologia de Alimentos*, v.32, n.1, p.151-155, 2012. http://www.scielo.br/pdf/cta/v32n1/aop_cta_5007.pdf

SCHUH, G.; GOTTARDI, R.; FERRARI FILHO, E.; ANTUNES, L.E.G.; DIONELLO, R.G. Efeitos de dois métodos de secagem sobre a qualidade físico-química de grãos de milho safrinha - RS, armazenados por 6 meses. *Semina: Ciências Agrárias*, v.32, n.1, p.235-244, 2011.http://www.uel.br/revistas/uel/index.php/semagrarias/article/view/3601/7190

SILVA, P.A.; DINIZ, K.A.; OLIVEIRA, J.A.; VON PINHO, E.V.R. Análise fisiológica e ultra-estrutural durante o desenvolvimento e a secagem de sementes de soja. *Revista Brasileira de Sementes*, v.29, n.2, p.15-22, 2007. http://www.scielo.br/pdf/rbs/v29n2/v29n2a03.pdf

SMANIOTTO, T.A.S.; RESENDE, O.; MARÇAL, K.A.F.; OLIVEIRA, D.E.C.; SIMON, G.A. Qualidade fisiológica das sementes de soja armazenadas em diferentes condições. *Revista Brasileira de Engenharia Agrícola e Ambiental*, v.18, n.4, p.446-453, 2014. http://www.agriambi.com.br/revista/v18n04/v18n04a13.pd f

SURKI, A.A.; SHARIFZADEH, F.; AFSHARI, R.T. Effect of drying conditions and harvest time on soybean seed viability and deterioration under different storage temperature. *African Journal of Agricultural Research*, v.7, n.36, p.5118-5127, 2012. http://www.academicjournals.org/journal/AJAR/article-full-text-pdf/D9AEAAA35317

TIECKER JUNIOR, A.; GUIMARÃES, L.E.; FERRARI FILHO, E.; CASTRO, B.; DEL PONTE, E.M.; DIONELLO, R.G. Qualidade físico-química de grãos de milho armazenados com diferentes umidades em ambientes hermético e não hermético. *Revista Brasileira de Milho e Sorgo*, v.13, n.2, p.174-186, 2014.http://rbms.cnpms.embrapa.br/index.php/ojs/article/view/484/pdf_93

ULLMANN, R.; RESENDE, O.; CHAVES, T.H.; OLIVEIRA, D.E.C.; COSTA, L.M. Qualidade fisiológica das sementes de sorgo sacarino submetidas à secagem em diferentes condições de ar. *Revista Brasileira de Engenharia Agrícola e Ambiental*, v.19, n.1, p.64-69, 2015. http://www.agriambi.com.br/revista/v19n01/v19n01a11.pdf

ULLMANN, R.; RESENDE, O.; SALES, J.F.; CHAVES, T.H. Qualidade das sementes de pinhão manso submetidas à secagem artificial. *Revista Ciência Agronômica*, v.41, n.3, p. 442-447, 2010. http://www.ccarevista.ufc.br/seer/index.php/ccarevista/article/view/911/463

VANZOLINI, S.; ARAKI, C.A.S.; SILVA, A.C.T.M.; NAKAGAWA, J. Teste de comprimento de plântula na avaliação da qualidade fisiológica de sementes de soja. *Revista Brasileira de Sementes*, v.29, n.2, p.90-96, 2007. http://www.scielo.br/pdf/rbs/v29n2/v29n2a12.pdf

VENANCIO, L.P.; LOPES, J.C.; MACIEL, K.S.; COLA, M.P.A. Teste do envelhecimento acelerado para avaliação da qualidade fisiológica de sementes de milho. *Enciclopédia Biosfera*, v.8, n.14, p.899- 906, 2012. http://www.conhecer.org.br/enciclop/2012a/agrarias/teste%20do.pdf

VIEIRA, B.G.T.L.; BARBOSA, G.F.; BARBOSA, R.M.; VIEIRA, R.D. Structural changes in soybean seed coat due to harvest time and storage. *Journal of Food, Agriculture & Environment*, v.11, n.1, p.625-628, 2013a. http://world-food.net/download/journals/2013-issue_1/2013-issue_1-agriculture/67.pdf

VIEIRA, R.D.; PANOBIANCO, M.; MARCOS-FILHO, J. Avaliação do potencial fisiológico de sementes. In: SEDIYAMA , T. (Ed.). Tecnologias de produção de sementes de soja. Londrina: ed. Mecenas, 2013b. p.109-127.

Germination and initial development of *Brachiaria brizantha* and *Brachiaria decumbens* on exposure to cadmium, lead and copper

Karine Sousa Carsten Borges[1], Raquel Custódio D'Avila[2], Mari Lúcia Campos[3*],
Cileide Maria Medeiros Coelho[4], David José Miquelluti[5], Natiele da Silva Galvan[6]

ABSTRACT – The objective was to evaluate the germination and initial development of three cultivars of *Brachiaria brizantha* and *Brachiaria decumbens* in the presence of trace elements of Cd, Pb and Cu and quantify the Cd and Cu contents in plant tissue of these species. First, seed germination occurred in towel paper containing cadmium, lead, copper and the control. We calculated the percentage of germination, germination speed index, vigor index and seedling length. Later, the seeds were germinated in soil pots contaminated with Cd and Cu, where the Soil-Plant Analysis Development index (SPAD index) was evaluated, as well as dry matter of shoot and root, and the Cd and Cu metal content in plant tissue was quantified. *B. decumbens* showed sensitivity to the elements studied, with decreases in all parameters. Cu was the element that caused more toxic effects on germination and early seedling development of the species studied. In contaminated soil, the species studied showed no differences in dry matter production of shoots, but the exposure to Cd and Cu caused reductions in SPAD index in all species. The highest Cd and Cu contents were found in the roots.

Index terms: inhibition, phytotoxicity, heavy metals.

Germinação e desenvolvimento inicial da *Brachiaria brizantha* e *Brachiaria decumbens* em exposição de cádmio, chumbo e cobre

RESUMO – O objetivo no trabalho foi avaliar a germinação e o desenvolvimento inicial de três cultivares de *Brachiaria brizantha* e a *Brachiaria decumbens* em presença dos elementos-traço Cd, Pb e Cu e quantificar os teores de Cd e Cu no tecido vegetal dessas espécies. A germinação das sementes ocorreu em papel germitest®, contendo cádmio, chumbo, cobre além da testemunha. Foram calculadas a porcentagem de germinação, índice de velocidade de germinação, índice de vigor e comprimento da plântula. Posteriormente, as sementes foram germinadas em vasos de solo contaminado com Cd e Cu, onde se avaliou o índice Soil Plant Analysis Development (SPAD), matéria seca da parte aérea e da raiz e quantificado o teor de metais Cd e Cu no tecido vegetal. A *B. decumbens* apresentou sensibilidade aos elementos estudados, apresentando decréscimos de todos os parâmetros avaliados. O Cu foi o elemento que mais causou efeitos tóxicos à germinação e desenvolvimento inicial das plântulas das espécies estudadas. Em solo contaminado, as espécies estudadas não apresentaram diferenças na produção de massa seca de parte aérea, porém a exposição de Cd e Cu ocasionou reduções do índice SPAD em todas as espécies. Os maiores teores de Cd e Cu foram encontrados nas raízes.

Termos para indexação: inibição, fitotoxidade, metais pesados.

Introduction

Human activity has increasingly raised trace element levels in the environment. Plants are the main entry for these elements into the food chain. Most trace elements are known as growth inhibitors and exert negative effects on plants, which may lead to broader phytotoxicity responses and decrease the yield and quality in agricultural crops (Gratão et al., 2005; Yang et al., 2010).

Some trace elements, such as Cd (cadmium) and Pb (lead), are considered toxic even in minute concentrations, causing deleterious effects on plants. Such deleterious

[1]Centro Universitário Leonardo da Vinci, Grupo UNIASSELVI, 89130-000 – Indaial, SC, Brasil.
[2]Fundação Municipal do Meio Ambiente, 88160-126 – Biguaçu, SC, Brasil.
[3]Departamento de Solos e Recursos Naturais, Universidade Estadual de Santa Catarina, 88520-000 – Lages, SC, Brasil.
[4]Departamento de Agronomia, Universidade Estadual de Santa Catarina, 88520-000 – Lages, SC, Brasil.
[5]Departamento de Solos e Recursos Naturais, Universidade Estadual de Santa Catarina, 88520-000 – Lages, SC, Brasil.
[6]Universidade Estadual de Santa Catarina, 88520-000 – Lages, SC, Brasil.
*Corresponding author <mari.lucia03@gmail.com>

effects can be observed on seedling growth, changes in the structure of chloroplasts, inhibition of photosynthesis, chlorosis, induced lipidic peroxidation, suppression of germination, reduction of the root system, disturbances in plant metabolic activities and reduction of plant biomass (Guimarães et al., 2008; Gill et al., 2013).

In addition, plants may exhibit indirect effects caused by these trace elements such as inhibition of water absorption and nutrient deficiency. Mineral nutrition disorders arise from deleterious effects caused by trace elements on the metabolism of essential elements, including calcium, magnesium, potassium, iron, zinc, manganese and coppe (Kabata-Pendias, 2011).

Cu (copper), on the other hand, plays a significant role in physiological processes such as photosynthesis and respiration, among others (Yruela, 2009; Karimi et al., 2012). However, in environmental conditions where copper is found in excess in the soil, plants may exhibit symptoms of toxicity which culminates in physiological disturbances inhibiting plant growth (Karimi, et al., 2012; Kabata-Pendias, 2011). Toxicity caused by Cu causes damage and disturbance in the integrity of the thylakoid membranes and photosynthesis impairment, which result in chlorosis or necrosis and inhibition of root and shoot growth (Yruela, 2009; Fidalgo et al., 2013).

In areas contaminated by trace elements, grasses are more promising in their establishment because they are relatively easy to develop, promote rapid growth in soil cover, improve soil physical and chemical structures, help in the cycling of nutrients and increase the soil organic matter content (Amaral et al., 2012).

Among grasses, genus *Brachiaria* spp. stands out due to high dry matter, easy cultivation and adaptation to different soils, allowing cultivation throughout the year, and low maintenance cost of the cultivated area (Lucena et al., 2010). However, few studies are carried out with this genus on trace element contamination (Gomes et al., 2011).

The present work aims to evaluate the germination and initial development of three cultivars of the species *Brachiaria brizantha* (cv. Piatã, Marandu and MG 5) and the species *Brachiaria decumbens* in the presence of trace elements of Cd, Pb and Cu, and quantify the content of Cd and Cu in its plant tissue.

Materials and Methods

The experiment was conducted in two distinct stages. In the first step, the experiment was conducted in a Seed Testing Laboratory. The second part of the experiment was carried out in a greenhouse and analyses were performed in a Laboratory of Environmental Survey.

Four different Brachiaria varieties were used, being three cultivars of *Brachiaria brizantha* (cv. Piatã, Marandu, MG 5)

and one of *Brachiaria decumbens*. The seeds used came from a commercial agriculture and livestock farm in the Brazilian municipality Lauro Müller, SC. Seeds purity was determined following the protocol suggested by Rules for Seed Testing (Brasil, 2009) (Table 1). Before all the tests, seeds were submitted to a sanitary treatment in a 3% solution of sodium hypochlorite for five minutes and afterwards they were washed in distilled water.

Table 1.   Characterization of the physical purity of seeds of *B. brizantha* (cv. Piatã, Marandu, MG 5) and *B. decumbens*.

Species	Percentage of purity %
B. brizantha (cv. Piatã)	71.5
B. brizantha (cv. Marandu)	55.6
B. brizantha (cv. MG 5)	73.5
B. decumbens	66.6

1st step – Germination test in the presence of Cd, Cu and Pb.

Seeds were placed on towel paper rolls for germination. In order to moisture the paper, 2.5 times the paper dry weight was used with solutions containing $Cd(NO_3)_2$, $Pb(NO_3)_2$ or $Cu(NO_3)_2$. The concentrations used followed the agricultural research values proposed by CONAMA (National Council for the Environment) Resolution no. 420 (2009) (3, 180 and 200 $mg.L^{-1}$ respectively), besides the control (0 $mg.L^{-1}$). Then the rolls were kept in a Mangelsdorf-type germinator at a temperature of 25 °C with a natural photoperiod. For the experimental unit, 50 seeds were considered, with four replicates for each treatment. The seeds were evaluated by the following parameters:

Germination Percentage (GP) – It was calculated by the number of normal seedlings (shoot and root) identified on the last day of the experiment, following the recommendations by the Rules for Seed Testing (Brasil, 2009).

Germination speed index (GSI) – Evaluations were performed every 24 hours until germination stabilization, which occurred on the tenth day. Seedlings were considered germinated when they reached root length greater than 5 mm and there was plumule rupture. To obtain the index, the formula proposed by Maguire (1962) was applied.

Length of shoot and main root – At the end of the germination test, the length of shoot and root of each seedling was measured in centimeters.

Vigor Index (VI) – Measurement of root length multiplied by the germination percentage (Dezfuli et al., 2008).

The germination percentage data were transformed to $arcsen\sqrt{x}/100$ and trace element content data were analyzed after logarithmic transformation, Y = log (X + 1). All data were submitted to analysis of variance (p ≤ 0.05) considering a factorial arrangement of the treatments (varieties and trace elements). When statistical significance was found, the Tukey's test (p ≤ 0.05) was used to verify the magnitude of the difference among treatments.

2nd step – Development in a greenhouse.

Plastic pots contained 0.5 kg of Haplic Cambisol (profile from 0 to 10 cm), collected in a natural environment in the Brazilian municipality of Lages, SC. The chemical characteristics of this soil are presented in Table 2.

Table 2.   Chemical characteristics and clay in the soil sample (Haplic Cambisol) used in the experiment.

pH H$_2$O	CTC$_{pH\ 7.0}$	Al^{+3}	Ca	Mg	K	P	Clay	CO^1
		Cmolc.dm^{-3}			mg.dm^{-3}		%	
5.0	35.2	56.1	3.3	1.0	52	0.6	34	1.4

Note: [1]Organic carbon.

For soil contamination, the concentrations of the elements also followed the values of agricultural research proposed by CONAMA (2009), 3 mg.kg^{-1} for Cd and 200 mg.kg^{-1} for Cu. Soil contamination occurred after soil drying and sieving in a 2-mm sieve, where each trace element represented a treatment. The soil, after being contaminated, was incubated for 30 days at humidity above field capacity to stabilize the chemical conditions.

A completely randomized design with four replications was used, with each pot receiving 10 seeds. For plants full development, soil moisture was maintained in field capacity and luminosity in natural conditions of intensity and photoperiod. The temperature remained between 15 and 25 °C and air moisture controlled between 70 and 95%.

Ten days after the plants emergence, thinning was carried out, keeping five plants in each pot, making up the experimental unit. On the 30th day, the leaves green intensity values were measured [Soil-Plant Analysis Development index (SPAD)], obtained by means of the SPAD 502 (Konica Minolta®, Tokyo, Japan) chlorophyll content portable meter in the period between 8-10 h am. The green intensity value considered was the average of readings carried out on five leaves having no physical damages or symptoms of pest and disease attack, randomly sampled in the five plants of each experimental unit.

After the plants development, these were collected and separated in shoot and root. The vegetative samples were washed with distilled water and dried in a forced-air circulation oven at a temperature of approximately 65 °C until reaching constant weight. After drying, the material was weighed to evaluate the shoot dry matter (SDM) and root dry matter (RDM) yields and processed in a plant tissue grinder. Quantification of the Cd and Cu elements in the plant was carried out in the shoot and root. Samples were submitted to acid digestion in Teflon® [polytetrafluoroethylene (PTFE)] tubes, according to the USEPA (The United States Environmental Protection Agency) 3051

method in a Multiwave 3000 (USEPA, 1994) microwave oven. In each battery a NIST 1573A Tomato Leaves reference sample was inserted, where 0.77 mg.kg^{-1} of Cd were obtained while what was expected was 1.52 ± 0.04 mg.kg^{-1} and 4.01 mg.kg^{-1} of Cu while what was expected was 4.70 mg.kg^{-1} ±.0.14 mg.kg^{-1}. The contents of the trace elements of Cd and Cu were quantified in a high resolution atomic absorption spectrometer and electrothermal atomization (CONTRAA 700 – Analytik Jena). Reading conditions were those indicated by the manufacturer. For the available soil contents, the Tedesco (1995) methodology was used. The contents were quantified in atomic absorption spectrophotometry (AA Perkin Elemer – A Analyst 100).

Data were submitted to the Shapiro-Wilk normality tests and the homogeneity of variance Levene's test, applying the required scale transformations when appropriate. Then the data were submitted to analysis of variance according to the completely randomized design in a factorial arrangement. The Tukey's averages comparison test was used at 5% probability to discriminate the effect of factor levels when this was the case.

Results and Discussion

Germination test in the presence of Cd, Cu and Pb.

For species *B. decumbens* there reductions in GP in exposure to all trace elements studied (Figure 1 A). The cv. Piatã, Marandu and MG 5 of *B. brizantha* have not presented difference in the germination percentage (GP) in exposure to Cd. In exposure to Pb, cultivars Piatã and Marandu have not presented GP differences either. However, cv. MG 5 has had 34% reduction in GP and *B. decumbens* 12% when comparing with the control.

Exposure to Cu has reduced GP in all species studied. However, it was for *B. decumbens* that Cu promoted the highest germination inhibition, because GP in exposure to Cu was only 1%, i.e., a reduction of 99% compared to the control.

The absorption of trace elements in brachiaria promotes oxidative stress, compromising the epidermis walls, which become thinner in relation to exodermis, resulting in the roots cellular degradation. In addition, the presence of trace elements accelerates maturation of the cell wall in mesoderm and endoderm (Gomes et al., 2011).

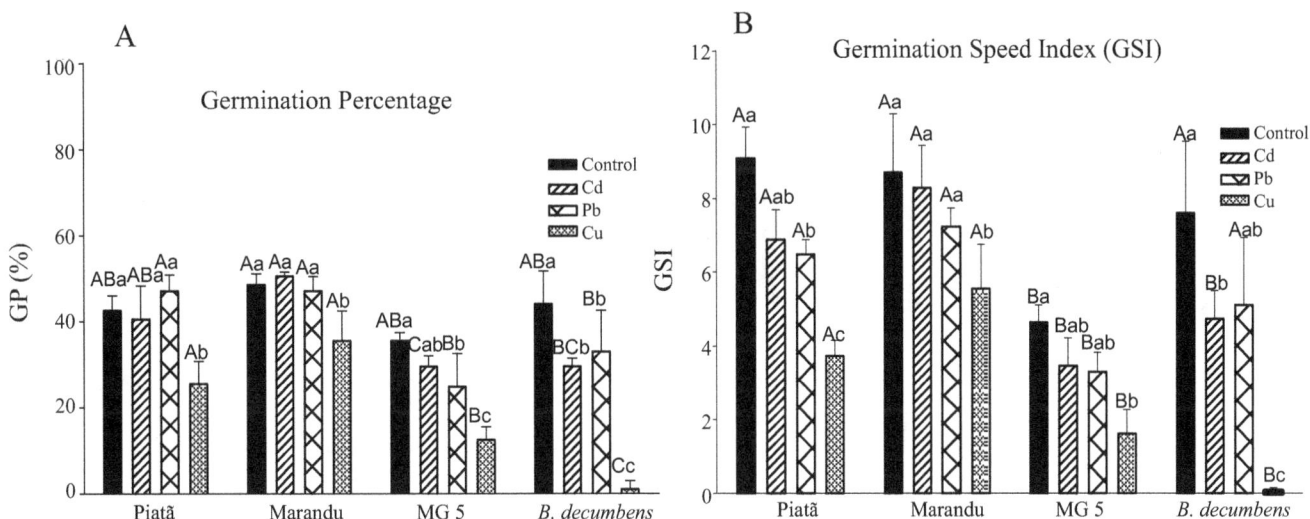

Figure 1. Germination percentage (GP) (A) and Germination speed index (GSI) (B) of species *Brachiaria brizantha* (cv. Piatã, Marandu and MG 5) and *Brachiaria decumbens* in the presence of Cd, Pb and Cu.

Note: Uppercase letters statistically compare species within the same treatment and lowercase letters compare treatments in the same variety.

The presence of trace elements may block the transport of water to the seeds due to the increase in the osmotic potential of the solution, resulting in a secondary effect caused by the low water absorption and not the element toxic effects on the embryo (Kranner and Colville, 2011). Nonetheless, Kalai et al., (2014) state that germination inhibition may occur due to a failure in mobilization of the endosperm caused by the decline of α-amylase, acid phosphatase activity and alkaline phosphatase, as well as a small modification of β-amylase activity resulting in a failure in the mobilization of Cd and Cu in the endosperm.

In exposure to Cd, cv. Piatã, Marandu and MG 5 of *B. brizantha* have not presented reductions in the germination speed index (GSI). However, in exposure to Pb, for cv. Piatã has caused reduction of GSI from 9 (control) to 6.5 (Pb). This reduction can be explained by germination delayed beginning on exposure to this element, which occurred two days after germination start in the control (Figure 1 B). As it occurred in GP, *B. decumbens* presented GSI reductions in exposure to Cd and Cu. In the exposure to Cu, all species studied had a germination delay of five days compared to that of the germination beginning without the presence of trace elements. Germination delay can occur by the protective role which integument can exert on seeds, being able to block and retain trace elements on its surface. However, when the trace element is absorbed, it is deposited in the endosperm, the organ responsible for providing nutrients to the germination process, thus being able to be translocated to the embryo (Sun and Luo, 2014). On the other hand, the GSI delay when exposed to Cu may be the result of a reduction in nitrogen availability in the embryonic axis. This is due to the inhibition of protein synthesis resulting in decreased availability of amino acids present in the endosperm (Karmous et al., 2012).

There was no significant difference in shoot length for cultivars *B. brizantha* (cv. Piatã, Marandu and MG 5) in the presence of Cd, Pb and Cu and for *B. decumbens* in the presence of trace elements Cd and Pb (Figure 2 A). However, in the presence of Cu, *B. decumbens* presented inhibition in shoot growth, which reduced from 5.5 to 1.0 cm. The low translocation of trace elements from the root to the shoot causes the shoot to be little influenced by the seedlings exposure to the trace elements (Fidalgo et al., 2013).

In the presence of Cd there was a reduction of root length of 2.5 and 1.0 cm for cv. Piatã of *B. brizantha* and in *B. decumbens* respectively (Figure 2 B). However, for cv. Marandu and MG 5 of *B. brizantha* there was no reduction in root length in the presence of Cd.

In exposure to Pb, cultivars Piatã, Marandu of *B. brizantha* and *B. decumbens* presented reductions in root length. As for exposure to Cu, the reductions in root length could be observed in cultivars Piatã, MG 5 of *B. brizantha* and *B. decumbens*. In cultivars Piatã and Marandu it was observed

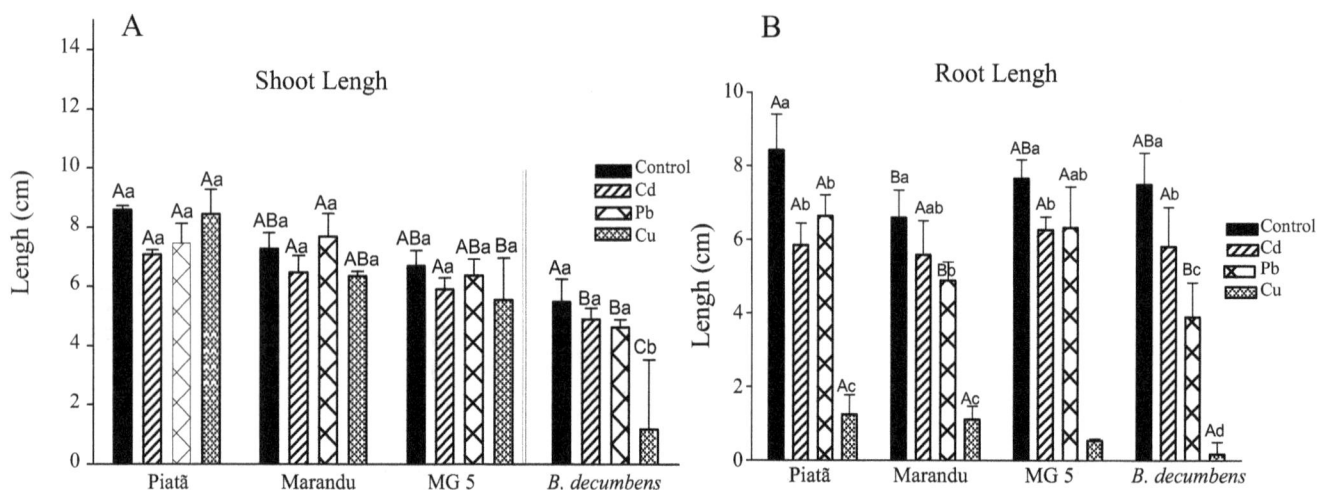

Figure 2. Shoot length (A) and root length (B) of seedlings of *Brachiaria brizantha* (cv. Piatã, Marandu and MG 5) and *Brachiaria decumbens* in the presence of Cd, Pb and Cu.

Note: Uppercase letters statistically compare varieties within the same treatment and lowercase letters compare treatments in the same variety.

that even when not presenting differences in GP, exposure to Pb caused reductions in root length. The greatest reduction of root length was observed in exposure to Cu, with growth not exceeding 1.5 cm in the species studied.

Blockade of enzyme activation and reduction or direct blockage of cell division and interference in mitosis formation may explain the reduction in root growth. Repression of protein synthesis, a DNA replication, can also block cell division (Moosavi et al., 2012).

When plants are exposed to excess Cu, this element can affect the metabolism of N by the reduction of nitrate reductase enzyme responsible for the root length, decrease of the amount of leaves, decrease of the plant biomass caused by the increase of Cu concentration in plant tissues (Xiong et al., 2006).

The presence of trace elements of Cd, Pb and Cu has caused reductions in the Vigor Index (VI) in all species studied (Figure 3). The decrease in VI, which is a result of the low germination percentage and/or decrease in root length, can be caused by the inhibition of mitosis and the synthesis reduction of the cell wall components and in the metabolism of polysaccharides (Heidari and Sarani, 2011). As with the data obtained here, Saderi and Zarinkamar (2012) have obtained germination reductions and low root growth of *Matricaria chamomilla* exposed to Cd and Pb. In seeds of *Populus alba* contaminated by Cd, Madejón et al. (2015) have not found differences in vigor of these seeds comparing with non-contaminated seeds. It is worth noting that the contamination tested by these authors was in seeds and not in the environment.

Germination is the most sensitive phase in a plant life cycle and there is no consistent test or measurable parameter valid for all possible conditions at the time of sowing, i.e., in

vitro germination does not refer to the plants field development conditions (Madejón et al., 2015). Germination tests serve as a preliminary parameter to seedlings development in exposure to trace elements (Di Salvatore et al., 2008). Thus, the experiment was conducted in a second step to evaluate these conditions.

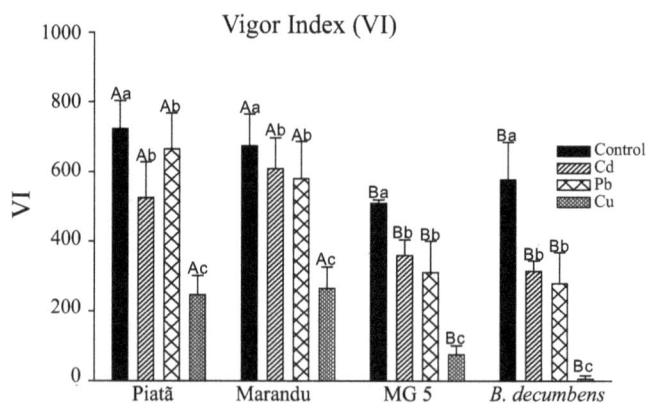

Figure 3. Vigor Index (VI) of seedlings of *B. brizantha* (cv. Piatã, Marandu and MG 5) and *B. decumbens* in the presence of Cd, Pb and Cu.

Note: Uppercase letters statistically compare varieties within the same treatment and lowercase letters compare treatments in the same variety.

Greenhouse

The root dry matter (RDM) results showed no interaction among the species studied and the presence or absence of Cd and Cu. By means of the test of means it was observed that there were no differences among the species studied nor for the treatments studied (Table 3).

There was an interaction between the shoot dry matter (SDM) and cultivars of *B. brizantha* and *B. decumbens*. Exposure to Cd has caused an increase of SDM from 1028.2 to 1457.8 mg for cv. MG 5 of *B. brizantha* (Table 3). The presence of Cu has increased SDM from 483.4 to 836.2 mg in cv. Marandu of *B. brizantha*, an increase of 57.8% when compared to the control. However, for cv. Piatã of *B. brizantha* and for *B. decumbens* there was no difference among treatments.

Table 3. Dry matter (mg.kg^{-1}) of the plant tissue of the species *B. brizantha* (cv. Piatã, Marandu and MG 5) and *B. decumbens* (Dec) in the absence and presence of trace elements of Cd and Cu.

Cultivars	Shoot (SDM)				Root (RDM)			
	Control	Cd	Cu	Mean	Control	Cd	Cu	Mean
Piatã	893.8 Ba	902.5 Aa	902.3 Ba	899.6	767.5	729.5	736.7	744.6 A
Marandu	483.4 Cb	836.2 Aa	891.3 Ba	737.0	751.3	681.8	922.4	785.2 A
MG5	1457.9 Aa	1028.2 Ab	1596.7 Aa	1361.0	776.4	805.8	850.2	810.8 A
Dec	1135.8 ABa	1180.8 Aa	1129.5 Ba	1148.7	569.5	697.3	773.8	680.2 A
Mean	992.7	987.0	1130.00		716.2 a	728.6 a	820.8 a	

Note: The representation of the uppercase letters statistically compare species within the same treatment and lowercase letters compare treatments within the same variety.

Andrade et al. (2014) have observed that exposure of *B. decumbens* to Ba and Pb has not resulted in differences in dry matter weight either. Exposure of brachiaria to trace elements may result in the production of phytochelatins and antioxidant compounds such as ascorbate and tocopherol, which minimizes the toxic effects of trace elements (Santos et al., 2011).

An indication of tolerance to plant exposure to trace elements is the dry matter weight proportionality between the absence and presence of these contaminants (Andrade et al., 2014; Tolentino et al., 2014). This has been observed in this study for *B. decumbens* and for cultivar MG 5.

There was a reduction in the SPAD index in all species studied in exposure to Cd and Cu (Figure 4). The negative effect caused by plant exposure to Cd and Cu is reflected in the indirect measure of the chlorophyll content by the SPAD index (Marchiol et al., 2004). The cv. Piatã of *B. brizantha* was the one which presented the highest reductions in the SPAD index from 27 in the absence of trace elements to 19 in the presence of Cd and 18 in the presence of Cu. Santos et al. (2013) have found reductions in the SPAD index in *Phaseolus vulgaris* L. in exposure to increasing doses of Cd. The uptake of Cd in plants can cause disorganization of chloroplast structures and reduction of chlorophyll biosynthesis, interfering with photosynthesis, respiration and water relations (Gill et al., 2013). The levels of chlorophyll, of the enzymes activity of the Calvin cycle and photosystem II (or water-plastoquinone oxidoreductase) are susceptible to trace elements. In combination, both directly and indirectly, these factors contribute to the reduction of carbon assimilation (Gomes et al., 2011).

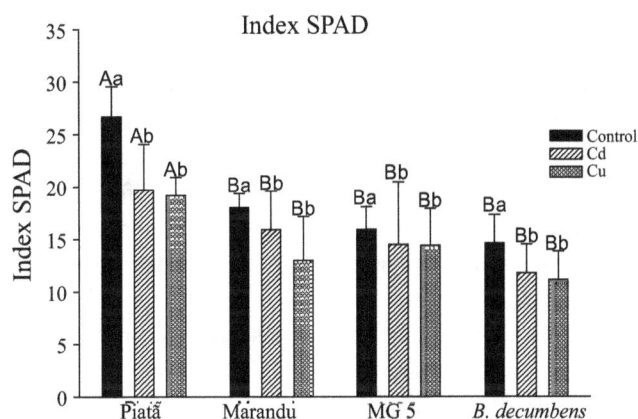

Figure 4. SPAD (Soil Plant Analysis Development) index of species *Brachiaria brizantha* (cv. Piatã, Marandu and MG 5) and *Brachiaria decumbens* in the presence of Cd and Cu.
Uppercase letters statistically compare varieties within the same treatment and lowercase letters compare treatments in the same variety.

The root tissue showed higher levels of Cd and Cu compared to the shoot for all species studied, as can be seen in Table 4. There was no statistical difference between the levels of Cd found in the roots of all cultivars of *B. brizantha* and *B. decumbens*. Cd contents in shoot have not differed among cv. Piatã, Marandu of *B. brizantha* and *B. decumbens*. As for cv. MG 5 of *B. brizantha* it presented the lowest content of Cd. Resistance to trace elements may be based on an exclusion mechanism in which the element accumulates in the roots and its translocation to the shoot is prevented (Rui et al., 2015).

The presence of trace elements may cause changes in size, shape and arrangement of cortical parenchyma in *B. decumbens* cells. These changes indicate an interference of these trace elements in the root maturation rate, possibly caused by the ability of these contaminants to disturb the plant hormonal balance. The cell walls thickening in the root may indicate a greater area supply for retention of the trace elements, reducing their translocation to the shoot (Gomes et al., 2011).

Table 4. Cd and Cu content in the plant tissue of species *B. brizantha* cv. Piatã, Marandu, MG 5 and *B. decumbens* (Dec) in the absence and presence of these elements.

	Cultivars	Shoot	Root	Shoot	Root
		Content of Cd mg.kg^{-1}		Content of Cu mg.kg^{-1}	
Absence	Piatã	0.40 Ab	1.19 Ab	0.94 Cb	8.22 Bb
	Marandu	1.30 Ab	1.47 Ab	5.88 Ab	12.26 Bb
	MG5	0.56 Ab	0.87 Bb	2.30 Bb	12.77 Ba
	Dec	0.86 Ab	1.44 Ab	5.40 Ab	14.20 Ab
Presence	Piatã	1.95 Aa	5.22 Aa	8.25 Aa	13.15 Ba
	Marandu	2.99 Aa	3.22 Aa	9.03 Aa	15.77 Ba
	MG5	1.43 Ba	3.35 Aa	5.32 Aa	12.96 Ba
	Dec	2.41 Aa	2.77 Aa	9.06 Aa	30.80 Aa

Note: The representation of the uppercase letters statistically compare species within the same treatment and lowercase letters compare treatments in the same variety.

Differences of accumulation in different cultivars of the same species were also verified by Rui and contributors (2015). These authors have observed that cultivars of species *Vicia sativa* L. have presented sensitivity differences to Cd. In another study with wheat, concentration differences of Cd, Pb, Zn, Cu, Ni and Cr were found in different cultivars cultivated in soils contaminated with domestic sewage sludge (Jamali et al,. 2009).

Santos et al. (2006) have observed an increase of up to 2.7 times in the absorption of Cd in *Brachiaria decumbens* after the application of phytochelines, showing that this species can also increase its remediation power in contaminated soils.

Contrary to germination observed in laboratory, the presence of a high concentration of Cu has not caused a phytotoxic effect for the plants evaluated in a greenhouse. This is possibly the result of Cu ability to be complex with organic matter, binding to the functional carboxylic and phenolic groups (Kim et al., 1999). Cu content available, observed in Table 5, is four times lower than the dose applied to the soil (200 mg.kg^{-1}). The same was not observed for Cd, where the content available obtained was of 2.8 mg.kg^{-1}, similar to the dose applied to the soil (3 mg.kg^{-1}).

Table 5. Available cadmium and copper (mg.kg^{-1}) content found in Haplic Cambisol in the absence and presence of contamination.

	Cd	Cu
Absence	0.83	1.64
Presence	2.88	52.97

In exposure to Cu, *B. decumbens* was the species with the highest Cu content in its root tissue and no phytotoxic effect was observed in plants of *B. decumbens* when in exposure to Cu. This result may be related to lower Cu availability in the soil and low Cu translocation for the shoot.

Conclusions

B. decumbens is more sensitive to the exposure of Cd, Pb and Cu.

B. decumbens presents germination inhibition, decrease in germination speed index and vigor index. These effects were less evident for cultivars Piatã, Marandu and MG 5 of *B. brizantha*.

The Vigor Index reduction measured by root length is more sensitive to the presence of Cd, Pb and Cu for the species studied. Shoot length is not affected in the presence of the same elements.

In the germination test, Cu is the trace element that caused the most damage to the germination and initial development of *B. brizantha* (cv. Piatã, Marandu and MG 5) and *B. decumbens* seedlings.

The exposure to Cu causes toxicity to all species studied.

There is no difference in root dry matter production among plants cultivated in the absence and presence of Cd and Cu in the soil.

The presence of Cd and Cu in the soil results in a reduction of the SPAD (Soil Plant Analysis Development) index for *B. brizantha* (cv. Piatã, Marandu and MG 5) and *B. decumbens*.

The highest contents of Cd and Cu are found in roots of all species studied.

B. decumbens presents the highest contents of Cu in its root.

References

AMARAL, C. S.; SILVA, E. B.; AMARAL, W. G.; NARDIS, B. O. Crescimento de *Brachiaria brizantha* pela adubação mineral e orgânica em rejeito estéril da mineração de quartizito. *Bioscience Journal*, v.28, n.1, p.130-141, 2012. http://www.seer.ufu.br/index.php/biosciencejournal

ANDRADE, A. F. M.; AMARAL SOBRINHO, N. M. B.; SANTOS, F. S.; MAGALHÃES, M. O. L.; TÓLON-BECERRA, A.; LIMA, L.S. EDTA-induced phytoextraction of lead and barium by brachiaria (*B. decumbens* cv. Basilisk) in soil contaminated by oil exploration drilling waste. *Acta Scientiarum. Agronomy*, v. 36, n. 4, p. 495-500, 2014. http://www.scielo.br/scielo.php?pid=S1807-86212014000400015&script=sci_arttext&tlng=pt

BRASIL. Ministério da Agricultura, Pecuária e Abastecimento. *Regras para análise de sementes*. Ministério da Agricultura, Pecuária e Abastecimento. Secretaria de Defesa Agropecuária. Brasília: MAPA/ACS, 2009. 395p. http://www.agricultura.gov.br/arq_editor/file/2946_regras_analise__sementes.pdf

CONAMA – Conselho Nacional do Meio Ambiente. Resolução 420, de 28 de dezembro de 2009. <http://www.mma.gov.br/port/conama/legiabre.cfm?codlegi=620> Accessed on: Feb, 26th, 2014.

DEZFULI, P. M.; SHARIFZADEH, P. F.; JANMOHAMMADI, M. Influence of priming techniques on seed germination behavior of maize inbred lines (*Zea mays* L.) ARPN. *Journal of Agricultural and Biological Science*, v.3, p.22-25, 2008. http://citeseerx.ist.psu.edu/viewdoc/download?doi=10.1.1.634.3942&rep=rep1&type=pdf

DI SALVATORE, M.; CARAFA, A.M.; CARRATÙ, G. Assessment of heavy metals phytotoxicity using seed germination and root elongation tests: A comparison of two growth substrates. *Chemosphere*, v.73, p.1461–1464, 2008. http://www.sciencedirect.com/science/article/pii/S0045653508009831

FIDALGO, F.; FIDALGO, F.; AZENHA, M.; SILVA, A. F.; SOUSA, A.; SANTIAGO, A.; FERRAZ, P.; TEIXEIRA, J. Copper-induced stress in *Solanum nigrum* L. and antioxidant defense system responses. *Food and Energy Security*, v.2, n.1, p.70–80, 2013. http://onlinelibrary.wiley.com/doi/10.1002/fes3.20/full

GILL, S.S.; HASANUZZAMAN, M.; NAHAR, K.; MACOVEI, A.; TUTEJA, N. Importance of nitric oxide in cadmium stress tolerance in crop plants. *Plant Physiology and Biochemistry*, v.63, p. 254-261, 2013. http://www.sciencedirect.com/science/article/pii/S0981942812003488

GOMES, M. P.; MARQUES, T. C. L. L. S. M.; NOGUEIRA, M. O. G.; CASTRO, E. M.; SOARES, A. M. Ecophysiological and anatomical changes due to uptake and accumulation of heavy metal in *Brachiaria decumbens*. *Scientia Agricola*, v.68, n.5, p.566-573, 2011. http://www.scielo.br/scielo.php?pid=S0103-90162011000500009&script=sci_arttext&tlng=pt

GRATÃO, P. L.; PRASAD, M.N.V.; CARDOSO, P.F.; LEA, P.J.; AZEVEDO, R.A. Phytoremediation: green technology for the clean up of toxic metals in the environment. *Brazilian Journal of Plant Physiology*, v.17, n.1, p.53-64, 2005. http://www.scielo.br/scielo.php?pid=S1677-04202005000100005&script=sci_arttext

GUIMARÃES, M. A.; SANTANA, T.A.; SILVA, E.V.; ZENZEN, I.L.; LOUREIRO, M. E. Toxicidade e tolerância ao cádmio às plantas. *Revista Trópica-Ciências Agrárias e Biológicas*, v.3, n.1, p.58-68, 2008. https://www.researchgate.net/profile/Marcelo_Guimaraes/publication/246044549_Cadmium_toxicity_and_tolerance_in_plants_(Toxicidade_e_tolerncia_ao_cdmio_em_plantas)/links/0046351d8d6eb6c6ee000000.pdf

HEIDARI, M.; SARANI, S. Effects of lead and cadmium on seed germination, seedling growth and antioxidant enzymes activities of mustard (*Sinapis arvensis* L.). *Agricultural and Biological Science*, v.6, p.44-47, 2011. https://www.researchgate.net/profile/Mostafa_Heidari/publication/267942856_Effects_of_lead_and_cadmium_on_seed_germination_seedling_growth_and_antioxidant_enzymes_activities_of_mustard_Sinapis_arvensis_L/links/5630fb0c08ae506cca676298.pdf

JAMALI, M. K.; KAZI, T. G.; ARAIM, M. B.; AFRIDI, H. I.; JALBANI, N.; KANDHRO, G. A.; SHAH, A. Q.; BAIG, J. A. Heavy metal accumulation in different varieties of wheat (*Triticum aestivum* L.) grown in soil amended with domestic sewage sludge. *Journal of Hazardous Materials*, v.164, n.2-3, p.1386-1391, 2009. http://www.sciencedirect.com/science/article/pii/S0304389408013915

KABATA-PENDIAS, A. Trace elements in soils and plants. 4 ed. Boca Raton: CRC Press, 2011. 534p.

KALAI, T.; KHAMASSI, K.; SILVA, J. A. T.; GOUIA, H.; BÉM-KAAB, L. B. Cadmium and cooper stress affect seedling grownth and enzymatic activities in germinating barley seeds. *Archives of Agronomy and Soil Science*, v.60, n.6, p.765-783, 2014. http://www.tandfonline.com/doi/abs/10.1080/03650340.2013.838001

KARIMI, P.; KHAVARI-NEJAD, R. A.; NIKNAM, V.; GHAHREMANINEJAD, F.; NAJAFI, F. The effects of excess copper on antioxidative enzymes, lipid peroxidation, proline, chlorophyll, and concentration of Mn, Fe, and Cu in *Astragalus neo-mobayenii*. *The Scientific World Journal*, v. 2012, p.1-6, 2012. https://www.hindawi.com/journals/tswj/2012/615670/abs/

KARMOUS, I.; KHADIJA, J.; CHAOUI, A.; EL FERJANI, E. Proteolytic activities in *Phaseolus vulgaris* cotyledons under copper stress. *Physiology and Molecular Biology of Plants*, v.18, n.4, p.337-343, 2012. http://link.springer.com/article/10.1007/s12298-012-0128-4

KIM, S. D.; MA, H.; ALLEN, H. E.; CHA, D. K. Influence of dissolved organic matter on the toxicity of copper to *Ceriodaphnia dubia*: Effect of complexation kinetics. *Environmental Toxicology and Chemistry*, v.18, n.11, p.2433-2437, 1999. http://onlinelibrary.wiley.com/doi/10.1002/etc.5620181108/full

KRANNER, I.; COLVILLE, L. *Metals and seeds*: Biochemical and molecular implications and their significance for seed germination. *Environmental and Experimental Botany*, v.72, p. 93-105, 2011. http://www.sciencedirect.com/science/article/pii/S0098847210001164

LUCENA, R. B.; PIEREZAN, F.; KOMMERS, G. D.; IRIGOYEN, L. F.; FIGHERA, R. A.; BARROS, C. S.L. Doenças de bovinos no Sul do Brasil: 6.706 casos. *Pesquisa Veterinária Brasileira*, v.30, n.5, p. 428-434, 2010. http://www.scielo.br/pdf/pvb/v30n5/a10v30n5

MADEJÓN, P.; CANTOS, M.; JIMÉNEZ-RAMOS, M. C.; MARAÑÓN, T.; MURILLO, J. M. Effects of soil contamination by trace elements on white poplar progeny: seed germination and seedling vigour. *Environmental Monitoring and Assessment*, v.187, n.11, p.663-674, 2015. http://link.springer.com/article/10.1007/s10661-015-4893-8

MAGUIRE, J. D. Speed of germination aid in selection and evaluation for seedling emergence and vigor. *Crop Science*, v. 2, n. 1, p. 176-177, 1962.

MARCHIOL, L.; ASSOLARI, S.; SACCO, P.; ZERBI, G. Phytoextraction of heavy metals by canola (*Brassica napus*) and radish (*Raphanus sativus*) grown on multicontaminated soil. *Environmental Pollution*, v.132, p.21-27, 2004. http://www.sciencedirect.com/science/article/pii/S0269749104001423

MOOSAVI, S. A.; GHARINEH, M. H.; AFSHARI, R. T.; EBRAHIMI, A. Effects of some heavy metals on seed germination characteristics of canola (*Brassica napus*),wheat (*Triticum aestivum*) and safflower (*Carthamus tinctorious*) to evaluate phytoremediation potential of these crops. *Journal of Agricultural Science*, v.4, n.9, p.11-19, 2012. http://www.ccsenet.org/journal/index.php/jas/article/view/15599

RUI, H. CHEN, C.; ZHANG, X.; ZHANG, F.; SHEN, Z. Cd-induced oxidative stress and lignification in the roots of two *Vicia sativa* L. varieties with different Cd tolerances. *Journal of Hazardous Materials*, v.15, p.30045-30050, 2015. http://www.sciencedirect.com/science/article/pii/S0304389415300455

SADERI, S. Z.; ZARINKAMAR, F. The effect of different Pb and Cd concentrations on seed germination and seedling growth of *Matricaria chamomilla*. *Advances in Environmetal Biology*, v.7, n.6, p.1940-1943, 2012. https://www.researchgate.net/profile/Zohreh_Saderi/publication/285588494_The_effect_of_different_Pb_and_Cd_concentrations_on_seed_germination_and_seedling_growth_of_Matricaria_chamomilla/links/56607a6308ae418a7866604e.pdf

SANTOS, F. S.; HERNÁNDEZ-ALLICA, J.; BECERRIL, J. M.; AMARAL-SOBRINHO, N.; MAZUR N.; GARBISU, C. Chelate-induced phytoextraction of metal polluted soils with *Brachiaria decumbens*. *Chemosphere*, v.65, n.1, p.43-50, 2006. http://www.sciencedirect.com/science/journal/00456535

SANTOS, F. S.; AMARAL SOBRINHO, N. M. B.; MAZUR, N.; GARBISU, C.; BARRUTIA, O.; BECERRIL, J. M. Resposta antioxidante, formação de fitoquelatinas e composição de pigmentos fotoprotetores em *Brachiaria decumbens* Stapf submetida à contaminação com Cd e Zn. *Química Nova*, v. 34, n. 1, p. 16-20, 2011. http://www.scielo.br/pdf/qn/v34n1/v34n1a04

SANTOS, A. P.; FAGAN, E. B.; TEIXEIRA, W. F.; SOARES, L. H.; REIS, M. R.; CORRÊIA, L. T. Influência de doses de cádmio na emergência e no crescimento do feijoeiro. *Revista do Centro Universitário de Patos de Minas*, v.4 p.1-8, 2013. http://revistaagrociencias.unipam.edu.br/documents/57126/179380/Influ%C3%AAncia+de+doses+de+c%C3%A1dmio.pdf

SUN, J. L.; LUO, L. Q. A study on distribution and chemical speciation of lead in corn seed germination by synchrotron radiation X-ray fluorescence and absorption near edge structure spectrometry. *Chinese Journal of Analytical Chemistry*, v.10, n.42, p.1447-1452, 2014. http://www.sciencedirect.com/science/article/pii/S187220401460774X

TEDESCO, M.J.; VOLKWEISS, S.J.; BOHNEN, H. Análises de solo, plantas e outros materiais. Porto Alegre: Departamento de solos, UFRGS. 118 p. (*Boletim Técnico*), 1995.

TOLENTINO, T.; BERTOLI, A.; CARVALHO, R.; BASTOS, A. R.; PIRES, M. Especiação do cádmio em *Brachiaria brizantha* e biodisponibilidade dos macro e micronutrientes. *Ciências Agrárias*, v.37, n.3, p.292-298, 2014. http://www.scielo.mec.pt/scielo.php?script=sci_arttext&pid=S0871-018X2014000300005

USEPA. U.S. Environmental Protection Agency. Microwave assisted acid digeston of sediments, sludges, soils and soils. Method 3051. Office of Solid Wast and Emergency Response, U.S. Government Printing Office, Washington, DC, 1994. <http://www.epa.gov.epaoswer/hazwaste/test/pdfs/3051.pdf> Acessed on Apr.4th, 2015.

XIONG, Z.; LIU, C.; GENG, B. Phytotoxic effects of copper on nitrogen metabolism and plant growth in *Brassica pekinensis* Rupr. *Ecotoxicology and Environmental Safety*, v.64, p.273-280, 2006. http://www.sciencedirect.com/science/article/pii/S0147651306000431

YANG, Y.; WEI, X.; LU, J.; YOU, J.; WANG, W.; SHI, R. Lead-induced phytotoxicity mechanism involved in seed germination and seedling growth of wheat (*Triticum aestivum* L.). *Ecotoxicology and Environmental Safety*, v.73, n.8, p.1982-1987, 2010. http://www.sciencedirect.com/science/article/pii/S014765131000254X

YRUELA, I. Copper in plants: acquisition, transport and interactions. *Functional Plant Biology*, v.36, p. 409-430, 2009. http://www.publish.csiro.au/fp/FP08288

Protective action of nitric oxide in sesame seeds submitted to water stress

Raquel Maria de Oliveira Pires[1*], Genaina Aparecida de Souza[2],
Denise Cunha Fernandes dos Santos Dias[3], Leonardo Araujo Oliveira[2],
Eduardo Euclydes de Lima e Borges[4]

ABSTRACT - The objective in this work was to investigate the effect of nitric oxide (NO) like protective agent in sesame seeds submitted to different osmotic potentials. The treatments, in total of eight, were water (control), water plus sodium nitroprusside (SNP) and the other treatments with PEG 6000 and PEG 6000 plus SNP: - 0.1 MPa, -0.1MPa +200 μM of SNP, 0.2 MPa, -0.2 MPa +200 μM of SNP, -0.3 MPa and -0.3 MPa, +200 μM of SNP. Were done the following determinations: germination, first count of germination, speed germination index, hypocotyl length, radicle length, dry mass of hypocotyl and radicle. It was quantified the activity of the antioxidative enzymes, superoxide dismutase, catalase, ascorbate peroxidase and total peroxidase. The experimental design was completely randomized with five replications. The water restriction reduced the germination of sesame seeds, however, the presence of nitric oxide (NO) due to the application of SNP, was beneficial, promoting increase in germination, vigor and seedlings. There was an increase of antioxidative enzymes activity in the period of 0 to 24 hours, demonstrating organization of antioxidative system in all long the time. The association of PEG 6000 to SNP, increased the activity of antioxidative enzymes, evidencing an efficient system of elimination of ROS formed during the exposition to water deficit.

Index terms: *Sesamum indicum* L., sodium nitroprusside, vigor, germination, antioxidative system.

Ação protetora do óxido nítrico em sementes de gergelim submetidas ao estresse hídrico

RESUMO - O objetivo neste trabalho foi investigar o efeito do óxido nítrico como agente protetor em sementes de gergelim submetidas a diferentes potenciais osmóticos. Os tratamentos, no total de oito, foram: água (controle), água acrescida de nitroprussiato de sódio (SNP) e os demais tratamentos referentes às concentrações de PEG e PEG acrescido de SNP: 0,1MPa, 0,1MPa +200 μM de SNP, -0,2 MPa, -0,2MPa +200 μM de SNP, -0,3 MPa e -0,3 MPa, +200 μM de SNP. Foram feitas as seguintes determinações: germinação, primeira contagem de germinação, índice de velocidade de germinação, comprimento de hipocótilo, radícula e massa seca de hipocótilo e radícula, além da quantificação da atividade das enzimas superóxido dismutase, catalase, ascorbato peroxidase e peroxidases totais. O delineamento estatístico utilizado foi o inteiramente casualizado com cinco repetições. A restrição hídrica reduz a germinação de sementes de gergelim, entretanto, a presença de óxido nítrico (ON) proporciona aumento na germinação e vigor. Há aumento da atividade das enzimas do sistema antioxidante no período de 0 a 24 horas, demonstrando organização do sistema antioxidante nas sementes de gergelim. A associação do PEG-6000 ao SNP aumenta a atividade das enzimas antioxidantes evidenciando um eficiente sistema de eliminação de ERO durante a exposição ao estresse hídrico.

Termos para indexação: *Sesamum indicum* L., nitroprussiato de sódio, vigor, germinação, sistema antioxidante.

Introduction

In the germination process, the first step in the events sequence that culminate in the resumption of embryo growth is the imbibition. The water absorbed by seeds is directly or indirectly involved in all other stages of the subsequent metabolism.

The seeds imbibition plays a key role in the germination process (Bewley et al., 2013). Between the various environmental factors that can influence this process, the availability of water is one of the most important. The water

[1]Departamento de Agricultura, UFLA, Caixa Postal 3037, 37200-000 – Lavras, MG, Brasil.

[2]Departamento de Biologia Vegetal, UFV, 36570-000 – Viçosa, MG, Brasil.

[3]Departamento de Fitotecnia, UFV, 36570-000 – Viçosa, MG, Brasil.
[4]Departamento de Engenharia Florestal, UFV, 36570-000 – Viçosa, MG,
*Corresponding author <raquel.oliveirapires@yahoo.com.br>

deficit occurrence in plants leads to the reduction of germination speed and the delaying of the seedlings development. Stronger restrictions come to harm the final germination percentage due to the delay or the non-occurrence of the metabolic processes necessary for germination (Rahimi, 2013). Prolonged exposure to water deficit generates several natural reactions in seeds. One of the consequences of this exhibition is the generation of reactive oxygen species (ROS), which leads to the oxidative stress, what means, there is an increase in the lipid peroxidation and a decrease in the activity of oxygen reactive species (ROS) sequestrant enzymes (Yao et al., 2012), affecting the structures and cell metabolism (Wang et al., 2009).

One of the defense mechanisms modulated by plants against oxidative stress is the antioxidant system that acts and constitutes an important primary defense against free radicals generated in seeds under stress conditions, such as superoxide dismutase (SOD), catalase (CAT), peroxidase (POX), and ascorbate peroxidase (APX) (Mittler, 2002).

Studies have shown that chemical compounds such as nitric oxide (NO), act in plants protection exposed to stress factors. Multi-functional molecule that acts in many physiological events, has among other functions, the cytoprotective by the ability to regulate the level and toxicity of ROS. Sodium nitroprusside (SNP) is currently the most used donor of nitrogen oxides that produce NO.

The donor use of NO in plants is being extensively explored in research to be an important signaling molecule present in many phases of plant development and essential element in the responses to biotic and abiotic stresses in plants. Studies show that NO is able to increase wheat seeds germination submitted to salt stress (Zheng et al., 2009), stress by heavy metals such as arsenic (Singh et al., 2013), copper (Hu et al., 2007), lead and cadmium (Kopyra and Gwóźdź, 2003), in addition to increasing the accumulation of dry matter in pumpkin seedlings under salt stress (Fan et al., 2013).

The NO action in the increase of plants tolerance to different types of stress has been strongly correlated with their ability to protect plant cell from oxidative damage. Seeds submitted to water stress and the NO influence as a protective molecule requires further studies.

Thus, the objective in this study was to investigate the NO effect as a protective agent and signaling in sesame seeds submitted to different osmotic potential, through evaluations of physiological and biochemical characteristics and antioxidant enzymes activity.

Material and Methods

Sesame seeds (*Sesamum indicum* L.) were used from the agricultural year 2013, from EMBRAPA Cotton. Initially, the preliminary tests were done to determine the concentrations of polyethylene glycol solution of molar mass 6000 (PEG 6000) and NO donor solution (sodium nitroprusside (SNP) to be studied. The amount of SNP was the one capable of reversing or mitigating the actions of the water stress. After that, the sesame seeds, in five replicates of 50, were put to germinate on a paper towel moistened with 3 mL of the solution, referent to the following treatments in Petri dishes: water (control), water plus sodium nitroprusside (SNP) and the other treatments with PEG 6000 and PEG 6000 plus SNP; - 0.1 MPa, -0.1 MPa +200 µM of SNP, 0.2 MPa, -0.2 MPa +200 µM of SNP, -0.3 MPa and -0.3 MPa, +200 µM of SNP, summing up to a total of eight treatments.

The seeds of each treatment were kept in B.O.D. regulated with alternate temperature of 20-30 °C, with the presence of constant light (Brasil, 2009). The following evaluations were done:

Germination percentage: in the sixth day after sowing, the percentage of normal seedlings was evaluated (Brasil, 2009).

First germination count: consisted of a record of the number of normal seedlings obtained in the third day after sowing, and the values were expressed in percentage (Brasil, 2009).

Germination speed index: daily counts were done of the number of seeds that issued a radicle higher than 1.0 mm and the index were calculated according to Nakagawa (1999).

Length of the hypocotyl and the radicle: the seeds of each treatment, in five replicates of 25, were sown in gerbox, following the methodology described above for the germination test (Pires et al., 2016). A measurement of the length of hypocotyls and the radicle was done in the seedlings classified as normal with a graduated ruler. The results were expressed in cm. seedling^{-1}.

Dry matter of the hypocotyl and the radicle: the seeds used to measure the length of the hypocotyl and the radicle were separated in hypocotyl and radicle, and later dried in an oven for 72 hours at 65 °C. The results were expressed in mg.seedling^{-1} (Brasil, 2009).

Activity of the main enzymes of the antioxidant system: determined using seeds soaked for 12 and 24 hours in water and in solutions of PEG and PEG increased by SNP. The enzyme extracts were obtained by maceration of 0.2 g of seeds in ice, followed by the addition of 2.0 mL of the following means of homogenizing: potassium phosphate buffer 0.1 M and pH 6.8, ethylenediaminetetraacetic acid (EDTA) 0.1 mm, phenylmethylsulfonyl fluoride (PMSF) 1 mm and polyvinylpolypyrrolidone (PVPP) 1% (p/v). After that, the extract was centrifuged at 15,000 g for 15 minutes at 4 °C and the supernatant was collected, where determinations were done on the activities of ascorbate peroxidase enzyme (APX), peroxidase (POX), catalase (CAT) and superoxide dismutase (SOD).

The activity of APX was determined by the addition of 200 μL of raw enzyme extract at 2.9 mL of reaction medium of ascorbic acid 10 mM and H_2O_2 10 mM in potassium phosphate buffer 100 mM, pH 6.0. A decrease in the absorbance was observed at 290 nm, at 25 °C, during the first minute of the reaction. The enzyme activity was calculated using the molar extinction coefficient of 2.8 $mM^{-1}.cm^{-1}$ and expressed in μmol $min^{-1}.mg^{-1}$ of protein (Sano et al., 2001).

The activity of POX was determined by the addition of 50 μL of raw enzyme extract at 2.97 mL of reaction medium composed of potassium phosphate buffer 100 mM and pH 6.8, pyrogallol 150 mM and hydrogen peroxide 125 mM (Kar and Mishra, 1976). The increase in the absorbance during the two first minutes of reaction at 420 nm at a constant temperature of 25 °C determined the production of purpurogallin. The enzyme activity was calculated using the molar extinction coefficient of 2.47 $mM^{-1}. cm^{-1}$ and expressed in μmol $min^{-1}. mg^{-1}$ of protein.

To quantify the CAT activity, was added in 30 μL of enzyme extract, 0.98 mL of sodium phosphate buffer 0.05 M, pH 6.8, H_2O_2 0.0125 mM dissolved in buffer. The enzymatic activity was determined by monitoring the absorbance decrease at 240 nm for 2 minutes, in intervals of 15 seconds and calculated based on the extinction factor of 36 $mM^{-1}. cm^{-1}$.

The activity of SOD was determined by the addition of 30 μL of raw enzyme extract at 2.95 mL of the reaction medium composed of sodium phosphate buffer 100 mM at pH 7.8, methionine 50 mM, p-nitro blue tetrazolium (NBT) 1 mM, EDTA 5 mM and riboflavin 100 mM. The reaction was conducted at 25 °C in a reaction chamber under lighting of a 15 W fluorescent lamp, kept inside a box internally coated with foil. After five minutes of exposure to the light, the lighting was interrupted and the blue formazan produced by NBT photo reduction was determined by the absorption at 560 nm in a spectrophotometer. A SOD unit was defined as the amount of necessary enzyme to inhibit the NBT photo reduction in 50%. The SOD activity was expressed in U min^{-1}. mg^{-1} protein. To determine the content of proteins, the method used was Bradford (1976) with a standard curve constructed with bovine serum albumin (BSA) as a reference protein.

For all determinations, the statistical design was entirely randomized with five replicates. The data was submitted to a variance analysis (ANAVA) with the help of the statistical program SISVAR® (Ferreira, 2011) and the averages obtained for the treatments were compared by the Tukey test at a 5% significance. The averages obtained in the treatments with and without SNP were compared by the F test at 5% of probability and for the enzyme determinations the Tukey test was also used, at 5% significance. The graphs were plotted with the help of Sigma Plot program.

Results and Discussion

Sesame seeds had 98.4% of germination under optimal conditions (water germination), occurring reduction in germination under water stress with values of 31.6% in lower osmotic potential (-0.3 MPa), 40% under osmotic intermediate potential (-0.2 MPa) and 66.8% in greater osmotic potential (-0.1 MPa) (Figure 1A). It was observed that the different values of water potential studied acted reducing germination. Despite this, the germination reached in the osmotic potential of -0.1 is above 60%, considered minimal for Brazilian marketing sesame seeds (Brasil, 2013). Reducing the percentage of seed germination in water deficit conditions is attributed to lower diffusion of water through tegument.

The SNP application does not affect the seeds germination in water, since there was no stress and the germination conditions were ideal (Figure 1), but allowed to obtain significant seed germination increase in treatments with PEG 6000 in the three concentrations tested (Figure 1A).

By the first germination count (FC) and germination speed index (GSI) results, it turns out that the water deficit in any of the tested potential reduced the germination speed (Figure 1B). There was marked reduction in germination speed with increasing of water stress, i.e., with decreased of osmotic potential of the PEG 6000 solutions. When PEG solutions were used, there was an increase in the germination speed (Figures 1B and 1C) using SNP. The water restriction effects on seed germination have been reported in several studies for different species. Ávila et al. (2007) found an accentuated reduction in canola (*Brassica napus*) seed germination when submitted to most negative osmotic potential. Pereira and Lopes (2011) observed germination, germination speed and performance drastic reduction of jatropha (*Jatropha curcas*) seedlings in the potential of -0.2 MPa and Teixeira et al. (2011) working with crambe (*Crambe abyssinica*) seeds verified significant reduction of germination and vigor in seeds submitted to osmotic potential more negative, without no formation of normal seedlings in lower potential at -0.6 MPa. The seed germination reduction when exposed to water stress is expected, since the solutes presence as PEG decreases the water absorption by the seeds.

Water restriction reduced the sesame seeds germination, however the NO presence due to the SNP application stimulated germination. The mechanism by which the NO stimulates germination has not been fully elucidated, but some studies show the NO effective effect in promoting seed germination. Deng and Song (2012) found the NO efficiency in promote germination when lettuce seeds were submitted to water stress, as well as Sarath et al. (2006) with *Panicum virgatum*. One hypothesis used by these authors is the NO direct effect on abscisic acid - ABA endogenous content. Sarath et al. (2006) demonstrated that the SNP application partially reverses the ABA inhibitory effects on germination, radicle elongation and the coleoptile emergence in *Panicum virgatum*.

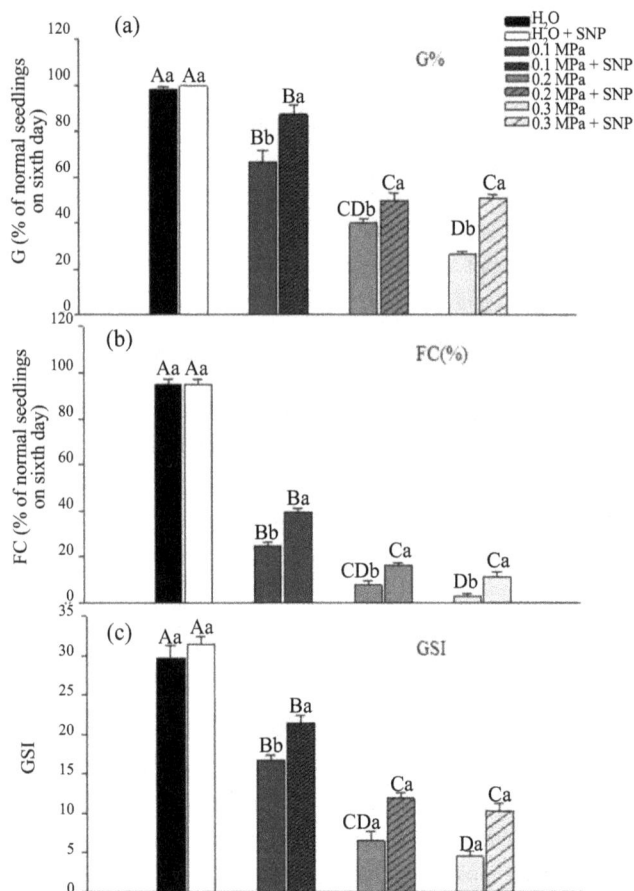

Figure 1. A- Germination (G%), B- first germination count (FC%) and C- germination speed index (GSI) of *Sesamum indicum* seeds submitted to imbibition in water (control), water + SNP, PEG 6000 solutions at -0.1 MPa, -0.2 MPa and -0.3 MPa added or not of SNP. Averages followed by the same upper case does not differ by Tukey test at 5%. Average of each treatment with and without SNP followed by the same lower case does not differ each by F test at 5%. The bars indicate the average standard deviation (n = 5).

Regarding the hypocotyl length (HL) (Figure 2A), it was observed reduction as the osmotic potential became more negative. In the potential of -0.3 MPa were obtained values for the HL 75% lower than those observed in seedlings grown under adequate conditions of water availability. In general, the root system was less compromised by water restriction than the shoots development (Figure 2B). These results can be explained by the fact that seedlings submitted to the most rigorous water stress, in general, tend to invest more biomass and develop larger root system as a survival strategy. The distribution of the root system in depth/length due water insufficiency is regarded as an indicator parameter of tolerance to drought and can confer adaptation in some species (Santos et al., 2015).

It was observed greater dry matter accumulation on seedlings of control treatment which presented dry mass of hypocotyl - DMH of 20.07 mg.seedling^{-1} (Figure 2C) and dry mass of radicle - DMR of 19.54 mg.seedling^{-1} (Figure 2D). When seeds were submitted to osmotic potential of -0.1 MPa there was reduction of 53% and 62% in DMH and DMR, respectively. In the potential of -0.2 MPa, the reduction was 69% and 74%, and in the potential -0.3 MPa the reduction of DMH and DMR in relation to control was 84% and 88%, respectively. This reduction of the seedlings' dry mass in function of water restriction can be explained by the lower speed of the physiological and biochemical processes or by difficulty of hydrolysis and mobilization of seed reserves (Bewley et al., 2013).

The NO application was able to reverse significantly the DMH and DMR reduction caused by water stress, except the DMH in the lower potential -0.3 MPa. Regarding DMH, reversal of reduction in the higher osmotic potential -0.1 MPa was 4.2 percentage points and 4.18 points at the intermediate potential -0.2 MPa. Regarding the DMR, reversal of reduction in the higher osmotic potential was 1.94 percentage points, 2.31 points at the intermediate potential and 3.02 in the lower potential (Figures 2C and 2D).

Adverse situations or biological dysfunctions lead to the formation of reactive oxygen species. Sesame seeds defense mechanism submitted to different concentrations of PEG were evaluated by means of the antioxidant enzymes activity (Tables 1 to 4). In general, it is observed increased activity of enzymes SOD, CAT, and APX with the progressive increase in the imbibition period from 0 to 24 hours. The POX, where treatments in the intervals 0 and 12 hours did not differ. These results indicate an apparent organization of the antioxidant system in sesame seeds over time. However, it was found that the period 0 h and 12 h did not differ statistically the treatments compared to the control for all enzymes, showing that these times are insufficient for organizing the antioxidant apparatus in sesame seeds submitted to stress by water deficit. Already in the 24 hours period, it is possible to observe differences in enzyme activity compared to control in practically all treatments.

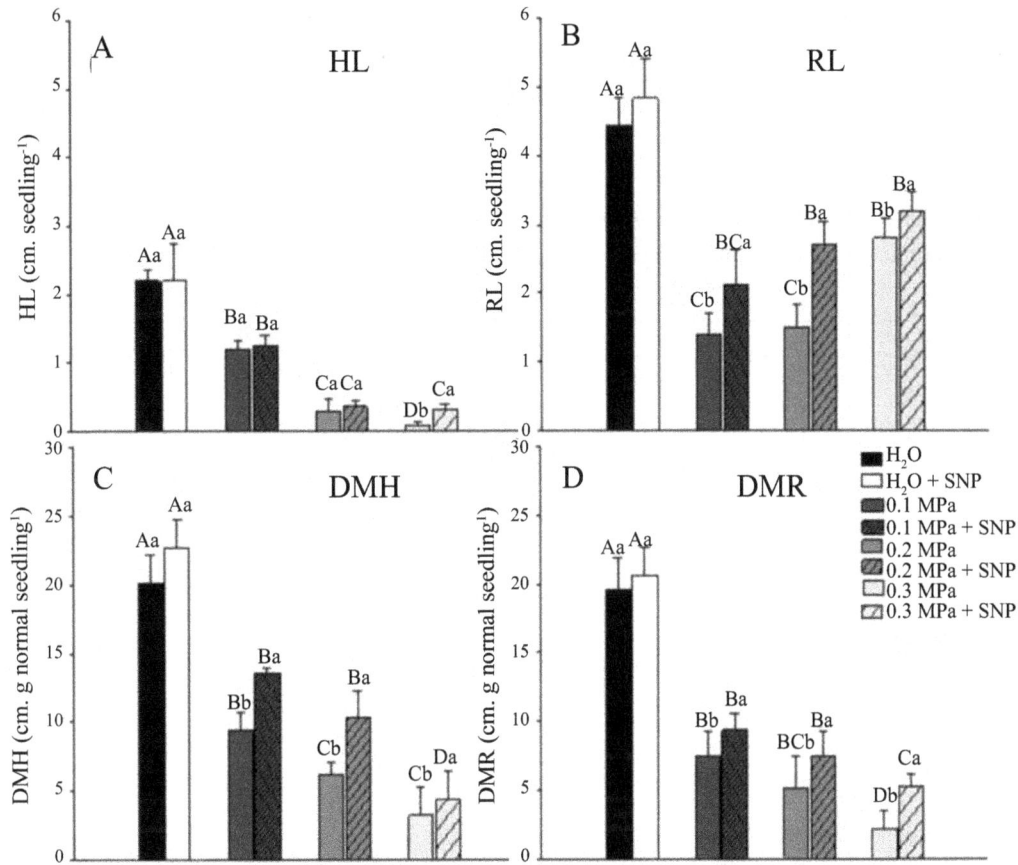

Figure 2. A- Hypocotyl length - HL (cm.normal seedling), B- Radicle lenght – RL (cm.normal seedling), C- dry mass of hypocotyl - DMH (cm.normal seedling), D- dry mass of radicle - DMR (cm.normal seedling) of *Sesamum indicum* seedlings submitted to imbibition with water (control), water + SNP, higher osmotic potential (-0.1 MPa), higher osmotic potential plus SNP (-0.1 MPa + SNP), osmotic intermediate potential (-0.2 MPa) plus SNP (-0,2 MPa + SNP), low osmotic potential (-0.3 MPa) and plus SNP (-0.3 MPa + SNP). Averages followed by the same upper case does not differ by Tukey test at 5%. Average of the treatment followed by the same lower case does not differ by F test at 5%. The bars indicate the average standard deviation (n = 5).

Table 1. Enzyme activity of superoxide dismutase (SOD) *Sesamum indicum* seeds after 0, 12 and 24 hours of imbition in solutions with different osmotic potentials supplemented or not of SNP.

	SOD (U min^{-1}. mg^{-1}. protein)		
Treatment	0 h	12 h	24 h
H_2O	0.38 ± 0.01 Ba	0.87 ± 0.01 Ba	1.09 ± 0.06 Ae
-0.1 MPa	0.38 ± 0.01 Ca	1.08 ± 0.03 Ba	1.22 ± 0.14 Ae
-0.1 MPa +SNP	0.38 ± 0.01 Ca	1.21 ± 0.14 Ba	2.18 ± 0.32 Ac
-0.2 MPa	0.38 ± 0.01 Ba	0.79 ± 0.03 Ba	1.69 ± 0.52 Ad
-0.2 MPa +SNP	0.38 ± 0.01 Ba	1.37 ± 0.06 Ba	2.68 ± 0.31 Ab
-0.3 MPa	0.38 ± 0.01 Ca	1.27 ± 0.30 Ba	1.75 ± 0.18 Ad
-0.3 MPa +SNP	0.38 ± 0.01 Ca	1.33 ± 0.32 Ba	3.29 ± 0.23 Aa
CV(%)	---------------	23.21	19.43

Averages followed by the same lower case in the column does not differ each other by Tukey test at 5%. Average followed by the same upper case in the line does not differ each other by Scott-Knott test at 5% probability. Average ± standard deviation.

It is worth remembering that in 0 h, there was no contact of the seeds with solutions PEG or PEG plus SNP, there was no

seeds contamination. Thus, it was calculated the values of each enzyme, assigning fixed value for all treatments in this interval.

Table 2. Enzyme activity of catalase (CAT) *Sesamum indicum* seeds after 0, 12 and 24 hours of imbibition in solutions with different osmotic potentials supplemented or not of SNP.

Treatment	CAT (μmol min^{-1}.mg^{-1}. protein)		
	0 h	12 h	24 h
H$_2$O	15.64 ± 1.24 Ca	29.51 ± 2.13 Ba	32.82 ± 2.56 Af
-0.1 MPa	15.64 ± 1.24 Ca	33.68 ± 1.97 Ba	42.88± 4.23 Ae
-0.1 MPa +SNP	15.64 ± 1.24 Ca	34.62 ± 1.89 Ba	50.31 ± 3.67 Ac
-0.2 MPa	15.64 ± 1.24 Ca	29.58 ± 2.31 Ba	49.53 ± 3.21 Ad
-0.2 MPa +SNP	15.64 ± 1.24 Ca	34.04 ± 3.14 Ba	55.44 ± 4.23 Ab
-0.3 MPa	15.64 ± 1.24 Ca	33.26 ± 3.18 Ba	49.13 ± 3.10 Ad
-0.3 MPa +SNP	15.64 ± 1.24 Ca	34.09 ± 2.33 Ba	63.26 ± 3.67 Aa
CV(%)	--------------	18.45	23.21

Averages followed by the same lower case in the column does not differ each other by Tukey test at 5% probability. Average followed by the same upper case in the line does not differ each other by Scott-Knott test at 5% probability. Average ± standard deviation.

Comparing the different treatments in the period 24 hours of imbibition (Tables 1 to 4), generally for all studied enzymes, there was higher enzyme activity in the most negative osmotic potential (-0.3 MPa) with lower activity in the potential 0 MPa (control). It is also observed increased activity of antioxidant enzymes in treatments with SNP application compared to the same treatment without SNP. It is assumed that the SNP application contributes to reducing ROS concentration during the sesame seeds germination process. According to Dong et al. (2014), NO promotes tolerance to water stress due to stimulate the increased activity of antioxidative enzymes, then providing greater protection against oxidative stress during the seed germination process.

In Table 3, there is the APX activity, at the time 0 h it was not observed enzyme activity (data not shown), however, there was catalase enzyme activity (CAT) (Table 2). The fact of the APX does not show activity in dry seeds is also reported by Bailly (2004) that justifies a higher catalase activity to supply APX absence, as both act dismuting superoxide in oxygen and hydrogen peroxide.

Already in the interval of 24 h after start of imbibition, similarly to what happened with SOD (Table 1), the peroxidase activity of ascorbate (APX) and peroxidase (POX) (Tables 3 and 4, respectively) increased the higher osmotic potential. It was also observed increased activity of both enzymes in treatments plus SNP.

It is interesting to observe that for all the analyzed enzymes, isolated treatments of -0.2 and -0.3 MPa did not differ, which may suggest that oxidative stress in both potential was similar. Similar results were found in the germination test, first germination count and germination speed index, where the two water potentials were significantly resembled in those parameters analyzed.

Table 3. Enzyme activity of ascorbate peroxidase (APX) in *Sesamum indicum* seeds after 12 and 24 hours of imbibition in solutions with different osmotic potentials supplemented or not of SNP.

Treatment	APX (μmol min^{-1}.mg^{-1}. protein)	
	12 h	24 h
H$_2$O	0,11 ±0.01 Ba	1.39 ±0.40 Af
-0.1 MPa	0.05 ±0.01 Ba	1.67 ±0.32 Ae
-0.1 MPa +SNP	0.13 ± 0.02 Ba	3.75 ± 0.21 Ac
-0.2 MPa	0.03 ±0.01 Ba	3.28 ± 0.19 Ad
-0.2 MPa +SNP	0.09 ±0.01 Ba	4.91 ± 0.31Ab
-0.3 MPa	0.06 ±0.01 Ba	3.59 ± 0.08Ad
-0.3 MPa +SNP	0.05 ±0.02 Ba	5.72 ± 0.08 Aa
CV(%)	21.93	15.38

Averages followed by the same lower case in the column does not differ each other by Tukey test at 5% probability. Average followed by the same upper case in the line does not differ each other by Scott-Knott test at 5% probability. Average ± standard deviation.

Table 4. Enzyme activity of peroxidase (POX) in *Sesamum indicum* seeds after 0, 12 and 24 hours of imbibition in solutions with different osmotic potentials supplemented or not of SNP.

Treatment	POX (μmol min^{-1}.mg^{-1}. protein)		
	0 h	12 h	24 h
H_2O	8.54± 1.03 Aa	12.14 ± 1.78 Aa	18.45 ± 2.34 Af
-0.1 MPa	8.54± 1.03 Ba	13.17 ± 1.23 Ba	31.39 ± 4.32 Ae
-0.1 MPa +SNP	8.54± 1.03 Ba	9.23 ± 1.29 Ba	52.99 ± 3.21 Ac
-0.2 MPa	8.54± 1.03 Ba	10.45 ± 2.01 Ba	44.39 ± 2.28 Ad
-0.2 MPa +SNP	8.54± 1.03 Ba	10.34 ± 1.45 Ba	59.12 ± 3.21 Ab
-0.3 MPa	8.54± 1.03 Ba	11.87 ± 2.65 Ba	48.45 ± 2.89 Ad
-0.3 MPa +SNP	8.54± 1.03 Ba	12.45 ± 3.12 Ba	65.45 ± 4.67 Aa
CV(%)	---------------	21.36	19.22

Averages followed by the same lower case in the column does not differ each other by Tukey test at 5% probability. Average followed by the same upper case in the line does not differ each other by Scott-Knott test at 5% probability. Average ± standard deviation.

Conclusions

The increase of PEG concentrations, reduces of germination and seed vigor, as well as initial seedling growth.

The nitric oxide presence due to the sodium nitroprusside application is beneficial, providing increases in germination and vigor.

The SNP increase the antioxidant enzymes activity, indicating an efficient removal system of reactive oxygen species formed during exposure to water deficit and allow increased tolerance of this species to the water restriction and stimulate seed germination.

Acknowledgements

The authors thank the Foundation for Research Support of the State of Minas Gerais (FAPEMIG), the Coordination for the Improvement of Higher Education Personnel (CAPES), and The National Council for Scientific and Technological Development (CNPq), for financial support and scholarships.

References

ÁVILA, M. R.; BRACCINI, A. L.; SCAPIM, C. A.; FAGLIARI, J. R.; SANTOS, J. L. Influência do estresse hídrico simulado com manitol na germinação de sementes e crescimento de plântulas de canola. *Revista Brasileira de Sementes*, v.29, n.1, p. 98-106, 2007. http://dx.doi.org/10.1590/S0101-31222007000100014.

BAILLY, C. Active oxygen species and antioxidants in seed biology. *Seed Science Research*, v.14, n.2, p. 93-107, 2004. http://dx.doi.org/10.1079/SSR2004159.

BEWLEY, J. D.; BRADFORD, K. J.; HILHORST, W. W. M.; NONOGAKI, H. *Seeds:* Physiology of development, germination and dormancy, 3ed. Springer, New York, 2013.

BRADFORD, M. M. A rapid and sensitive method for the quantitation of microgram quantities of protein utilizing the principle of protein-dye binding. *Analytical Biochemistry*, v.72, p.248-254, 1976. http://www.sciencedirect.com/science/article/pii/0003269776905273

BRASIL. *Instrução Normativa* n° 45, de 17 de Setembro de 2013. *Legislação de sementes e mudas*. Brasília: 2013. http://www.abrasem.com.br/wp-content/uploads/2012/10/Instru%C3%A7%C3%A3o-Normativa-n%C2%BA-45-de-17-de-Setembro-de-2013-Padr%C3%B5es-de-Identidade-e-Qualiidade-Prod-e-Comerc-de-Sementes-Grandes-Culturas-Republica%C3%A7%C3%A3o-DOU-20.09.13.pdf

BRASIL. Ministério da Agricultura, Pecuária e Abastecimento. *Regras para análise de sementes*. Ministério da Agricultura, Pecuária e Abastecimento. Secretaria de Defesa Agropecuária. Brasília: MAPA/ACS, 2009. 395p. http://www.agricultura.gov.br/arq_editor/file/2946_regras_analise__sementes.pdf

DENG, Z.; SONG, S. Sodium nitroprusside, ferricyanide, nitrite and nitrate decrease the thermo-dormancy of lettuce seed germination in a nitric oxide-dependent manner in light. *South African Journal of Botany*, v.78, p.139-146, 2012. http://www.sciencedirect.com/science/article/pii/S0254629911000998

DONG, Y. J.; JINC, S. S.; LIU, S.; XU, L. L.; KONG, J. Effects of exogenous nitric oxide on growth of cotton seedlings under NaCl stress. *Journal of Soil Science and Plant Nutrition*, v.14, n.1, p.1-13, 2014. http://dx.doi.org/10.4067/S0718-95162014005000001.

FAN, H.; DU, C.; GUO, S. Nitric oxide enhaces salt tolerance in cucumber seedlings by regulating free polyamine content. *Environmental and Experimental Botany*, v.86, p.52-59, 2013. https://www.researchgate.net/publication/251507365_Nitric_oxide_enhances_salt_tolerance_in_cucumber_seedlings_by_regulating_free_polyamine_content

FERREIRA, D. F. Sisvar: A computer statistical analysis system. *Ciência e Agrotecnologia*, v.35, n.6, p. 1039-1042, 2011. http://www.scielo.br/pdf/cagro/v35n6/a01v35n6

HU, K.; HU, L.; LI, Y.; ZHANG, F.; ZHANG, H. Protective roles of nitric oxide on germination and antioxidante metabolism in wheat seeds under copper stress. *Plant Growth Regulation*, v. 53, n. 3, p. 173-183, 2007. http://link.springer.com/article/10.1007%2Fs10725-007-9216-9

KAR, M.; MISHRA, D. Catalase, peroxidase and polyphenoloxidase activities during rice leaf senescence. *Plant Physiology*, v. 57, p.315-319, 1976. http://dx.doi.org/10.1104/pp.57.2.315.

KOPYRA, M.; GWÓŹDŹ, E. A. Nitric oxide stimulates seed germination and counteracts the inhibitory effect of heavy metals and salinity on root growth of *Lupinus luteus*. *Plant Physiology and Biochemistry*, v.41, p.1011-1017, 2003. http://dx.doi:10.1016/j.plaphy.2003.09.003.

MITTLER, R. Oxidative stress, antioxidants and stress tolerance. *Trends in Plant Science*, v.7, p.405-410, 2002. http://www.cell.com/trends/plant-science/abstract/S1360-1385(02)02312-9?_returnURL=http%3A%2F%2Flinkinghub.elsevier.com%2Fretrieve%2Fpii%2FS1360138502023129%3Fshowall%3Dtrue

NAKAGAWA, J. Testes de vigor baseados no desempenho das plântulas. In: KRZYZANOSKI, F.C.; VIEIRA, R.D.; FRANÇA-NETO, J.B. (Ed.). *Vigor de sementes:* conceitos e testes. Londrina: *ABRATES*, 1999. https://books.google.com.br/books?id=lpyRGwAACAAJ

PEREIRA, M. D.; LOPES, J. C. Germinação e desenvolvimento de plântulas de pinhão manso sob condições de estresse hídrico simulado. *Semina: Ciências Agrárias*, v. 32, n.4, p.1837-1842, 2011. http://www.uel.br/revistas/uel/index.php/semagrarias/article/view/5224/9144

PIRES, R. M. O.; SOUZA, G. A.; CARDOSO, A. A.; DIAS, D. C. F. S.; BORGES, E. E. L. Action of nitric oxide in sesame seeds (*Sesamum indicum* L.) submitted to stress by cadmium. *Journal of Seed Science,* v.38, n.1, p.22-29, 2016. http://www.scielo.br/scielo.php?script=sci_arttext&pid=S2317-15372016000100022

RAHIMI, A. Seed priming improves the germination performance of cumin (*Cuminum syminum* L.) under temperature and water stress. *Industrial Crop and Products*, v.42, n.1, p.454-460, 2013. https://www.researchgate.net/publication/245541582_Seed_priming_improves_the_germination_performance_of_cumin_Cuminum_syminum_L_under_temperature_and_water_stress

SANO, S.; UEDA, M.; KITAJIMA, S.; TAKEDA, T.; SHIGEOKA, S.; KURANO, N.; MIYACHI, S.; MIYAKE, C; YOKOTA, A. Characterization of ascorbate peroxidases from unicellular red alga *Galdieria partita.* *Plant Cell and Physiology*, v.42, p.433-440, 2001. http://oxfordindex.oup.com/view/10.1093/pcp/pce054.

SANTOS, H. O.; PIRES, R. M. O.; VON PINHO, E. V. R.; RESENDE, E. S.; SILVA, V. F.; CARVALHO, M. L. M. Proteins expression and germination of maize seeds submitted to saline stress. *African Journal of Agricultural Research*, v.10, n.44, p.4102-4107, 2015. http://www.academicjournals.org/journal/AJAR/article-full-text-pdf/3A8355256021

SARATH, G.; BETHKE, P. C.; JONES, R.; BAIRD, L. M.; HOU, G.; MITCHELL, R. B. Nitric oxide accelerates seed germination in warm-season grasses. *Planta*, v.223, p.1154-1164, 2006. http://dx.doi 10.1007/s00425-005-0162-3.

SINGH, V. P.; SRIVASTAVA, P. K.; PRASAD, S. M. Nitric oxide alleviates arsenic-induced toxic effects in ridged *Luffa* seedlings. *Plant Physiology and Biochemistry*, v.71, p. 155-163, 2013. https://www.ncbi.nlm.nih.gov/pubmed/23917073

TEIXEIRA, R. N.; TOLEDO, M. Z.; FERREIRA, G.; CAVARIANI, C.; JASPER, S.P. Germinação e vigor de sementes de crambe sob estresse hídrico. *Irriga*, v.16, n. 1, p. 42-51, 2011. http://revistas.fca.unesp.br/index.php/irriga/article/view/190/99

WANG, W.; KIM, Y.; LEE, H.; KIM, K.; DENG, X.; KWAK, S. Analysis of antioxidant enzyme activity during germination of alfalfa under salt and drought stress. *Plant Physiology and Biochemistry*, v.47, n.7, p.570-577, 2009. https://www.ncbi.nlm.nih.gov/pubmed/19318268

YAO, Z.; LIU, L.; GAO, F.; RAMPITSCH, C.; REINECKE, D. M.; OZGA, J. A.; AYELE, B. T. Developmental and seed aging mediated regulation of antioxidative genes and differential expression of proteins during pre- and post-germinative phases in pea. *Journal of Plant Physiology*, v.169, n.15, p.1477-1488, 2012. https://www.researchgate.net/publication/228087046_Developmental_and_seed_aging_mediated_regulation_of_antioxidative_genes_and_differential_expression_of_proteins_during_pre-_and_post-germinative_phases_in_pea

ZHENG, C.; JIANG, D.; LIU, F.; DAI, T.; LIU, W.; JING, Q.; CAO, W. Exogenous nitric oxide improves seed germination in wheat against mitochondrial oxidative damage induced by high salinity. *Environmental and Experimental Botany*, v.67, n.1, p.222-227, 2009. http://www.sciencedirect.com/science/article/pii/S0098847209000896

Physiological changes in osmo and hydroprimed cucumber seeds germinated in biosaline water

Janete Rodrigues Matias[1], Renata Conduru Ribeiro[1], Carlos Alberto Aragão[2], Gherman Garcia Leal Araújo[1], Bárbara França Dantas[1*]

ABSTRACT- Biosaline agriculture has been a viable alternative for agricultural production in regions with lack of good quality water. To enable the cultivation of vegetables in high electrical conductivities, seed priming has been used to increase tolerance to stress caused by use of brackish water. We aimed to evaluate the efficiency of osmo and hydropriming in cucumber seeds cv. Caipira germinated biosaline in water, regarding germination and biochemical changes during the germinative process. The experimental design was totally randomized, with four replications of 50 seeds or 10 seedlings, in a factorial scheme 6x3; with six priming conditions (control, osmopriming during 24 and 48 hours, hydropriming with 1, 2 and 3 cycles of hydration-dehydration) and three biosaline water (fish farming biosaline wastewater) concentrations in distilled water (0, 50 e 100%). We evaluated the kinetics and percentage of germination; germinative metabolism and activity of antioxidant enzymes. According to the results, one hydropriming cycle is faster and more efficient to improve the performance of cucumber seedling in biosaline water and this can be used in substrate for germinating seeds of cucumber cv. Caipira.

Index terms: water reuse, Curcubitaceae, seedling development.

Modificações fisiológicas em sementes de pepino osmo e hidrocondicionadas germinadas em água biossalina

RESUMO - A agricultura biossalina tem sido uma alternativa viável para a produção agrícola em regiões com escassez de água de boa qualidade. Para possibilitar o cultivo de hortaliças em condutividades elétricas elevadas, o condicionamento fisiológico de sementes tem sido utilizado para aumentar a tolerância destas ao estresse causado pela utilização de água salobra. Objetivou-se avaliar a eficiência do osmo e hidrocondicionamento em sementes de pepino cv. Caipira em água biossalina, quanto à germinação e as modificações bioquímicas durante o processo germinativo. O delineamento foi inteiramente casualizado, com quatro repetições de 50 sementes ou 10 plântulas, em esquema fatorial 6x3; sendo seis condições de condicionamento fisiológico (controle, osmocondicionamento durante 24 e 48 horas; hidrocondicionamento com 1, 2 e 3 ciclos de hidratação e secagem) e três concentrações de água biossalina (água residuária de piscicultura) em água destilada (0, 50 e 100%). Foram avaliados: porcentagem de germinação; crescimento inicial de plântulas; metabolismo germinativo e atividade de enzimas antioxidantes. De acordo com os resultados obtidos pode-se concluir que um ciclo de hidrocondicionamento é mais rápido e eficiente para melhorar o desempenho das plântulas de pepino em água biossalina, que pode ser utilizada no substrato de germinação de sementes de pepino cv. Caipira.

Termos para indexação: reuso de água, Curcubitacea, desenvolvimento de plântulas.

Introduction

Worldwide, soil salinity is a problem for agriculture in more than 8 million hectares; wich is 6% for the total agricultural area (Bot et al., 2000).The most serious salinity problems are held in the irrigated arid and semi-arid regions of the world, such as Northeastern Semiarid Region of Brazil. Soil salinity is also a serious problem in areas where groundwater of high salt content is used for irrigation (FAO, 2014). Because of the serious problems due to water scarcity, the search for alternatives which aims at better utilization of water resources has become extremely important. As an example, Israel has overcome this problem in an economically viable manner, which is the biosaline agriculture, enabling coexistence with water restriction (Rocha et al., 2010).

Biosaline agriculture aims to seek alternative use of salty water

[1]Embrapa Semiárido, Caixa Postal 23, 56302-970 - Petrolina, PE, Brasil.
[2]Departamento de Tecnologia e Ciências Sociais Universidade do Estado da Bahia, 48900-000 - Juazeiro, BA, Brasil.
*Corresponding author <barbara.dantas@embrapa.br>

with efficient crop development and has been effective in salt affected regions, with use of sea water, brackish groundwater or wastewater (Al-Said et al., 2012). Although biosaline agriculture may be an alternative to crop production, in glycophytes, wich are most of crop species, it causes yield decrease by submitting seeds, transplants and growing plant to salt stress.

A way to increase seeds tolerance to abiotic stress, such as salt stress, is seed priming through various techniques of osmo or hydropriming used effectively in different crops (Sanchéz et al., 1997; Aragão et al., 2002; Azevedo Neto et al., 2005).

The contact of the seed with polyethylene glycol (PEG) solution during osmopriming, allows the occurance of all preparatory processes for seed germination, but prevents radicle emergence (Heydecker et al., 1975). Apparently, imposition of osmotic stress to seeds can induce tolerance to other stresses, including drought and salt tolerance. Primed seeds tend to germinate faster, since these have already started imbibition, requiring less water to complete germination (Mohammadi and Amiri, 2010).

At hydropriming, the amount of water absorbed by seeds is controlled by the length of time it remains in contact with a moist substrate (Farooq et al., 2010) or high relative humidity atmosphere, and subsequent drying. This technique results in increased seed germination and seedling emergence during re-imbibition (Ashraf and Foolad, 2005).

Within this approach, the aim of this work was to evaluate the effect of osmo and hydropriming in cucumber seeds germination (*Cucumis sativus* L. - Cucurbitaceae) cultivar Caipira, in fish farming biosaline wastewater.

Material and Methods

This study was carried out in 2012, with cucumber seeds (*Cucumis sativus* L. – Cucurbitaceae) cultivar Caipira. The experimental design was completely randomized with four replications of 50 seeds in a factorial scheme 6x3, regarding six priming conditions (control, osmopriming for 24 and 48 hours, 1, 2 and 3 cycles of hydropriming) and three biosaline water dilutions in distilled water (0, 50 and 100% of biosaline water).

Osmopriming: seeds were placed in plastic boxes (gerboxes) on substrate blotter paper, soaked in -1.0 MPa PEG

6000 solution (Villela et al., 1991), at a volume equivalent to 2.5 times the substrate paper weight and kept in germinator for 24 and 48 hours at 25 °C. After these periods, seeds were rinsed in running tap water to remove PEG solution, and dried at laboratory environment.

Hydropriming: was performed on gerboxes adapted with aluminum sieve with 40 mL of distilled water under the sieve and on it seeds were distributed and maintained for 8 h under laboratory conditions, characterizing hydration of seeds. For dehydration, water was replaced by 32 g silica, equivalent to 40 mL volume, for 12 hours, completing one cycle hydropriming. Three cycles of hydropriming were performed. Before and after each cycle, seeds water content was evaluated (Brasil, 2009).

Biosaline water (BW): after osmo and hydropriming, seeds were submitted to germination test in three different dilutions of BW in distilled water (0, 50 and 100%), obtained from tanks used for raising Nile tilapia (*Oreochromis* sp). The chemical characteristic of BW can be observed in Table 1.

Germination and vigor variables accessed: seeds were evaluated for water content and percentage of normal and abnormal seedlings (Brasil, 2009). Seedlings were evaluated for shoot and root length and dry weight (Nakagawa, 1999).

Evaluation of germinative metabolism: embryo, cotyledons, shoots and roots were collected from seedlings with 2 and 4 days, and frozen at -80 ° C until extraction. Total protein (TP); total soluble sugars (TSS); reducing sugars (RS); total amino acids (AA); catalase (CAT, E.C. 1.11.1.6) and ascorbate peroxidase (APX, EC 1.11.1.11) activities were quantified according to Ribeiro-Reis et al. (2012).

Statistical analysis: all data were subjected to analysis of variance and means were compared by Scott-Knott test at 5% probability.

Results and Discussion

During hydropriming cycles, cucumber seeds water content ranged from 15% after hydration to 5% after dehydration. Water content of osmoprimed seeds reached values around 27% and decreased to 7%, after drying, at both periods of osmopriming.

Table 1. Chemical characteristics of biosaline water (BW).

Cations					Anions					pH	CE
Ca^{2+}	Mg^{2+}	Na^+	K^+	Σ^+	CO_3^{2-}	HCO_3^-	SO_4^{2-}	Cl^-	Σ^-		
mmol$_c$/L											dS.m^{-1}
9.57	22.0	20.2	0.54	52.4	0	2	1.9	56	56.9	6.74	4.94

Provided by the Laboratory of Soil, Water and Plants - Embrapa Tropical Semiarid.

The germination process of cucumber cv. Caipira seeds was not affected by increasing salt concentration in germination medium. This also occurred in watermelon seeds cv. Crimson Sweet when subjected to domestic saline wastewater (Mota et al., 2011) or in squash and pumpkin seeds with fish farming brackish wastewater (Silva et al., 2014).

Although, a 4.94 dS.m^{-1} electrical conductivity was formerly considered damaging to Cucurbitaceae (Maas and Hoffman, 1977), recent literature showed cucumber seeds are tolerant up to 22 dS.m^{-1} (Torres et al., 2000) and over 90% of cv. Caipira seeds germinate at 14 dS.m^{-1} (Matias et al., 2014).

Low electrical conductivities can induce an increased germination percentage and decreased number of abnormal seedlings by reducing seed damage by imbibition (Matthews and Powell, 2006). Thus, cucumber cv. Caipira seeds subjected to 50% BW showed a higher percentage of normal seedlings when compared to seeds germinated in distilled water and subjected to a 100% BW. Furthermore, 100% BW induced an increase in cucumber seedlings root dry mass. In the highest salinity (100% BW), there was dry mass accumulation in shoots compared to roots, emphasizing that salinity does not interfere with the growth of these seedlings (Table 2).

In general, cycles of hydropriming increased normal seedlings percentage when seeds germinated in distilled water (Table 2). Agreeing with these findings, one cycle of hydropriming was efficient to speed up and increase germination in cucumber cultivars Hatuey-1 and Japonés (Sanchez et al., 1997). In bean cv. Carioca seeds, hydropriming also was favorable, providing a faster germination (Aragão et al., 2002). Additionally, two and three cycles of hydropriming increased shoots dry mass in 100% BW, compared to other priming treatments (Table 2).

When cucumber seeds were primed for 24 hours, seedlings had higher shoot length growth when subjected to biosaline water, compared to other priming treatments. Evaluating shoot: root length ratio, all priming treatments increased shoot growth, when compared to unprimed seedlings. On the other hand, seeds hydro and osmopriming inhibited growth in root length of cucumber seedlings (Table 2).

Table 2. Germination and vigor of cucumber seedlings cv. Caipira germinated in different percentages of water biosaline (BW) and subjected to osmopriming for 24 and 48 hours (OP) and three cycles hydropriming (HP).

BW	Control	24h OP	48h OP	1Cycle HP	2Cycle HP	3Cycle HP
(%)			Normal seedlings (%)			
0	81.0 cB	79.5 bB	69.5 cC	90.0 aA	90.0 aA	91.0 aA
50	94.0 aA	92.0 aA	93.5 aA	91.5 aA	94. 0aA	92.5 aA
100	88.5 bA	89.5 aA	75.5 bC	83.5 bB	82.0 bB	76.0 bC
			CV(%) = 3.23			
			Abnormal seedlings (%)			
0	19.0 aB	20.5 aB	30.5 aA	10.0 bC	10.0 bC	9.0 bC
50	6.0 cA	8.0 bA	6.0 cA	8.5 bA	6.0 cA	7.5 bA
100	11.5 bC	10.5 bC	24.5 bA	16.5 aB	18.0 aB	24.0 aA
			CV(%) = 20.34			
			Shoot length (cm)			
0	5.89 aB	6.69 bA	4.95 bB	5.66 aB	7.70 aA	7.36 aA
50	6.64 aB	8.85 aA	6.58 aB	6.16 aB	6.16 bB	6.73 aB
100	6.22 aA	6.37 bA	6.95 aA	6.28 aA	6.43 bA	6.04 bA
			CV(%) = 9.76			
			Root length (cm)			
0	9.76 aA	4.33 cC	4.56 bC	7.73 bB	7.69 aB	7.55 aB
50	9.80 aA	7.55 aB	6.32 aC	8.96 aA	7.79 aB	6.88 aC
100	8.28 bA	6.59 bB	6.30 aB	7.83 bA	6.75 bB	5.78 bB
			CV(%) = 8.93			
			Shoot dry mass (mg)			
0	91.90 bC	91.75 cC	107.50 aB	96.75 bC	117.25 aA	123.50 bA
50	110.25 aC	126.75 aB	116.00 aC	116.00 aC	129.00 aB	148.00 aA
100	100.00 bB	112.75bA	117.25 aA	108.00 aB	117.25 aA	118.25 bA
			CV(%) =7.63			
			Root dry mass (mg)			
0	36.17 bA	44.25 aA	39.50 aA	29.00 bB	26.75 bB	21,80 cB
50	28.50 cC	26.75 bC	24.75 bC	40.75 aB	40.75 aB	52.50 aA
100	47.50 aA	27.75 bB	22.50 bB	27.00 bB	25.00 bB	30.50 bB
			CV(%) = 13.62			

Means followed by the same capital letter on the line, and lower case letter in the column do not differ statistically from each other. Scott-Knott test at 5% probability level was applied.

Although fish farming effluents induce root growth in seedlings of cucurbits (Medeiros et al., 2010), cucumber seedlings that developed in BW with electrical conductivity of 4.9 dS.m^{-1}, showed decreased root length (Table 2), as did gherkin seedlings in electrical conductivities greater than 2.15 dS.m^{-1} (Oliveira et al., 2013).

Substrate salinity initially changes absorption of water and nutrients, as well as, membrane permeability. By analyzing plant characteristics, one can realize that tolerance to salinity varies with stage of development in which the plant is in and if stress is imposed on a cell, a tissue or an organ of the individual. Cucumber seeds and seedlings showed different degrees of tolerance between organs and developmental stages (Table 2).

Conus et al. (2009) described salinity tolerance as the ability of plants to avoid, through an osmotic regulation, excessive amount of salt of substrate from reaching the plasma and to tolerate toxic and osmotic effects associated with increased salt concentration. Under stress conditions when seeds are primed, germination performance in adverse situation is favored, since seeds have already started soaking with lower osmotic potential and require less water to complete germination (Carvalho and Nakagawa, 2012). Due to a possible osmotic adjustment, which is a reduction of cell osmotic potential by accumulating compatible organic solutes with metabolism, seed priming has been considered an important enhancer of tolerance to salt and drought stress in plants (Azevedo Neto et al., 2005).

Accumulation of organic solutes, compatible with plant metabolism is an important mechanism of salt tolerance in plants. This mechanism promotes reduction of cell osmotic potential, thereby allowing an osmotic adjustment to the stress condition (Azevedo Neto et al., 2005), which may have occurred in cucumber cv. Caipira seedlings invigorated with hydropriming and subjected to salinity conditions during post-germinative growth (Tables 2-6).

Total protein (TP) in cotyledons, of cucumber seedlings with two days after sowing (DAS), increased significantly when they were not preconditioned and germinated in solutions of 50% BW (Table 3). This result was also observed in the hypocotyl-root axis. In 2 DAS cotyledons, under ideal conditions, the cycles of hydropriming caused a significant increase in TP concentration.

Mobilization of TP to the seedlings axis was affected by pre-treatments only in ideal germination conditions (0% BW). Under these conditions, hydropriming induced a higher content in shoot TP, while osmopriming induced accumulation in roots TP content (Table 3). BW reduced TP degradation in cucumber cotyledons (Table 3).

Table 3. Content of total protein (µg.g^{-1} of fresh weight) in cucumber seedlings cv. Caipira of 2 and 4 days after sowing (DAS), germinated in different percentages of biosaline water (BW) and subjected to osmopriming for 24 and 48 hours (OP) and three cycles hydropriming (HP).

BW	Control	24h OP	48h OP	1Cycle HP	2Cycle HP	3Cycle HP
(%)			Cotyledons			
0	92.26 cD	136.96 aC	157.92 aB	189.82 aA	198.51 aA	192.62 aA
50	248.30 aA	156.49 aB	173.99 aB	161.20 bB	181.72 aB	176.43 aB
100	148.40 bA	97.62 bB	105.48 bB	157.68 bA	139.52 bA	96.07 bB
			CV(%) = 8.34			
			Hypocotyl-root axis 2 DAS			
0	50.48 bA	65.54 aA	65.77 aA	86.79 aA	85.29 aA	64.70 aA
50	87.79 aA	66.25 aA	85.95 aA	85.42 aA	100.13 aA	72.62 aA
100	47.32 bA	59.34 aA	57.86 aA	60.65 aA	63.98 aA	58.63 aA
			CV(%) = 23.34			
			Cotyledons 4 DAS			
0	56.43 aB	58.45 aB	36.84 aB	90.36 aA	52.98 aB	94.40aA
50	69.23 aA	78.45 aA	68.21 aA	53.28 bA	70.36 aA	76.37 aA
100	68.51 aB	71.43 aB	54.17 aB	95.65 aA	63.93 aB	52.20 bB
			CV(%) = 16.49			
			Shoot 4 DAS			
0	6.26 aB	9.05 aA	8.65 aA	5.68 aB	7.26 aB	5.40 aB
50	5.08 aA	5.86 bA	6.08 bA	6.04 aA	4.91 aA	5.41 aA
100	5.95 aA	6.73 bA	4.42 bA	7.23 aA	5.40 aA	4.23 aA
			CV(%) = 21.27			
			Root 4 DAS			
0	5.34 aB	4.48 aB	6.44 aB	14.09 aA	12.89 aA	10.78 aA
50	5.55 aA	7.72 aA	7.65 aA	11.04 aA	6.98 bA	8.11 aA
100	7.85 aA	6.58 aA	10.29 aA	11.57 aA	4.83 bA	6.75 aA
			CV% = 33.11			

Means followed by the same capital letter on the line, and lower case letter in the column do not differ statistically from each other. The Scott-Knott test at 5% probability level was applied.

Both 2 and 4 DAS cotyledons of seedlings which were osmoprimed and germinated in distilled water showed a lower content of amino acids (AA) compared to other pretreatments. This content increased when osmoprimed seedlings were subjected to biosaline water (Table 4). During germination process, synthesis of new proteins relies on provision of adequate amino acids derived from breakdown of storage proteins (Kim et al., 2011). Furthermore, unprimed 2 DAS seedlings germinated under ideal conditions, showed lower AA content in embryonic axis than other treatments, different from what occurred in cotyledons (Table 4).

Seedlings with 4 DAS showed the accumulation of amino acids when grown from seeds primed for 24 hours. Seedlings not osmoprimed presented increased AA content, both in shoots and roots, as the concentration of BW increased, reaching values higher than twice in distilled water (Table 4). BW had positive effect on mobilization of seed reserves, which favored germination and seedling performance, leading to a better development in BW (Tables 2-6).

Table 4. Content of total aminoacids ($\mu mol.g^{-1}$ of fresh weight) in cucumber seedlings cv. Caipira of 2 and 4 days after sowing (DAS), germinated in different percentages of biosaline water (BW) and subjected to osmopriming for 24 and 48 hours (OP) and three cycles hydropriming (HP).

BW (%)	Control	24h OP	48h OP	1Cycle HP	2Cycle HP	3Cycle HP
			Cotyledons			
0	153.07 aA	120.25 bB	131.35 bB	163.43 aA	153.39 aA	145.83 aA
50	155.44 aA	155.95 aA	171.41 aA	158.66 aA	156.52 aA	151.99 aA
100	165.32 aA	144.27 aA	166.89 aA	159.65 aA	157.51 aA	73.61 bB
			CV(%) = 9.29			
			Hypocotyl-root axis 2 DAS			
0	12.24 bB	68.02 aA	23.33 bB	46.64 bA	33.39 bB	55.27 aA
50	98.32 aA	56.18 aB	65.06 aB	109.23 aA	91.83 aA	70.73 aB
100	24.26 bA	27.96 bA	25.33 bA	25.00 cA	27.72 bA	31.50 bA
			CV(%) = 19.83			
			Cotyledons 4 DAS			
0	114.24 aA	84.88 bB	73.37 bB	133.74 aA	121.40 aA	140.32 aA
50	95.41 aB	116.79 aA	124.77 aA	88.83 bB	94.01 bB	66.21 bC
100	109.47 aA	106.18 aA	72.21 bB	107.75 bA	93.19 bA	73.45 bB
			CV(%) = 10.43			
			Shoot 4 DAS			
0	40.63 bC	25.66 bD	35.94 aC	98.53 aA	82.17 aB	45.81 aC
50	56.51 aA	26.15 bB	43.09 aA	48.03 bA	26.81 bB	19.82 bB
100	59.71 aA	55.10 aA	33.47 aB	21.79 cB	26.32 bB	20.72 bB
			CV(%) = 14.68			
			Root 4 DAS			
0	7.46 cD	12.72 bD	29.94 aB	48.28 aA	20.56 aC	25.66 aB
50	19.02 bA	20.81 aA	15.54 bA	24.48 bA	22.51 aA	22.62 aA
100	37.8 aA	23.28 aB	18.95 bB	24.59 bB	21.71 aB	19.79 aB
			CV(%) = 13.96			

Means followed by the same capital letter on the line, and lower case letter in the column do not differ statistically from each other. The Scott-Knott test at 5% probability level was applied.

Among the compounds predominant in seed reserves, carbohydrates act as an energy source in early seedling development (Carvalho and Nakagawa, 2012). Germination in BW, diluted or not, induced a higher content of reducing sugars (RS) in cotyledons of 2 DAS seedlings, except for those which were osmoprimed for 24 hours. Moreover, hypocotyl-root axis in this response was reversed for most pretreatments, in which 4 DAS cotyledons showed lower values RS compared to 2 DAS seedlings. Hydropriming RS resulted in higher levels in cotyledons of seedlings subjected to distilled water, whereas those maintained in 50% BW showed a high cotyledon RS content in all pretreatments. One cycle of hydropriming caused an increase in RS content in 2 DAS cotyledons and in 4 DAS shoots (Table 5).

Total soluble sugar (TSS) in cotyledons was lower in 2 DAS seedlings hydroprimed for one and three cycles and grown in 50% BW solution (Table 6). Development on 50% BW solution induced an increase in TSS levels in cotyledons and hypocotyl-root axis of unprimed seedlings. TSS content in 4 DAS cotyledons was lower compared to 2 DAS cotyledons. The level of TSS was higher in 4 DAS cotyledons of hydroprimed seedlings that developed in distilled water.

In pure BW, all pre-treatments induced increased levels of TSS, with the most significant increase after one cycle of hydropriming (Table 6).

Table 5. Content of total soluble sugars (μmol.g⁻¹ of fresh weight) in cucumber seedlings cv. Caipira of 2 and 4 days after sowing (DAS), germinated in different percentages of biosaline water (BW) and subjected to osmopriming for 24 and 48 hours (OP) and three cycles hydropriming (HP).

BW	Control	24h OP	48h OP	1Cycle HP	2Cycle HP	3Cycle HP
(%)			Cotyledons			
0	86.42 aB	74.65 bB	74.80 bB	103.89 bA	100.54 aA	110.80 aA
50	91.13 aA	96.37 aA	84.60 bA	98.12 bA	100.39 aA	87.79 bA
100	62.73 bC	103.13 aB	102.68 aB	121.05 aA	105.94 aB	103.21 aB
			CV(%) = 6.21			
			Hypocotyl-root axis 2 DAS			
0	106.59 aA	42.78 bC	71.69 bB	110.19 aA	53.21 bC	44.17 aC
50	72.83 bB	65.54 aB	95.95 aA	62.25 bB	104.99 aA	58.14 aB
100	114.43 aA	47.41 bB	52.04 cB	48.19 bB	49.32 bB	49.49 aB
			CV(%) = 9.86			
			Cotyledons 4 DAS			
0	60.22 aA	52.40 aA	59.77 aA	74.88 aA	108.98 aA	101.84 aA
50	91.96 aA	93.03 aA	84.22 aA	70.02 aA	73.06 aA	74.19 aA
100	109.21 aA	101.46 aA	84.98 aA	91.68 aA	51.41 aA	53.31 aA
			CV(%) = 38.68			
			Shoot 4 DAS			
0	52.93 bC	49.51 cC	128.49 aA	83.53 aB	113.00 aA	89.76 aB
50	63.18 bC	141.63 aA	97.36 aB	89.79 aB	131.84 aA	29.16 bD
100	138.44 aA	105.11 bA	116.27 aA	102.14 aA	97.96 aA	80.57 aA
			CV(%) =16.92			
			Root 4 DAS			
0	88.96 bB	303.69 aA	87.73 bB	85.46 bB	98.44 bB	139.33 bB
50	121.31 bC	193.00 bB	68.06 bD	304.98 aA	248.39 aB	222.17 aB
100	235.41 aB	195.43 bB	141.43 aC	288.63 aA	82.60 bD	214.90 aB
			CV(%) = 13.24			

Means followed by the same capital letter on the line, and lower case letter in the column do not differ statistically from each other. The Scott-Knott test at 5% probability level was applied.

Reserve mobilization during germination, which provides energy for formation of the new plant, is a determining factor in seedling vigor produced by the end of the process (Carvalho and Nakagawa, 2012). Thus, during imposition of any stress, mobilization of storage substances of seeds should adjust so that normal, healthy and vigorous seedlings can be formed. In plants subjected to salt stress, accumulation of soluble sugars is a common process in the cells. Sugars have a role in osmotic adjustment, but also have indirect effects of protection as in stabilization of proteins (Bianchi et al., 1991). Priming, by imposing a moderate osmotic stress to seeds, induces the concentration of osmolites, such as sugars, to keep the cell's osmotic potential compatible with maintenance of stability of cell proteins, minimizing the loss of enzymatic activity or membrane integrity which occurs when there is water or saline stress (Gonçalves et al., 2007).

Catalase (CAT) prevents formation of reactive compounds, converting hydrogen peroxide into water and oxygen, which are non-reactive oxygen species (Nakada et al., 2011).The activity of CAT in the hypocotyl-root axis of 2 DAS seedlings was not affected by either methods of priming imposed on cucumber seeds (Table 7). This means these seeds showed no changes in oxidative stress, and therefore respiratory capacity, energy supply (ATP) and assimilates for seed germination may not be affected (Demirkaya et al., 2010). In 2 DAS cucumber seedlings cotyledons, CAT activity was higher than other treatments only for those who were osmoprimed for 48 hours or hydroprimed for two cycles and germinated in distilled water (Table 7), indicating that in 2 DAS cotyledons BW caused a mild salt stress, which was not alleviated by priming.

In other tissues and developmental stages differences in

CAT activity were hardly observed (data not shown), indicating no increase in reactive oxygen species (ROS) or damage to cell membranes which could influence seedling performance.

The activity of ascorbate peroxidase (APX) in 2 DAS cotyledons was lower in seedlings subjected to 100% BW. Hydropriming in cotyledons induced a reduction in APX activity when subjected to BW. In 2 DAS seedlings hypocotyl-root axis, APX activity was low in relation to cotyledons. In hypocotyl-root axis of osmoprimed seedlings, APX activity was lower compared to non primed seedlings developed in 50% BW (Table 7) and at 4 DAS the activity was null (data not shown). The decrease of APX activity when seeds were hydroprimed, may have been caused by decreased oxidative stress, resulting in an insufficient amount of ROS to induce the activity of this enzyme (Ribeiro-Reis et al., 2012).

Table 6. Content of reducing sugars ($\mu mol.g^{-1}$ of fresh weight) in cucumber seedlings cv. Caipira of 2 and 4 days after sowing (DAS), germinated in different percentages of biosaline water (BW) and subjected to osmopriming for 24 and 48 hours (OP) and three cycles hydropriming (HP).

BW	Control	24h OP	48h OP	1Cycle HP	2Cycle HP	3Cycle HP
(%)			Cotyledons			
0	88.96 bB	303.69 aA	87.73 bB	85.46 bB	98.44 bB	139.33 bB
50	121.31 bC	193.00 bB	68.06 bD	304.98 aA	248.39 aB	222.17 aB
100	235.41 aB	195.43 bB	141.43 aC	288.63 aA	82.60 bD	214.90 aB
			CV(%) = 13.24			
			Hypocotyl-root axis 2 DAS			
0	106.59 aA	42.78 bC	71.69 bB	110.19 aA	53.21 bC	44.17 aC
50	72.83 bB	65.54 aB	95.95 aA	62.25 bB	104.99 aA	58.14 aB
100	114.43 aA	47.41 bB	52.04 cB	48.19 bB	49.32 bB	49.49 aB
			CV(%) = 9.86			
			Cotyledons 4 DAS			
0	72.30 aB	55.46 bB	72.16 aB	106.82 aA	121.66 aA	112.62 aA
50	56.02 aC	95.53 aA	100.63 aA	122.44 aA	83.37 bB	102.79 aA
100	74.24 aA	80.99 aA	91.67 aA	89.16aA	95.44 bA	78.69 bA
			CV(%) = 13.43			
			Shoot 4 DAS			
0	47.93 aC	9.20 cE	8.64 bE	89.12 aA	75.23 aB	28.07 aD
50	48.15 aB	40.36 aC	32.18 aC	78.95 aA	55.07 bB	29.46 aC
100	30.19 bB	23.70 bB	25.82 aB	81.57 aA	23.57 cB	30.58 aB
			CV(%) = 12.17			
			Root 4 DAS			
0	70.08 aB	87.30 aA	28.68 bD	67.79 bB	51.91 aC	24.18 cD
50	53.86 bC	26.73 bD	35.64 bD	81.85 aA	65.01 aB	85.79 aA
100	54.42 bB	79.99 aA	74.11 aA	54.12 bB	68.66 aA	50.74 bB
			CV(%) = 12.60			

Means followed by the same capital letter on the line, and lower case letter in the column do not differ statistically from each other. The Scott-Knott test at 5% probability level was applied.

Considering antioxidative enzymes play vital roles in growth and development of early embryos, an increase in activity of these enzymes might result in vigorous growth and good crop establishment. Studies have shown priming affects activity of oxidative enzymes in the germination of seeds of different species (Ashraf and Foolad, 2005). However, not all cultures respond satisfactorily when developing in saline conditions, thus antioxidant defense mechanisms may provide a strategy to increase salt tolerance (Gonçalves et al., 2007).

In the present study, cucumber seedlings have grown satisfactorily in conditions that would be detrimental for the crop. When subjected to priming treatments, cucumber seeds showed higher tolerance to mild salt stress induced by the biosaline water, reflecting on seedling vigor. This study evaluated the germination and early seedling performance of cucumber for only eight days according to Brasil (2009). In order to check effects of priming treatments and biosaline water on transplants or adult plants, further studies should be conducted over a longer period of time.

Table 7. Catalase- CAT activity (μmol H_2O_2 min^{-1}.mg $protein^{-1}$) and ascorbate peroxidase- APX activity (ηmol ascorbate.min^{-1}. mg $proteín^{-1}$) in cucumber seedlings cv. Caipira of 2 days after sowing (DAS), germinated in different percentages of biosaline water (BW) and subjected to osmopriming for 24 and 48 hours (OP) and three cycles hydropriming (HP).

BW (%)	Control	24h OP	48h OP	1Cycle HP	2Cycle HP	3Cycle HP
			Catalase activity			
			Cotyledons			
0	21.76 aB	24.82 aB	56.67 aA	28.49 aB	43.92 aA	27.58 aB
50	29.54 aA	22.84 aA	36.20 bA	31.19 aA	48.11 aA	19.79 aA
100	33.30 aA	17.79 aA	18.71 bA	25.98 aA	17.67 bA	11.15 aA
			CV(%) = 37.35			
			Hypocotyl-root axis 2 DAS			
0	1.91 aA	2.13 aA	1.48 aA	2.54 aA	2.75 aA	2.59 aA
50	2.16 aA	2.16 aA	2.24 aA	1.75 aA	1.80 aA	2.04 aA
100	2.96 aA	1.65 aA	2.61 aA	2.29 aA	2.56 aA	2.52 aA
			CV(%) = 20.97			
			Ascorbate peroxidase activity			
			Cotyledons			
0	46.57 aA	48.51 aA	51.15 aA	59.49 aA	58.12 aA	65.67 aA
50	40.17 aA	40.33 aA	40.34 aA	18.33 bB	14.79 bB	19.87 bB
100	12.18 bA	8.21 bA	18.24 bA	12.52 bA	7.98 bA	7.15 bA
			CV(%) = 29.79			
			Hypocotyl-root axis 2 DAS			
0	1.23 bA	1.33 aA	1.54 aA	1.28 aA	1.13 bA	1.29 aA
50	1.86 aA	1.39 aB	1.13 bC	0.88 bC	0.82 cC	1.02 bC
100	1.50 bA	1.34 aA	1.33 bA	1.42 aA	1.44 aA	1.47 aA
			CV(%) = 10.73			

Means followed by the same capital letter on the line, and lower case letter in the column do not differ statistically from each other. The Scott-Knott test at 5% probability level was applied. Data were transformed by the equation $X = \sqrt{x}$.

Conclusions

The cucumber seedlings were tolerant to salinity up to 4.9 $dS.m^{-1}$ imposed by biosaline water (BW).The BW may be used in germinating seeds of cucumber cv. Caipira and may be a viable alternative use for the production of seedlings through new studies.

One cycle of hydropriming can improve the overall performance of cucumber seedlings.

References

AL-SAID, F. A.; ASHFAQ , M.; AL-BARHI, M.; HANJRA, M.A.; KHAN, I. A. Water productivity of vegetables under modern irrigation methods in Oman. *Irrigation and Drainage*, v.61, n.4, p. 477-489, 2012. http://onlinelibrary.wiley.com/doi/10.1002/ird.1644/full.

ARAGÃO, C.A.; DANTAS, B.F.; ALVES, E.; CORRÊA, M.R. Sementes de feijão submetidas a ciclos e períodos de hidratação-secagem. *Scientia Agricola*, v.59, n.1, p.87-92, 2002. http://www.scielo.br/pdf/sa/v59n1/8079.pdf

ASHRAF, M.; FOOLAD, M.R. Pre-sowing seed treatment - a shotgun approach to improve germination, plant growth, and crop yield under saline and non-saline conditions. *Advances in Agronomy*, v.88, p.223-271, 2005. http://ac.els-cdn.com/S006521130588006X/1-s2.0-S006521130588006X-main.pdf?_tid=3fac8852-47b7-11e3-904a-00000aab0f6b&acdnat=1383833992_e16d08759dccfa0e527a9dd6345849e4

AZEVEDO NETO, A.D.; PRISCO, J.T.; ENEAS-FILHO, J.; MEDEIROS, J.V.R.; GOMES-FILHO, E. Hydrogen peroxide pretreatment induces salt-stress acclimation in maize plants. *Journal of Plant Physiology*, v.162, p.1114-1122, 2005. http://ac.els-cdn.com/S0176161705000726/1-s2.0-S0176161705000726-main.pdf?_tid=9bc0d724-4875-11e3-9b0d-00000aacb35f&acdnat=1383915751_9f120673cdc3ff3f5f5c07d869edaae9

BIANCHI, G.; GAMBA, A.; MURELLI, C.; SALAMINI, F.; BARTELS, D. Novel carbohydrate metabolism in the resurrection plant *Cratero stigma plantagineum*. *The Plant Journal*, v.1, p. 355-359, 1991. http://onlinelibrary.wiley.com/doi/ 10.1046/j.1365-313X.1991.t01-11-00999.x/pdf

BOT, A.; NACHTERGAELE, F.; YOUNG, A. *Land resource potential and constraints at regional and country levels*. Rome: FAO, 2000. (FAO.World Soil Resources Report, 90). ftp://ftp.fao.org/agl/agll/docs/wsr.pdf

BRASIL. Ministério da Agricultura, Pecuária e Abastecimento. *Regras para análise de sementes*. Ministério da Agricultura, Pecuária e Abastecimento. Secretária de Defesa Agropecuária. Brasília: Mapa/ACS, 2009. 395p. http://www.agricultura.gov.br/arq_editor/file/2946_regras_analise__sementes.pdf

CARVALHO, N. M.; NAKAGAWA, J. *Sementes:* ciência, tecnologia e produção. 5.ed. Jaboticabal: FUNEP, 2012. 590p.

CONUS, L.A.; CARDOSO, P.C.; VENTUROSO, L.R.; SCALON, S.P.Q. Germinação de sementes e vigor de plântulas de milho submetidas ao stress e salino induzido por diferentes sais. *Revista Brasileira de Sementes*, v.31, n.4, p.67-74, 2009. http://www.scielo.br/pdf/rbs/v31n4/08.pdf

DEMIRKAYA, M.; DIETZ, K.J.; SIVRITEPE, H.O. Changes in antioxidant enzymes during aging of onion seeds. *Notula e Botanica e Hortiagrobotanici*, v.38, n.1, p.49-52, 2010. http://www.notulaebotanicae.ro/index.php/nbha/article/view/4575/4417

FAO - Food and Agriculture Organization of the United Nations, 2014. *Soil salinity management*. http://www.fao.org/tc/exact/sustainable-agriculture-platform-pilot-website/soil-salinity-management/en/. Accessed on May, 17th, 2014.

FAROOQ, M.; WAHID, A.; AHMAD, N.; ASAD, S.A. Comparative efficacy of surface drying and re-drying seed priming in rice: changes in emergence, seedling growth and associated metabolic events. *Paddy Water Environmental*, v.08, n.01, p.15-22, 2010. http://link.springer.com/article/10.1007/s10333-009-0170-1#page-1

GONÇALVES, J.F.; BECKER, A.G.; CARGNELUTTI, D.; TABALDI, L.A.; PEREIRA, L.B.; BATTISTI, V.; SPANEVELLO, R.M.; MORSCH, V. M.; NICOLOSO, F.T.; SCHETINGER, M. R.C. Cadmium toxicity causes oxidative stress and induces response of the antioxidant system in cucumber seedlings. *Brazilian Journal Plant Physiology*, v.19, n.3, p.223-232, 2007. http://www.scielo.br/scielo.php?script=sci_arttext&pid=S1677-04202007000300006&lng=en&nrm=iso

HEYDECKER, W.; HIGGING, J.; TURNER, Y.J. Invigoration of seeds. *Seed Science and Technology*, v.3, p.881-888, 1975.

KIM, H.T.; CHOI, U.K.; RYU, H.S.; LEE, S.J.; KWON, O.S. Mobilization of storage proteins in soybean seed (*Glycine max* L.) during germination and seedling growth. *Biochimica et Biophysica Acta*, v.1814, p.1178-1187, 2011. http://www.ncbi.nlm.nih.gov/pmc/articles/PMC1877431/pdf/amjpathol00113-0109.pdf8

MAAS, E.V., HOFFMAN, G.J. Crop salt tolerance, current assessment. *Journal of the Irrigation and Drainage Division*, v.103, n.2, p.115-134, 1977. http://www.waterrights.ca.gov/baydelta/docs/southerndeltasalinity/hist_exhibits/1977bdh_p2ex1.pdf

MATIAS, J.R.; SILVA, T.C.F.S.; OLIVEIRA, G.M.; ARAGÃO, C.A.; DANTAS, B.F. Germinação de sementes de pepino cv. Caipira em condições de estresse hídrico e salino. *Revista Sodebras*, v.10, n.110, 2014. (in press)

MATTHEWS, S.; POWELL, A.A. Electrical conductivity vigour test: physiological basic and use. *ISTA News Bulletin*, n.131, p.32-35, 2006. https://www.seedtest.org/upload/cms/user/STI131April2006.pdf

MEDEIROS, D.C.; MARQUES, L.F.; DANTAS, M.R.S.; MOREIRA, J.N.; AZEVEDO, C.M.S.B. Produção de mudas de meloeiro com efluente de piscicultura em diferentes tipos de substratos e bandejas. *Revista Brasileira de Agroecologia*, v.5, n.2, p. 65-71, 2010. http://www.aba-agroecologia.org.br/ojs2/index.php/rbagroecologia/article/view/9619/pdf

MOHAMMADI, G.R.; AMIRI, F.The effect of priming on seed performance of canola (*Brassica napus* L.) under drought stress. *American Eurasian Journal of Agricultural & Environmental Sciences*, v.2, n.9, p.202-207, 2010. http://www.idosi.org/aejaes/jaes9%282%29/16.pdf

MOTA, A.F.; ALMEIDA, J.P.N.; SANTOS, J.S.; AZEVEDO, J.; GURGEL, M.T. Desenvolvimento inicial de mudas de melancia 'Crimson sweet' irrigadas com águas residuárias. *Revista Verde*, v.6, n.2, p.98–104, 2011. http://www.gvaa.com.br/revista/index.php/RVADS/article/viewFile/628/pdf_175

NAKADA, P.G.; OLIVEIRA, J.A.; MELO, L.C.; GOMES, L.A.A.; VON PINHO, E.V.R. Desempenho fisiológico e bioquímico de sementes de pepino nos diferentes estádios de maturação, *Revista Brasileira de Sementes*, v.33, n.1,p.113–122, 2011. http://www.scielo.br/pdf/rbs/v33n1/13.pdf

NAKAGAWA, J. Testes de vigor baseados no desempenho das plântulas. In: KRZYZANOSKI, F.C.; VIEIRA, R.D.; FRANÇA-NETO, J.B. (Ed.). *Vigor de sementes*: conceitos e testes. Londrina: ABRATES, 1999. p.2.1-2.24.

OLIVEIRA, F.N.; TORRES, S.B.; BENEDITO, C.P.; MARINHO, J.C. Comportamento de três cultivares de maxixe sob condições salinas. *Semina: Ciências Agrárias*, v.34, n.6, p.2753-2762, 2013. http://www.uel.br/revistas/uel/index.php/semagrarias/article/view/12015/pdf_122

RIBEIRO-REIS, R.C.; DANTAS, B.F.; PELACANI, C.R. Mobilization of reserves and germination of seeds of *Erythrina velutina* Willd.(Leguminosae-Papilionoideae) under different osmotic potentials. *Revista Brasileira de Sementes*, v.34, n.4, p.580-588, 2012. http://www.scielo.br/scielo.php?script=sci_arttext&pid=S0101-31222012000400008&lng=en&nrm=iso

ROCHA, F.A.; SILVA, J.O.; BARROS, F.M. Reuso de águas residuárias na agricultura: a experiência israelense e brasileira. *Enciclopédia Biosfera*, v.6, n.11, p.1-9, 2010. http://www.conhecer.org.br/enciclop/2010c/reuso%20de%20aguas.pdf.

SANCHÉZ, J.A.; CALVO, E.; ORTA, R.; MURILO, Z. Tratamientos pré-germinativos de hidratación–deshidratación pra semillas de pepino (*Cucumis sativus* L.). *Acta Botánica Mexicana*, n.38, p.13-20, 1997. http://www.redalyc.org/pdf/574/57403803.pdf

SILVA, J.E.S.B; BARBOSA, L.G.; SILVA, F.Z.; SILVA, T.B.; MATIAS, J.R.; ARAGÃO, C.A.; DANTAS, B.F. Produção de mudas de moranga e abóboras irrigadas com água biossalina. *Scientia Plena*, v.10, n.10, p.109906-1 - 109906-9, 2014. http://www.scientiaplena.org.br/sp/article/view/1949/1042

TORRES, S.B.; VIEIRA, E.L.; MARCOS-FILHO, J. Efeitos da salinidade na germinação e no desenvolvimento de plântulas de pepino. Revista Brasileira de Sementes, v.22, n.2, p.39-44, 2000.

VILLELA, F.A.; DONI FILHO, L.; SEQUEIRA, E.L. Tabela de potencial osmótico em função da concentração de polietileno glicol 6000 e da temperatura. *Pesquisa Agropecuária Brasileira*, v.26, n.11-12, p.1957-1968, 1991. http://seer.sct.embrapa.br/index.php/pab/article/view/3549/882

Action of nitric oxide in sesame seeds (*Sesamum indicum* L.) submitted to stress by cadmium

Raquel Maria de Oliveira Pires[1*], Genaina Aparecida de Souza[2],
Amanda Ávila Cardoso[2], Denise Cunha Fernandes dos Santos
Dias[3], Eduardo Euclydes de Lima e Borges[4]

ABSTRACT - The objective of this paper was to evaluate the effect of nitric oxide (NO) as a protecting agent of sesame seeds submitted to different concentrations of cadmium. The treatments were: water (control), water increased by sodium nitroprusside (SNP) and other treatments regarding the concentrations of cadmium increased by SNP. The following determinations were done: germination, first count of germination, germination speed index, length of hypocotyl and radicle and dry matter of hypocotyl and radicle, besides quantification of enzyme activities, superoxide dismutase, catalase, ascorbate peroxidase and total peroxidases. The statistical design was entirely randomized with five replicates. The data was submitted to a variance analysis and the averages obtained for the treatments were compared by the Tukey test at 5% significance. The averages obtained in the treatments with and without SNP were compared by the F test at 5% probability. The NO due to the application of SNP was beneficial, providing an increase in germination, vigor and growth of seedlings. There was a progressive increase of the antioxidant enzymes activity in the period of 0 to 24 hours, showing an organization of the antioxidant system in the sesame seeds throughout germination time.

Index terms: SNP, vigor, germination, antioxidant system.

Ação protetora do óxido nítrico em sementes de gergelim (*Sesamum indicum* L.) submetidas ao estresse por cádmio

RESUMO - Objetivou-se avaliar o efeito do óxido nítrico (ON) como agente protetor em sementes de gergelim submetidas a diferentes concentrações de cádmio. Os tratamentos foram: água (controle), água acrescida de nitroprussiato de sódio (SNP) e os demais tratamentos referentes às concentrações de cádmio e cádmio acrescido de SNP. Foram feitas as seguintes determinações: germinação, primeira contagem de germinação, índice de velocidade de germinação, comprimento de hipocótilo e radícula e massa seca de hipocótilo e radícula, além da quantificação da atividade das enzimas, superóxido dismutase, catalase, ascorbato peroxidase e peroxidases totais. O delineamento estatístico utilizado foi o inteiramente casualizado com cinco repetições. Os dados foram submetidos à análise de variância e as médias obtidas para os tratamentos foram comparadas pelo teste de Tukey a 5% de significância. As médias obtidas nos tratamentos com e sem SNP foram comparadas pelo teste F a 5% de probabilidade. O ON devido à aplicação de SNP foi benéfico, proporcionando aumento na germinação, vigor e crescimento de plântulas. Houve aumento progressivo da atividade das enzimas antioxidativas no período de 0 a 24 horas, demonstrando organização do sistema antioxidante nas sementes de gergelim com o decorrer do tempo de germinação.

Termos para indexação: SNP, vigor, germinação, sistema antioxidativo.

Introduction

Problems related to the pollution and its harmful effects on living organisms are highly important. The heavy metals, currently mentioned as trace elements or persistent organic pollutants are increasingly assuming a prominent role in the concern with the environment (Wani et al., 2012). Among these metals, cadmium (Cd) stands out for easily accumulating in living organisms (Bridgen et al., 2000).

The accumulation of these contaminating elements in plants

[1]Departamento de Agricultura, UFLA, Caixa Postal 3037, 37200-000 – Lavras, MG, Brasil.
[2]Departamento de Biologia Vegetal, UFV, 36570-000 – Viçosa, MG, Brasil.
[3]Departamento de Fitotecnia, UFV, 36570-000 – Viçosa, MG, Brasil.
[4]Departamento de Engenharia Florestal, UFV, 36570-000 – Viçosa, MG, Brasil.
*Corresponding author < raquel.oliveirapires@yahoo.com.br>

may provoke several physiological damages, causing growth disorders (John et al., 2009), and structural and ultrastructural alterations in the plants (Kasim, 2006). In addition, high levels of cadmium may affect germination and the initial growth of seedlings, besides the oxidative stress and lipid peroxidation in plant tissues where they are found (Zhang et al., 2007).

With the intensification of the seeds metabolism during the germination process, there is usually production of reactive species of oxygen (EROs) that can be increased in the presence of stressing agents, such as cadmium. However, antioxidant enzyme systems act and are an important primary defense against the free radicals generated in seeds in stress conditions, such as superoxide dismutase (SOD), catalase (CAT), peroxidases (POX), and ascorbate peroxidase (APX).

Studies show that chemical compounds, such as nitric oxide (NO) act in the protection of plants exposed to stress factors. The multifunctional molecule that acts in several physiological events is also cytoprotective, due to its capacity of regulating the level and toxicity of EROs. Sodium nitroprusside (SNP) is currently the most used donor of nitrogen oxides that produce NO.

Some results show the efficiency of NO in promoting germination. Seeds of *Plathymenia reticulata,* submitted to accelerated aging, had their germination increased when SNP was applied (Pereira et al., 2010a). In studies with seeds of yellow lupine (*Lupinus luteus*), Kopyra and Gwózdz (2003) reported that SNP had a considerable effect on the promotion of germination in stress conditions, caused by the presence of lead and cadmium, indicating the efficiency of NO against the negative impact of the heavy metals on germination.

The introduction of crops of sesame in regions of Brazil is highly interesting, once it presents several nutritional and socio-economic advantages. However, studies related to the germination physiology of sesame seeds are important in order to favor the correct handling of the species and to get to know its tolerance and/or sensitivity to heavy metals, frequently found in environments where there was human intervention. The objective of this paper was to investigate the effect of NO as a protecting agent of sesame seeds submitted to different concentrations of cadmium, through evaluations of physiological characteristics and activity of the antioxidant enzymes.

Material and Methods

The research was carried out in the Forest Seeds Testing Laboratory of the Forest Science Department, at the Federal University of Viçosa. Sesame seeds *(Sesamum indicum)* harvested in 2013 were used.

Initially, preliminary tests were done to determine the concentrations of cadmium chloride ($CdCl_2$), and the NO donor solution (Sodium nitroprusside - SNP) to be studied. The concentrations of cadmium were established so that they would interfere negatively in the germination of seeds, without killing them. The amount of SNP was the one capable of reversing or mitigating the actions of the metal. After that, the sesame seeds, in five replicates of 50, were put to germinate on a paper towel moistened with 3 mL of the solution, referent to the following treatments in Petri dishes: water (control), water increased by SNP and other treatments regarding the concentrations of cadmium and cadmium increased by SNP: 800 µM of $CdCl_2$, 800 µM of $CdCl_2$ +200 µM of SNP, 600 µM of $CdCl_2$, 600 µM of $CdCl_2$ +200 µM of SNP, 400 µM of $CdCl_2$ and 400 µM of $CdCl_2$ +200 µM of SNP, summing up to a total of eight treatments. The seeds of each treatment were kept in B.O.D. regulated with alternate temperature of 20-30 °C, with the presence of constant light (Brasil, 2009). The following evaluations were done:

Germination percentage: in the sixth day after sowing, the percentage of normal seedlings was evaluated.

First germination count: consisted of a record of the number of normal seedlings obtained in the third day after sowing, and the values were expressed in percentage.

Germination speed index: daily counts were done of the number of seeds that issued a radicle higher than 1.0 mm and were calculated according to Nakagawa (1999).

Length of the hypocotyl and the radicle: the seeds of each treatment, in five replicates of 25, were sown in gerbox, following the methodology described above for the germination test. A measurement of the length of hypocotyl was done in the seedlings classified as normal with the help of a graduated ruler. The results were expressed in cm. seedling^{-1}.

Dry matter of the hypocotyl and the radicle: the seeds used to measure the length of the hypocotyl and the radicle were separated in hypocotyl and radicle, and later dried in an oven for 72 hours at 65 °C . The results were expressed in mg.seedling^{-1}.

Activity of the main enzymes of the antioxidant system: determined by using the seeds soaked for 12 and 24 hours in water and in solutions of cadmium and cadmium increased by SNP. The enzyme extracts were obtained by maceration of 0.2 g of seeds in ice, followed by the addition of 2.0 mL of the following means of homogenizing: potassium phosphate buffer 0.1 M and pH 6.8, ethylenediaminetetraacetic acid (EDTA) 0.1 mm, phenylmethylsulfonyl fluoride (pmsf) 1 mm and polyvinylpolypyrrolidone (pvpp) 1% (p/v). After that, the extract was centrifuged at 15,000 g for 15 minutes at 4 °C and the supernatant was collected, where determinations were done on the activities of ascorbate peroxidase enzyme (APX), peroxidase (POX), catalase (CAT) and superoxide dismutase (SOD).

The activity of peroxidase ascorbate (APX) was determined by the addition of 200 μL of raw enzyme extract at 2.9 mL of reaction medium of ascorbic acid 10 mM and H_2O_2 10 mM in potassium phosphate buffer 100 mM, pH 6.0. A decrease in the absorbance was observed at 290 nm, at 25 °C, during the first minute of the reaction. The enzyme activity was calculated using the molar extinction coefficient of 2.8 $mM^{-1}.cm^{-1}$ and expressed in μmol $min^{-1}.mg^{-1}$ of protein (Nakano and Asada, 1981).

The activity of peroxidase (POX) was determined by the addition of 50 μL of raw enzyme extract at 2.97 mL of reaction medium composed of potassium phosphate buffer 100 mM and pH 6.8, pyrogallol 150 mM and hydrogen peroxide 125 mM (Kar and Mishra, 1976). The increase in the absorbance during the two first minutes of reaction at 420 nm at a constant temperature of 25 °C determined the production of purpurogallin. The enzyme activity was calculated using the molar extinction coefficient of 2.47 $mM^{-1}.cm^{-1}$ and expressed in μmol $min^{-1}.mg^{-1}$ of protein.

To quantify the CAT activity, 0.98 mL of sodium phosphate buffer 0.05 M pH 6.8, H_2O_2 0.0125 mM, dissolved in an adapted buffer of Madhusudhan et al. (2003) was added to a 30 μL enzyme extract. The enzyme activity was determined by the follow up of the absorbance drop at 240 nm, for 2 minutes, in intervals of 15 seconds, and calculated based on the extinction factor of 36 $mM^{-1}.cm^{-1}$.

The activity of superoxide dismutase (SOD) was determined by the addition of 30 μL of raw enzyme extract at 2.95 mL of the reaction medium composed of sodium phosphate buffer 100 mM at pH 7.8, methionine 50 mM, p-nitro blue tetrazolium (NBT) 1 mM, EDTA 5 mM and riboflavin 100 mM. The reaction was conducted at 25 °C in a reaction chamber under lighting of a 15 W fluorescent lamp, kept inside a box internally coated with foil. After five minutes of exposure to the light, the lighting was interrupted and the blue formazan produced by NBT photo reduction was determined by the absorption at 560 nm in a spectrophotometer. A SOD unit was defined as the amount of necessary enzyme to inhibit the NBT photo reduction in 50%. The SOD activity was expressed in U $min^{-1}.mg^{-1}$ protein. To determine the content of proteins, the method used was Bradford (1976) with a standard curve constructed with bovine serum albumin (BSA) as a reference protein.

For all determinations, the statistical design was entirely randomized with five replicates. The data was submitted to a variance analysis and the averages obtained for the treatments were compared by the Tukey test at a 5% significance. The averages obtained in the treatments with and without SNP were compared by the F test at 5% probability and for the enzyme determinations the Tukey test was also used, at 5% significance.

Results and Discussion

The sesame seeds had their germination inhibited under stress by cadmium, with germination of 98.4% in great conditions (germination in water), 36.4% in high concentration of cadmium (800 μM of $CdCl_2$), 42.8% in intermediate concentration (600 μM of $CdCl_2$) and 56% in low concentration (400 μM of $CdCl_2$) (Figure 1A). Comparing the values of germination obtained in water and in greater cadmium concentration, a sharp reduction of 62 percentage points was seen.

The application of SNP did not affect the germination of seeds in water, which was more or less expected, since the germination conditions were ideal (Figure 1A). The application of SNP in all treatments with Cd allowed the significant increase of germination regarding the SNP treatments. Therefore, in high, intermediate and low concentrations, SNP reversed the damaged caused by cadmium, providing an increase in germination of 13.2, 19.6 and 8.4 percentage points, respectively (Figure 1A).

Through the vigor, germination first count (FC) and germination speed index (GSI) tests, it is seen that the cadmium, in any of the tested concentrations, affected negatively the performance of the seeds, reducing germination speed (Figures 1B and 1C). In optimum conditions, the FC was 94.8%, and the GSI was 29.72, causing a significant drop under stress by Cd, with values of 24.4% and 9.99 for FC and GSI, respectively, in a greater concentration of cadmium, 31.6% and 10.35 for the intermediate concentration and 30.8% and 10.9 for the lowest concentration.

After the application of SNP, there was a significant increase in the values of FC in relation to the treatments without SNP, for all tested concentrations of cadmium (Figure 1B). On the other hand, the application of SNP did not cause a significant increase in the germination speed (GSI) when comparing to the treatments with cadmium (Figure 1C). This behavior may be related to a possible time of response of the SNP action in restoring or minimizing the damaging actions of the Cd.

Chugh and Sawhney (1996) associate the drastic effect of cadmium in germination of seeds to the reduction of the activity of α and β amylases, which compromises respiration, causing inhibition of the growth of the embryonic axis and the radicle. Moreover, the phytotoxic effect of cadmium promotes a disorder in the development and cellular differentiation, resulting in abnormal seedlings and decreasing the percentage of normal seedlings in germination (Rossi and Lima, 2001).

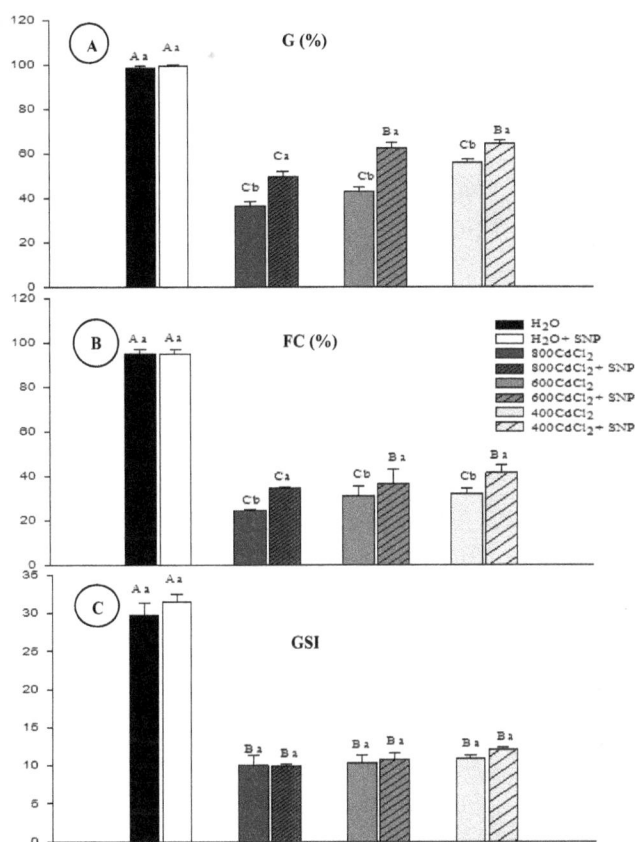

Figure 1.   A- Germination (G%) of normal seedlings on the sixth day; B- Germination First Count (FC%), done on the sixth day; and C- Germination Speed Index (GSI) of sesame seeds in the tests conducted in substrate moistened with water (control), water + SNP, cadmium solution 800 µM, and 800 µM of $CdCl_2$ + SNP, 600 µM of $CdCl_2$, and 600 µM of $CdCl_2$ + SNP, 400 µM of $CdCl_2$ and 400 µM of $CdCl_2$ +SNP.

*Averages followed by the same uppercase letter do not differ among each other by the Tukey test at 5%. **Averages of each treatment with and without SNP followed by the same lowercase letter do not differ from each other by the F test at 5% probability. The bars correspond to the average standard deviation (n = 5).

According to Kopyra and Gwózdz (2003), NO stimulates germination in situations of high concentrations of heavy metals. Beligni and Lamattina (2002) reported that NO acts as an inductor of the germination process, increasing germination in the treatments with SNP. Exposure of seeds to the heavy metal caused an increase of the reactive species of oxygen produced due to some biological dysfunction caused by cadmium, such as the interference in the action of channel proteins or its negative action in the enzyme activity (Hasan et al., 2009). The use of SNP as a NO donor increased germination, probably due to its regulation capacity or elimination of these EROs, reducing the oxidative stress

and recovering germination partially. It is highlighted that the action of SNP is efficient in increasing germination, but not in restoring it before the application of cadmium.

The radicle is the smallest mechanic resistance point and it is the first region to be in contact with the solution, being one of the main entryways of toxic metals. It is possible that this is the first organ to suffer damage because of these elements, and the other organs are harmed after the transportation of these metals, or as a consequence of its damaging effects in the radicle itself. A significant reduction was seen along the length of the radicle when the seeds were treated in a cadmium solution in relation to the seeds soaked in water (Figure 2B). It is observed that the development of the radicle was more compromised by the concentrations of the metal than the aerial part (Figures 2A, 2B). It is noted that there was no formation of roots in most part of the treated seedlings in relation to the control (Figures 3A and 3B, C and D). Accioly et al. (2004), working with seeds of *Eucalyptus camaldulensis* observed a greater content of Cd on the radicle in relation to the aerial part, being an indication of the non-translocation of this metal.

Long exposures to the radicle system of the plants to Cd lead to the manifestation of a set of symptoms that express the continuous effect of this ion on the growth of the radicular system, among them, the darkening of the radicle (Figure 3D). These same symptoms of toxicity by Cd were observed in species of *Eucalyptus maculata* and *Eucalyptus urophylla* after a week of exposure to the stressing element, in which the radicles presented a smaller development and darkening in concentrations of 180 µM of Cd (Soares et al., 2005).

A greater accumulation of the dry matter in the control seedlings that presented MSH of 20.07 mg.seedling-1 and 19.54 mg. seedling-1 of MSR (Figures 2C and 2D, respectively). In the highest concentration of cadmium, there was a reduction of 58.4% and 83% in the MSH and MSR regarding the control, respectively. The application of NO, through SNP, NO donor molecule, was able to partially reverse the reduction of MSR, only in the smallest concentration of cadmium.

Regarding the defense mechanism of the sesame seeds, a greater activity of the SOD, CAT, APX and POX enzyme (Tables 1 to 4, respectively) with a progressive increase of the soaking period of 0 to 24 hours, for basically all treatments. These results indicate an apparent organization of the antioxidant system in the sesame seeds as time goes by. However, it was seen that the period between 0 h and 12 h did not statistically differ the treatments in relation to the control for all the enzymes, with the exception of SOD, showing that these times are not enough for the organization of the antioxidant apparatus in sesame seeds submitted to stress by cadmium, or that possibly the toxic effect of cadmium was not expressive until the 12-hour period. In the 24-hour period, it is possible to see the difference in the activity of the enzymes in relation to the control in basically all treatments. In the initial time (0 h), there was no contact of the seeds with the solutions of cadmium or cadmium increased by SNP, so there is no contamination of the seeds. Therefore, the values of each enzyme were calculated, attributing a fixed value for all treatments in this time.

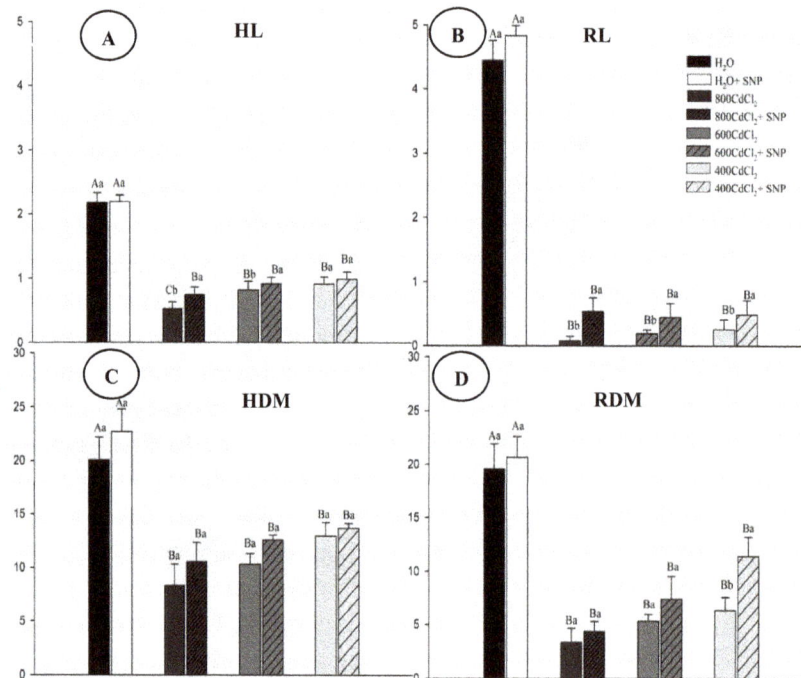

Figure 2.   A- Hypocotyl length- HL (cm/normal seedling^{-1}); B- radicle length- RL (cm/normal seedling^{-1}); C- hypocotyl dry matter- HDM (mg normal seedling^{-1}); D- radicle dry matter- RDM (mg normal seedling^{-1}) of seedlings of *Sesamum indicum* in the tests conducted in substrate moistened with water (control), water + SNP, cadmium solution 800 μM, and 800 μM of CdCl$_2$ + SNP, 600 μM of CdCl$_2$ and 600 μM of CdCl$_2$ + SNP, 400 μM of CdCl$_2$ and 400 μM of CdCl$_2$ + SNP.
*Averages followed by the same uppercase letter do not differ from each other by the Scott-Knott test at 5%. **Averages of each treatment with and without SNP followed by the same lowercase letter do not differ from each other by the F test at 5% probability.

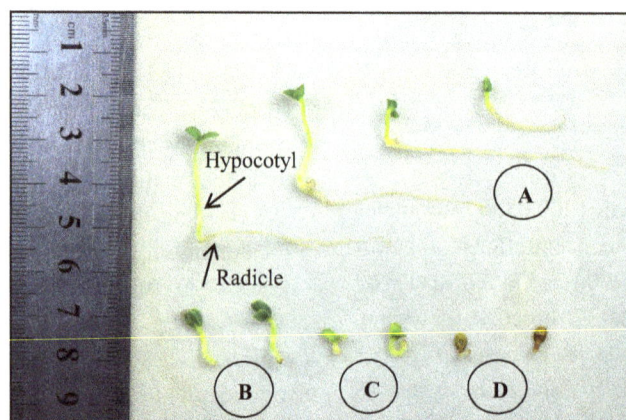

Figure 3. A- The normal seedlings of *S. indicum* obtained in the seeds germinated in water on the sixth day; B- abnormal seedlings of *S. indicum* with total inhibition of the radicle and partial of the hypocotyl obtained in the germinated seeds in solutions of cadmium on the sixth day; C- abnormal seedlings of *S. indicum* with total inhibition of radicle and total of the hypocotyl obtained from seeds germinated in solutions of cadmium on the sixth day; and D- abnormal seedlings of *S. indicum* darkened by the toxic effect of the stressing element on the sixth day of germination.

For all the enzymes in the interval of 24 hours, it is seen that the application of Cd solutions stimulated the antioxidant system of the seeds, and this increase is more expressive in higher concentrations of the solution. However, a higher activity of the enzymes may be seen in the treatments increased by SNP, which suggests detoxifying activity of these molecules in seeds submitted to stress by cadmium (Tables 1 to 4, respectively).

SOD is considered the first antioxidant line of defense against EROs (Valko et al., 2006), which explains the increase of the activity in relation to the control in the first 12 hours of soaking (Table 1).

Table 1. Activity of the enzyme superoxide dismutase (SOD) in seeds of *S. indicum* after 0, 12 and 24 hours of soaking in different concentrations of cadmium added or not by SNP.

	SOD (U min^{-1}.mg^{-1}.protein)		
Treatment	0 h	12 h	2 4h
Test	0.38 ± 0.01 Ba	0.58 ± 0.02 Ba	1.16 ± 0.02 Ba
800 CdCl$_2$	0.38 ± 0.01 Ba	3.13 ± 0.01 Ba	4.25 ± 0.05 Ae
800 CdCl$_2$ + SNP	0.38 ± 0.01 Ba	0.69 ± 0.02 Bd	8.21 ± 0.07 Aa
600 CdCl$_2$	0.38 ± 0.01 Ca	2.40 ± 0.01 Bb	4.93 ± 0.01 Ad
600 CdCl$_2$ + SNP	0.38 ± 0.01 Ba	0.91 ± 0.05 Bd	7.39 ± 0.01 Ab
400 CdCl$_2$	0.38 ± 0.01 Ca	1.89 ± 0.03 Bc	2.64 ± 0.01 Af
400 CdCl$_2$ + SNP	0.38 ± 0.01 Ba	0.92 ± 0.08 Bd	5.83 ± 0.06 Ac
VC (%)	------	22.97	20.45

*Averages followed by the same lowercase letter in the column do not differ among each other by the Tukey test at 5% probability. **Averages followed by the same uppercase letter in the row do not differ from each other by the Scott Knott test at 5% probability. Average ± standard deviation.

Table 2. Activity of the catalase enzyme (CAT) in seeds of *S. indicum* after 0, 12 and 24 hours of soaking in different concentrations of cadmium added or not by SNP.

	CAT (µmol.min^{-1}. g^{-1}.protein)		
Treatment	0 h	12 h	24 h
Test	15.64 ± 1.24 Ca	22.51 ± 3.11 Ba	35.44 ± 3.45 Ae
800 CdCl$_2$	15.64 ± 1.24 Ca	23.96 ± 3.67 Ba	29.33 ± 2.78 Af
800 CdCl$_2$ + SNP	15.64 ± 1.24 Ca	24.08 ± 2.43 Ba	52.19 ± 4.12 Ac
600 CdCl$_2$	15.64 ± 1.24 Ca	26.50 ± 2.12 Ba	37.56 ± 3.28 Ae
600 CdCl$_2$ + SNP	15.64 ± 1.24 Ca	26.76 ± 2.18 Ba	58.23 ± 2.13 Ab
400 CdCl$_2$	15.64 ± 1.24 Ca	27.59 ± 3.12 Ba	48.69 ± 2.17 Ad
400 CdCl$_2$ + SNP	15.64 ± 1.24 Ca	27.82 ± 1.56 Ba	64.90 ± 1.76 Aa
VC (%)	------	17.14	18.34

*Averages followed by the same lowercase letter in the column do not differ among each other by the Tukey test at 5% probability. **Averages followed by the same uppercase letter in the row do not differ from each other by the Scott Knott test at 5% probability. Average ± standard deviation.

Table 3. Activity of the ascorbate peroxidase enzyme (CAT) in seeds of *S. indicum* after 12 and 24 hours in different concentrations of Cd and Cd added by SNP.

	APX (µmol.min^{-1}. g^{-1}.protein)	
Treatment	12 h	24 h
Test without soaking	0.12 ± 0.01 Ba	1.06 ± 0.02 Ag
800 CdCl$_2$	0.09 ± 0.02 Ba	4.71 ± 0.01 Ac
800 CdCl$_2$ + SNP	0.06 ± 0.01 Ba	7.73 ± 0.14 Aa
600 CdCl$_2$	0.16 ± 0.02 Ba	3.30 ± 0.02 Ad
600 CdCl$_2$ + SNP	0.09 ± 0.02 Ba	6.76 ± 0.23 Ab
400 CdCl$_2$	0.20 ± 0.04 Ba	2.12 ± 0.22 Af
400 CdCl$_2$ + SNP	0.11 ± 0.04 Ba	2.87 ± 0.78 Ae
VC (%)	19.43	21.76

*Averages followed by the same lowercase letter in the column do not differ among each other by the Tukey test at 5% probability. **Averages followed by the same uppercase letter in the row do not differ from each other by the Scott Knott test at 5% probability. Average ± standard deviation

CAT is responsible for removing the hydrogen peroxide present in high concentrations in peroxisomes, protecting the cells from oxidation damage (Kibinza et al., 2011). On Table 2, an increase is seen in the activity of CAT in relation to the control in all treatments of 24 hours; however, in opposition to what happened to the SOD, there was a decrease in the activity of the enzyme in higher concentrations of the metal, as well as in the treatments with SNP. Several factors can affect the catalase activity, such as the stressing toxic agent, its used concentration, the time of exposure and the plant species, which, to a certain extent, can make this enzyme activity be submitted to great variations (Pereira et al., 2010b). This decrease in the CAT activity can also be related to the inhibition of the synthesis of the enzyme in the presence of cadmium.

The APX acts against the reactive intermediate of oxygen,

degrading the H_2O_2 in water in the presence of ascorbate, specific donor of electrons (Noctor and Foyer, 1998). Given the importance of this enzyme in the antioxidant defense of the plants, the increase of its activity has been presented by several species when exposed to different toxic agents (Moller et al., 2007). On Table 3, APX activity can be verified. At 0 h, there was no enzyme activity (data not presented) possibly due to some methodology mistake during the determination of its activity. In the interval of 24 hours after the beginning of soaking, similarly to what happened to the SOD (Table 1), the activity of peroxidase of ascorbate (APX) and peroxidase (POX) (Tables 3 and 4, respectively) increase in the higher concentrations of cadmium. It was also possible to see greater activity in both the enzymes in the treatments increased by SNP.

Table 4.   Activity of the peroxidase enzyme (POX) in seeds of *S. indicum* after 0, 12 and 24 hours of soaking in different concentrations of cadmium added or not by SNP.

Treatment	POX (μmol.min^{-1}. g^{-1}.protein)		
	0 h	12 h	24 h
Test	8.54 ± 1.03 Ba	11.67 ± 0.12 Ba	20.18 ± 0.24 Ag
800 CdCl$_2$	8.54 ± 1.03 Ca	20.16 ± 0.32 Ba	30.73 ± 0.13 Ad
800 CdCl$_2$ + SNP	8.54 ± 1.03 Ca	19.17 ± 0.27 Ba	46.73 ± 0.16 Aa
600 CdCl$_2$	8.54 ± 1.03 Ca	16.72 ± 0.12 Ba	25.29 ± 0.17 Ae
600 CdCl$_2$ + SNP	8.54 ± 1.03 Ca	18.17 ± 0.31 Ba	42.73 ± 0.17 Ab
400 CdCl$_2$	8.54 ± 1.03 Ba	12.78 ± 0.27 Ba	23.83 ± 0.17 Af
400 CdCl$_2$ + SNP	8.54 ± 1.03 Ca	18.69 ± 0.25 Ba	34.83 ± 0.37 Ac
VC (%)	--------	16.56	21.32

*Averages followed by the same lowercase letter in the column do not differ among each other by the Tukey test at 5% probability. **Averages followed by the same uppercase letter in the row do not differ from each other by the Scott Knott test at 5% probability. Average ± standard deviation

As mentioned by Moller et al. (2007), the action of these toxic agents such as the Cd affects the action of enzymes of oxidative stress. Consequently, these elements can interfere in the cellular metabolism during germination, causing damages that can compromise the vigor and the quality of the seeds. We believe that more efforts must be done in the sense of evaluating the complexity of these damages and finding a way to mitigate them, once the contamination of soils by toxic elements is an ever-increasing concern for agriculture.

Conclusions

The increase in the concentration of cadmium in the soaking solution reduces germination and vigor of the sesame seeds and the initial growth of the seedlings, showing a possible toxic effect of this element to the seeds. The SNP seems to partially reverse the damage caused by the heavy metal.

Mitigation of a damaging action of cadmium by the use of SNP occurs due to the increase of the activity of the antioxidant enzymes, showing an elimination system of the species reactive of oxygen which occurs after the application of the NO donor in response to exposure to heavy metal.

Acknowledgments

We thank Fundação de Amparo à Pesquisa do Estado de Minas Gerais (Foundation of Research Support of the State of Minas Gerais - FAPEMIG), the Coordination of the Higher Education Personnel Training (CAPES) and the National Board of Technological and Scientific Development (CNPq) for the financial support and grant of scholarships.

References

ACCIOLY, A. M. A.; SIQUEIRA, J. O.; CURI, N.; MOREIRA, F. M. S. Amenização do calcário na toxidez de zinco e cádmio para mudas de *Eucalyptus camaldulensis* cultivadas em solo contaminado. *Revista Brasileira de Ciência do Solo*, v.28, n.4, p.775-783, 2004. http://www.scielo.br/pdf/rbcs/v28n4/21800.pdf

BELIGNI, M. V.; LAMATTINA, L. Nitric oxide interferes with plant photo-oxidative stress by detoxifying reactive oxygen species. *Plant Cell Environment*, v.25, p.737-74, 2002. http://onlinelibrary.wiley.com/doi/10.1046/j.1365-3040.2002.00857.x/full

BRADFORD, M. M. A rapid and sensitive method for the quantitation of microgram quantities of protein utilizing the principle of protein-dye binding. *Analytical Biochemistry*, v.72, p.248-254, 1976. http://ac.els-cdn.com/0003269776905273/1-s2.0-0003269776905273-main.pdf?_tid=cb19ee2e-c37b-11e5-83d6-00000aab0f26&acdnat=1453737396_8c39b9bfbe5174426b1f9e0387d7c13d

BRASIL. Ministério da Agricultura, Pecuária e Abastecimento. *Regras para análise de sementes*. Ministério da Agricultura, Pecuária e Abastecimento. Secretaria de Defesa Agropecuária. Brasília: MAPA-ACS, 2009. 395p. http://www.agricultura.gov.br/arq_editor/file/2946_regras_analise_sementes.pdf

BRIDGEN, K.; STRINGER, R. LABUSKA, I. Poluição por organoclorados e metais pesados, associada ao fundidor de ferro da Gerdau em Sapucaia do Sul, Brasil, Rio Grande do Sul, Greenpeace, 2000.

CHUGH, L. K.; SAWHNEY, S. K. Effect of cadmium on germination, amylases and rate of respiration of germinating pea seeds. *Environmental Pollution*, v.92, p.1-5, 1996. http://www.sciencedirect.com/science/article/pii/0269749195000933

HASAN, S.A.; FARIDUDDIN, Q.; ALI, B.; HAYAT, S.; AHMAD, A. Cadmium: Toxicity and tolerance in plants. *Journal of Environmental Biology*, v.30, n.2, p.165-174, 2009. https://www.researchgate.net/publication/41396456_Cadmium_toxicity_and_tolerance_in_plants_J_Environ_Biol

JOHN, R.; AHMAD, P.; GADGIL, K.; SHARMA, S. Heavy metal toxicity: Effect on plant growth, biochemical parameters and metal accumulation by *Brassica juncea* L. *International Journal of Plant Production*, v.3, n.3, p.65-76, 2009. https://www.researchgate.net/publication/237724355_Heavy_metal_toxicity_effect_on_plant_growth_biochemical_parameters_and_metal_accumulation_by_Brassica_juncea_L_Int_J_Plant_Prod_366-75

KAR, M.; MISHRA, D. Catalase, peroxidase and polyphenol oxidase activities during rice leaf senescence. *Plant Physiology*, v.57, p.315–319, 1976. http://www.plantphysiol.org/content/57/2/315.long

KASIM, W. A. Changes induced by copper and cadmium stress in the anatomy and grain yield of *Sorghum bicolor* (L.) Moench. *International Journal of Agriculture and Biology*, v.8, n.1, p.123-128, 2006. http://www.fspublishers.org/published_papers/64929_..pdf

KIBINZA S.; BAZINA J.; BAILLY C.; FARRANT J. M.; CORBINEAUA O.; BOUTEAUA H. Catalase is a key enzyme in seed recovery from ageing during priming. *Plant Science*, v.181, p. 309-315. 2011. https://www.researchgate.net/publication/51496832_Catalase_is_a_key_enzyme_in_seed_recovery_from_ageing_during_priming_Plant_Sci

KOPYRA, M.; GWÓZDZ, E.A. Nitric oxide stimulates seed germination and counteracts the inhibitory effect of heavy metals and salinity on root growth of *Lupinus luteus*. *Plant Physiology Biochemistry*, v.41, p.1011-1017, 2003. http://www.sciencedirect.com/science/article/pii/S098194280300175X

MADHUSUDHAN, R.; ISHIKAWA, T.; SAWA, Y.; SHIGEOKA S.; SHIBATA, H. Characterization of an ascorbate peroxidase in plastids of tobacco BY-2 cells. *Physiologia Plantarum*, v.117, p.550-557, 2003. http://onlinelibrary.wiley.com/doi/10.1034/j.1399-3054.2003.00066.x/epdf

MOLLER, I. M.; JENSEN, P. E.; HANSSON, A. Oxidative modifications to cellular components in plants. *Annual Review of Plant Biology*, v.58, p.459-481, 2007. http://www.annualreviews.org/doi/pdf/10.1146/annurev.arplant.58.032806.103946

NAKAGAWA, J. Testes de vigor baseados no desempenho das plântulas. In: KRZYZANOSKI, F.C.; VIEIRA, R.D.; FRANÇA-NETO, J.B (Ed.). *Vigor de sementes:* conceitos e testes. ABRATES, Londrina, p.2-1–2.21, 1999.

NAKANO, Y.; ASADA, K. Hydrogen peroxide is scavenged by ascorbato specific peroxidase in spinach chloroplasts. *Plant and Cell Physiology*, v.22, p.867-880, 1981. http://pcp.oxfordjournals.org/content/22/5/867.full.pdf+html

NOCTOR, G; FOYER, C. H. Ascorbate and Glutathione: Keeping Active Oxygen Under Control. *Annual Review of Plant Physiology and Plant Molecular Biology*, v.49, p.249–279, 1998. http://www.annualreviews.org/doi/pdf/10.1146/annurev.arplant.49.1.249

PEREIRA, B.L.C.; BORGES, E.E.L.; OLIVEIRA, A.C.; LEITE, H.G.; GONÇALVES, J.F.C. Influência do óxido nítrico na germinação de sementes de *Plathymenia reticulata* Benth com baixo vigor. *Scientia Florestalis*, v.38, n.88, p.629-636, 2010a. http://www.ipef.br/publicacoes/scientia/nr88/cap09.pdf

PEREIRA, F. J.; MAGALHÃES, P. C.; SOUZA, T. C.; CASTRO, E. M.; ALVES, J. D. Atividade do sistema antioxidante e desenvolvimento de aerênquima em raízes de milho 'Saracura'. *Pesquisa Agropecuária Brasileira*, v.45, p.450-456, 2010b. http://www.scielo.br/pdf/pab/v45n5/03.pdf.

ROSSI, C.; LIMA, G. P. P. Cádmio e a atividade de peroxidase duante a germinação de sementes de feijoeiro. *Scientia Agricola*, v.58, n.1, p.197-199, 2001. http://www.scielo.br/pdf/sa/v58n1/a30v58n1.pdf

SOARES, C. R. F. S.; SIQUEIRA, J. O.; CARVALHO, J. G.; MOREIRA, F. M. S. Fitotoxidez de cádmio para *Eucalyptus maculata* e *E. urophylla* em solução nutritiva. *Revista Árvore*, v.29, n.2, p.175-183, 2005. http://www.scielo.br/pdf/rarv/v29n2/a01v29n2.pdf

VALKO, M.; RHODES, C.J.; MONCOL, J. Free radicals, metals and antioxidants in oxidative stress-induced cancer. *Chemico Biological Interaction*, v.160, p.1-40, 2006. http://ac.els-cdn.com/S0009279705004333/1-s2.0-S0009279705004333-main.pdf?_tid=cb6f97a8-c37f-11e5-beeb-00000aacb361&acdnat=1453739114_ab3e2366f88da46b7e4c1d505f6b29e0

WANI, P. A.; KHAN, M. S.; ZAIDI, A. Toxic effects of heavy metals on germination and physiological processes of plants. In: ZAIDI, A,; WANI, P.A.; KHAN, M.S. Toxicity of heavy metals to legumes and bioremediation. *Springer*, p.45–66, 2012. http://link.springer.com/chapter/10.1007%2F978-3-7091-0730-0_3

ZHANG, F. Q.; WANG, Y. S.; LOU, Z. P.; DONG, J. D. Effect of heavy metal stress on antioxidant enzymes and lipid peroxidation in leaves and roots of two mangrove plant seedlings (*Kandelia candel* and *Bruguiera gymnorrhiza*). *Chemosphere*, v.67, p. 44–50, 2007. http://ac.els-cdn.com/S004565350601335X/1-s2.0-S004565350601335X-main.pdf?_tid=0e07c766-c380-11e5-ac0d-00000aab0f6c&acdnat=1453739226_252a42fe2b03c7c3a33fc7a109719340.

Tetrazolium test for viability estimation of *Eugenia involucrata* DC. and *Eugenia pyriformis* Cambess. seeds

Fernanda Bernardo Cripa[1*], Laura Cristiane Nascimento de Freitas[1], Andrieli Cristine Grings[1], Michele Fernanda Bortolini[1]

ABSTRACT - The study aimed to adapt the tetrazolium test methodology to assess the viability of seed of *Eugenia involucrata* DC. (cherry) and *E. pyriformis* Cambess. (uvaia) freshly harvested and stored. Three lots of seeds of both species were used, being I (freshly harvested), II (stored for 15 days) and III (stored for 30 days). Seeds (lot I) of both species were immersed in distilled water for 24 hours and submitted to four preparation methods: whole seeds without seed coat and with ¾ of seed coat; longitudinally cut with and without seed coat (immersed in 0.1% tetrazolium for 4 hours). The most suitable preparation method was applied to the remaining lots, where three tetrazolium concentrations were tested: 0.075, 0.1 and 0.5% and three immersion time periods: 2, 4 and 6 hours. The tetrazolium test results were compared to the germination results. For the conduction of the tetrazolium test in seeds of both species we recommend soaking in distilled water by 24 hours and cut lengthwise on seeds with the seed coat. For cherry, the seeds should be immersed in 0.5% tetrazolium solution for 2 hours. For uvaia seeds, additional studies are recommended.

Index terms: viability test, cherry, uvaia, forest species.

Teste de tetrazólio para a estimativa de viabilidade de sementes de *Eugenia involucrata* DC. e *Eugenia pyriformis* Cambess.

RESUMO - O estudo teve como objetivo adequar a metodologia do teste de tetrazólio para avaliar a viabilidade das sementes de *Eugenia involucrata* DC. (cereja) e *E. pyriformis* Cambess. (uvaia) recém-colhidas e armazenadas. Foram empregados três lotes de sementes para ambas as espécies, sendo o I (recém-colhidas), II (armazenadas durante 15 dias) e III (armazenadas durante 30 dias). As sementes (lote I) de ambas as espécies, foram imersas em água destilada por 24 horas e submetidas a quatro métodos de preparo: sementes inteiras sem tegumento e com ¾ do tegumento; sementes cortadas longitudinalmente com tegumento e sem tegumento (imersas em solução de tetrazólio a 0,1% por 4 h). O método de preparo mais adequado foi aplicado aos demais lotes, utilizando-se três concentrações de tetrazólio: 0,075; 0,1 e 0,5% e três tempos de imersão: 2, 4 e 6 h. Os resultados do teste de tetrazólio foram comparados com os de germinação. Para a condução do teste de tetrazólio em sementes de ambas as espécies recomenda-se a embebição em água destilada por 24 horas e corte longitudinal das sementes com o tegumento. Para cereja, recomenda-se a imersão das sementes em solução de tetrazólio a 0,5% por 2 horas. Para uvaia, recomendam-se estudos adicionais.

Termos para indexação: teste de viabilidade, cereja, uvaia, espécie florestal.

Introduction

Eugenia involucrata DC. (cherry) and *Eugenia pyriformis* Cambess. (uvaia) are species that can be used in degraded areas' recovery, being of great importance in the Brazilian flora. *E. involucrata*, popularly known as cherry is a fruit species that can be found from Minas Gerais to Rio Grande do Sul states, whose fruits ripen from October to December. The *E. pyriformis*, known as uvaia can be seen from Sao Paulo to Rio Grande do Sul states and its fruits ripen from September to January (Lorenzi et al., 2006). According to Maluf et al. (2003), Eugenia species have potential for areas' recovery, because of the fauna attraction that acts as a seeds disperser.

Seeds are structures capable of maintaining viability until environmental conditions are favorable for the beginning of a new generation. However, they cannot preserve its vital functions indefinitely (Marcos-Filho, 2005). Therefore, the seeds storage practice management becomes essential to

[1]Pontifícia Universidade Católica do Paraná, Escola de Saúde e Biociências, 85802-532 - Toledo, PR, Brasil.
*Corresponding author <fernandacripa@hotmail.com>

maintain their physiological quality and ensure the vigor and viability maintenance in the period between sowing and harvest (Azevedo et al., 2003).

The Eugenia species have recalcitrant seeds (Carvalho et al., 2006), sensitive to dehydration, which causes problems in germination as well as storage potential. Therefore, studies on seeds storage are necessary, since the feasibility is under direct influence of time and storage conditions (Fonseca and Freire, 2003). Some studies regarding the seeds storage of Eugenia species have been reproduced to verify the ideal time and storage conditions, as reported by Barbosa et al. (1990); Andrade and Ferreira (2000); Scalon et al. (2004); Delgado and Barbedo (2007); Justo et al. (2007); among others.

The official parameter usually used to evaluate the seeds physiological quality is the germination test, which is performed under controlled conditions, and being ideal for germination (Brasil, 2009). The germination test acts as a support for all the other analyzes (Sena et al., 2010), however, there are some limitations to its conduction, such as the results' delay and their possible modification by fungi presence. Therefore, the development of methods aiming to obtain quick and reliable information of *E. involucrata* and *E. pyriformis* seeds physiological quality may help in decision making regarding their use, as well as, in the viability estimation of stored seeds.

Among tests with fast results, there is the tetrazolium test based on the dehydrogenize enzymes activity that reduces salt at 2,3,5-triphenyl tetrazolium chloride in the seeds living tissues, generating a compound called trifenilformazan of red color and not diffusible, indicating that a respiratory activity exists, and cell and tissues viability (França-Neto et al., 1998). This test can be used to speed up the seed's quality evaluation, allowing decisions making related to the seed management (Pinto et al., 2008).

However, the tetrazolium test methodology is not widespread among forest species, although remaining a good alternative, for many of them has a long germination period, which is a limiting factor for the germination test realization (Fogaça et al., 2011). Given this, the methodology for the tetrazolium test has been studied for several forest species as the *Peltophorum dubium* (Sprengel) Taubert, (Oliveira et al., 2005) and the *Eugenia pleurantha* O. Berg. (Masetto et al., 2009), as well as, assessing the seeds' stored quality as the *Amburana cearenses* (Allemão) A. C. Smith. (Guedes et al., 2010).

Given the above, this study aims to propose a faster method to evaluate *Eugenia involucrata* DC. and *Eugenia pyriformis* Cambess. seeds viability, by the tetrazolium test, evaluating the applicability of this test in seeds stored for 15 and 30 days.

Materials and Methods

The experiment was conducted at the Biotechnology Laboratory of the Pontifical Catholic University of Paraná - Toledo Campus, where *Eugenia involucrata* DC. fruits (cherry) were used and *E. pyriformis* Cambess. (uvaia), harvested in October and December 2011, respectively. For both species, the collection was made from five previously selected arrays in the western region of Paraná, Brazil.

The fruits were sent to the laboratory, where they were pulped by friction in a sieve under running water and seeds left to dry in the shade, in laboratory environment for 24 hours. The freshly harvested seeds (lot I) were then separated into different lots according to the storage period: seeds stored for 15 days (lot II) and seeds stored for 30 days (lot III). The storage of lots II and III took place in refrigerator at a temperature around 10 °C, in sealed and drilled polyethylene bags (10 holes of approximately 0.5 mm) (Barbedo et al., 1998).

For both species the assessment of lot I started with the moisture content determination, using four replicates of 5 g of seeds, in aluminum containers with known weight, and dried seeds in greenhouse at 105 ± 3 °C, for 24 hours, according to the methodology of Brasil (2009). For lots II and III of both species the moisture degree was also determined, when using seeds on remaining tests, enabling to verify whether there was seeds moisture loss during storage.

For lot I germination test, for both cherry and uvaia, three different substrates were evaluated, namely: autoclaved sand, three sheets of towel paper and fine vermiculite, arranged in gerbox.

The sand and vermiculite substrates were previously moistened with distilled water until saturation, the towel paper was moistened with distilled water 2.5 times its dry weight (Brasil, 2009).

The seeds of both species underwent disinfection with 2% sodium hypochlorite for 5 minutes, followed by washing in running water and then placed onto substrates. The test was conducted in a germination chamber (BOD), at a 30 °C constant temperature and 12 hours photo period (Barbedo et al., 1998).

The germination test evaluation was performed by daily counting of germinated seeds until germination stabilization, which occurred 48 days after sowing. Were considered germinated seeds, those showing primary root equal to or higher than 2.0 mm.

The average germination rate (v), calculated according to Labouriau (1983) was determined. The experimental design was completely randomized with four replications of 25 seeds per experimental unit for each substrate.

The same procedure was performed for stored uvaia

and cherry seeds lots (lots II and III) once, after 15 and 30 storage days of these lots, the lot I germination test had not yet been completed, making it impossible to indicate the best germination test substrate.

To obtain the germination test results of lot I of both species, the treatments' variances for homogeneity were tested by the Bartlett's test, and those being homogeneous were submitted to the variance analysis (F test). The treatments average values were compared using the Tukey test at 5% probability level to find out what would be the test's best substrate.

Therefore, after determining the most suitable substrate for lot I germination, for each species under study, the same was done to calculate the stored seeds lots average comparison (lots II and III).

Regarding the tetrazolium test, preliminary tests were performed with lot I seeds, both cherry and uvaia, to verify which would have the most appropriate conditions for seeds preparation and staining of both species.

For that purpose, a 24 hours immersion in distilled water was used, followed by the following treatments: whole seeds without seed coat; whole seeds with ¾ of seed coat; longitudinally cut seeds with seed coat and without seed coat. Each treatment seeds were immersed in 0.1% tetrazolium for 4 hours at 30 °C temperature in the dark (Masetto et al., 2009). During each repetition preparation, they were kept immersed in distilled water, until the entire sample was prepared (Brasil, 2009).

After the described above preparations and still using lot I, the distilled water was drained and tetrazolium solution was added at 0.075; 0.1 and 0.5% concentrations for 2, 4 and 6 hours. The seeds were kept in a BOD chamber at 30 °C in the dark (Brasil, 2009; Masetto et al., 2009). After the immersion time, the seeds were washed in running water and kept submerged in distilled water until the analysis time. After determining the appropriate methodologies for the tetrazolium test for each studied species, these have been applied for the stored seeds lots (lots II and III).

The evaluation was done with a stereoscopic microscope magnification aid (4 x) and the result was expressed in viability percentage, according to the different staining patterns observed in the seeds' internal structures (Masetto et al., 2009).

Regarding the tetrazolium test, the experimental design used was completely randomized in a factorial design 3 x 3 (three concentrations and three different immersion periods of seeds in tetrazolium solution), using per experimental unit, four replicates of 25 seeds for each treatment.

The results were submitted initially to the Bartlett test, where treatments variances were tested for homogeneity and those being homogeneous, submitted to the variance analysis (F test), and their averages compared using the Tukey test at 5% probability.

To confirm the tetrazolium test reliability, its result in viability percentage was used for the germination percentages comparison, using a completely randomized design in a factorial 2 x 3 (two viability tests, germination test and tetrazolium test, for the three lots at different storage times).

Results and Discussion

The average speed germination and percentage average results for both species are shown in Table 1. As germination percentage the higher vermiculite and sand substrates averages were observed, being vermiculite 94% (cherry) and 85% (uvaia) and sand 83% (cherry) and 78% (uvaia), values that did not significantly differ for both species.

Table 1. Germination and average speed germination percentages of *Eugenia involucrata* DC. (cherry) and *Eugenia pyriformis* (uvaia) freshly harvested seeds (lot I).

Substrate	Germination (%)[1]		Average speed (seeds/days)	
	Cherry	Uvaia	Cherry	Uvaia
Sand	83 ab*	78 ab	0.065 a	0.065 a
Paper	54 b	57 b	0.072 a	0.043 b
Vermiculite	94 a	85 a	0.072 a	0.071 a
CV(%)	16.45	15.69	12.90	11.43

*Averages followed by the same letter in the column are not statistically different as per Tukey's test (p > 0.05).
[1]Data transformed by arcsine $\sqrt{x}/100$.

A significant difference in germination percentage was observed among vermiculite and paper substrates only, the paper values being inferior compared to the vermiculite ones. The germination average speed values obtained for cherry seeds were statistically the same for the three substrates. Regarding uvaia, the paper substrate significantly differed from the remaining substrates, showing the lowest germination speed (0.043 germinated seeds per day).

As for the germination tests performed in laboratory, the most frequently substrates used are sand and paper (Brasil, 2009). However, it was observed that for both cherry and uvaia seeds, the best results were obtained with vermiculite, followed by the sand substrate.

The substrate adopted in lots II and III germination test for both species was vermiculite, which proved to be ideal providing good conditions for the germination test of both tested species. Vermiculite is a substrate commonly used in laboratories for seed germination analysis, presenting advantages such as: easy to get, economic viability, low density, chemical composition and particle size uniformity, porosity and water holding capacity (Martins et al., 2012).

The best conditions for the seeds staining preparation was the longitudinal cut realization with or without removing the seed coat, which allowed getting adequate tissue staining of cherry and uvaia seeds.

The main purpose of preconditioning and seed preparation is to facilitate the tetrazolium solution penetration through the tissues (Mendes et al., 2009), therefore, as preconditioning was used in this work, the soaking in distilled water for 24 hours.

Before staining, the seed preparation method with the longitudinal section without removing the seed coat was chosen, once the removal procedure is laborious and time-consuming. It is worth mentioning that results for the longitudinal cut without removing the seed coat were similar to those obtained with the removal, therefore both methods can be used.

Regarding cherry seeds it could be observed that the best results were obtained with 2 hours soaking in 0.5% tetrazolium solution, with 87% of viable seeds which was significantly higher than the values obtained in 0.075 and 0.1% concentrations (Table 2).

Table 2.   Percentage viability (lot I) of *Eugenia involucrata* DC. (cherry) and *Eugenia pyriformis* Cambess. (uvaia) freshly harvested seeds, evaluated by the tetrazolium test solution at different concentrations and coloring times.

Eugenia involucrate DC. (cherry)				
	Time			
Concentration	2 hours	4 hours	6 hours	Average
0.075%	0 Cb*	30 Ba	0 Bb	10
0.1%	57 Ba	24 Bb	60 Aa	47
0.5%	87 Aa	47 Ab	8 Bc	47
Average	48	34	23	
CV (%)	31.04			
Eugenia pyriformis Cambess. (uvaia)				
	Time			
Concentration	2 hours	4 hours	6 hours	Average
0.075%	0 Ab*	3 Bb	49 Ba	17
0.1%	1 Ab	6 ABb	51 Ba	19
0.5%	9 Ab	16 Ab	83 Aa	36
Average	3.33	8.33	61	
CV (%)	25.18			

*Averages followed by the same uppercase in the column and lowercase in the row do not differ among themselves as per Tukey's test (p > 0.05).

It appears that the 4 and 6 hours staining time, the highest viability values were obtained at 0.5% (47%) and 0.1% (60%) concentrations, respectively. However, these values were below the observed at 0.5% (87%), which obtained the highest viability of cherry seeds.

For the tetrazolium test, the soaking in 0.5% for 2 hours was chosen as the best procedure to be adopted in cherry seeds (Table 2), once it allowed maximum viability, a shorter time (2 hours), and similar results to the germination in vermiculite (Table 1). To affirm the tetrazolium test reliability, its results should be similar to those obtained with the germination test. Also, at this condition a more homogeneous coloring of the seed internal tissues was observed, facilitating the results' interpretation. Therefore, the tetrazolium concentration and immersion time used on the cherry tests lots II and III (stored seeds) was 0.5% of tetrazolium solution for 2 hours.

Regarding uvaia seeds, the best results were obtained with 6 hours soaking in 0.5% tetrazolium concentration (83% of viable seeds), significantly differing from the remaining concentrations, 0.075 and 0.1% (Table 2). At this concentration (0.5%) a significant difference between 2 and 4 hours of soaking in the tetrazolium solution was also found (9 and 16%, respectively).

The method applied to uvaia lots II and III (stored seeds) was the seeds soaking in 0.5% tetrazolium solution for 6 hours, once these conditions provided maximum viability and similar results to germination in vermiculite (84%) (Table 1).

According to Fogaça et al. (2006), the seeds preparation, the tetrazolium solution concentration and the staining time, are specific to each species. Therefore, after the methodology definition (seeds preparation, tetrazolium concentration and immersion time in the solution), of each studied species, classes of viable and non-viable seeds were established, according to the tissues' staining intensity observed from the whole seed interior.

For cherry seeds, two viable seeds classes and three non-viable seed classes (Figure 1) were established. For uvaia seeds it was not possible to establish feasibility classes, only the distinction between viable and non-viable seeds was made (Figure 2).

To interpret the tetrazolium test, according to França-Neto et al. (1998), the analyst must know the seed structures and have experience and capacity for critical judgment. It is important to mention that the uvaia seeds staining were weak (Figure 2), making it difficult to assess them, differently from the cherry seeds situation (Figure 1), demonstrating that the staining pattern can be different for each species and depends primarily on the concentration used. The uvaia seeds weaker staining also indicates greater resistance to the tetrazolium solution penetration for this species, suggesting further studies with higher tetrazolium concentrations, which could aid in the viability classification.

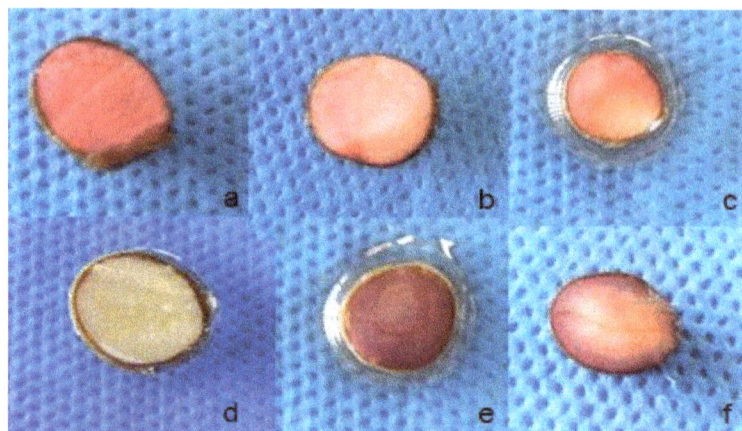

Figure 1. Viability categories found in tetrazolium test of *Eugenia involucrata* DC. (cherry) seeds. (a and b) viable seeds with homogeneous and firm pinkish; (c) viable seed with pink color in more than 50% of embryonic tissue; (d) unviable seed with milky white opaque coloration; (e) unviable seed with more than 50% of embryonic tissue with bright red color; (f) unviable seed with less than 50% of embryonic tissue with pink color, with edges necrosis and intense red coloration tissue.

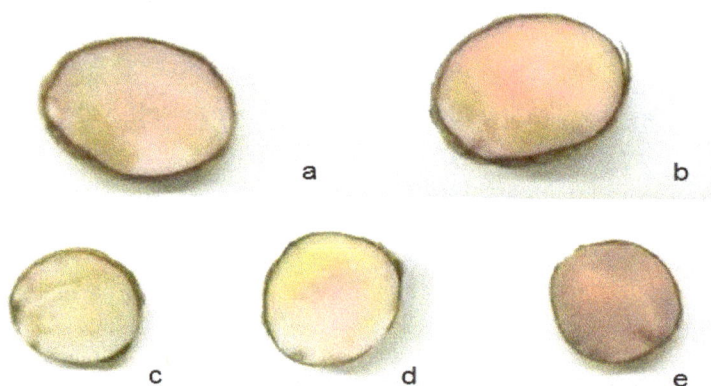

Figure 2. Viability categories found in tetrazolium test of *Eugenia pyriformis* Cambess. (uvaia) seeds. (a and b) viable seeds with pink color in more than 50% of embryonic tissue; (c, d, e) unviable seeds.

According to Ferreira et al. (2001), to assess the tetrazolium test reliability, its result must be compared with the germination test result, and obtained viability results should be similar for both. To assess the tetrazolium test reliability for the three lots of both species, the variance analysis was used, in which no significant interaction was found between the test's factors (germination or tetrazolium) and storage time (lots).

Once there was no significant difference between germination and tetrazolium tests in cherry and uvaia seeds, the tetrazolium test applicability was confirmed for the studied species.

For cherry, the general averages values for the three lots showed 83% of germination and 77% of viable seeds per the tetrazolium test (Table 3). For the cherry seeds storage time, based on the germination and tetrazolium tests results, a higher vigor in lot I was observed (freshly harvested), being

significantly larger than lot III only.

Concerning the uvaia seeds storage time, based on the germination and tetrazolium tests results, the higher vigor was observed on freshly harvested seeds (lot I) and stored for 15 days (lot II), which did not significantly differ from each other, but were significantly higher than those stored for 30 days (lot III) (Table 3).

When assessing the cherry and uvaia lots physiological quality, it was found that lots stored for 30 days, showed greater quality loss (Table 3). This possibly happened because these are recalcitrant seeds, characteristic that can be observed in the moisture content determination results (Table 4).

It is difficult to identify the specific causes for the intolerance to desiccation in recalcitrant seeds, which can be a result of a combination of factors (Ferreira and Borghetti, 2004).

Table 3. Average values of the germination and tetrazolium tests on *Eugenia involucrata* DC. (cherry) and *Eugenia pyriformis* Cambess. (uvaia) seeds, at different storage time periods.

Eugenia involucrata DC. (cherry)			
Storage time period	Germination test (%)	Tetrazolium test (%)	Average
Lot I (freshly harvested)	94	87	90.5 A
Lot II (15 days)	84	76	80.0 AB
Lot III (30 days)	72	67	69.5 B
Average	83.3 a*	76.6 a	
CV (%)	12.04		
Eugenia pyriformis Cambess. (uvaia)			
Storage time period	Germination test (%)	Tetrazolium test (%)	Average
Lot I (freshly harvested)	85	83	84.0 A
Lot II (15 days)	77	77	77.0 A
Lot III (30 days)	57	50	53.5 B
Average	73.0 a*	70 a	
CV (%)	13.28		

*Averages followed by the same uppercase in the column and lowercase in the row do not differ among themselves as per Tukey's test (p > 0.05).

Table 4. Moisture content average values of cherry and uvaia seeds at different storage time periods.

Storage period	Cherry	Uvaia
Lot I (freshly harvested)	47.4 %	44.2 %
Lot II (15 days)	46.3 %	44.0 %
Lot III (30 days)	43.0 %	43.5 %

Even without the complete information on the viability loss cause of recalcitrant seeds, when dehydrated beyond their limits, according to Marcos-Filho (2005), the membrane integrity seems to be the main injuries target. These seeds critical moisture level varies according to the species as per Delgado and Barbedo (2007), the *Eugenia* seeds losing viability at water contents of 15 to 20%

For *Eugenia involucrata* lots (cherry) and *Eugenia pyriformis* (uvaia), high water contents were observed, a characteristic of recalcitrant seeds (Table 4), indicating that there is no desiccation phase at the final development, that they are disconnected from the parent plant with a high water content, being intolerant to desiccation and also to prolonged storage time (Kerbauy, 2008)

Therefore, according to the conditions under which this work was developed, the *Eugenia involucrata* DC. seeds (cherry) and *E. pyriformis* Cambess. (uvaia) can be stored for a 15 days period without significant viability loss. Being aware that the germination test for both species took on average 48 days to be finalized and considering that most forest species require a long time to germinate and, that the delay of the germination test implementation is considered a limitation to it (Fogaça et al., 2006), the tetrazolium test is recommended

to speed up the decision making regarding the cherry and uvaia seeds use, which are seen as recalcitrant seeds and have a short life time (Scalon et al., 2012).

Conclusions

The tetrazolium test can be used to evaluate the feasibility of *Eugenia involucrata* DC. (cherry) and *E. pyriformis* Cambess. (uvaia) species.

To execute the tetrazolium test, on both species seeds, the soaking in distilled water for 24 hours is recommended, followed by the seed longitudinal cut with the seed coat. For cherry, it is recommended to soak the seeds in 0.5% tetrazolium solution for 2 hours. For uvaia seeds, the use of this concentration for 6 hours did not allow to obtain an optimal staining pattern in living tissues, indicating the need of additional studies.

Acknowledgements

To all, who assisted in this work, as well as the Araucaria Foundation, for its financial support.

References

ANDRADE, R.N.B.; FERREIRA, A.G. Germinação e armazenamento de sementes de uvaia (*Eugenia pyriformis* Camb.) Myrtaceae. *Revista Brasileira de Sementes*, v.22, n.2, p.118-125, 2000. http://www.lume.ufrgs.br/bitstream/handle/10183/23264/000293585.pdf?sequence=1&locale=pt_BR

AZEVEDO, M.R.Q.A.; GOUVEIA, J.P.G.; TROVÃO, D.M.M.; QUEIROGA, V.P. Influência das embalagens e condições de armazenamento no vigor de sementes de gergelim. *Revista Brasileira de Engenharia Agrícola e Ambiental*, v.7, n.3, p. 519-524, 2003. http://www.scielo.br/pdf/rbeaa/v7n3/v7n3a19.pdf

BARBOSA, J.M.; BARBOSA, L.M.; SILVA, T.S.; FERREIRA, D.L. Influência de substratos e temperaturas na germinação de sementes de duas frutíferas silvestres. *Revista Brasileira de Sementes*, v.12, n.2, p. 66-73, 1990. http://www.abrates.org.br/revista/artigos/1990/v12n2/artigo07.pdf

BARBEDO, C.J.; KOHAMA, S.; MALUF, A.M.; BILIA, D.A.C. Germinação e armazenamento de diásporos de cerejeira (*Eugenia involucrata* DC. - MYRTACEAE) em função do teor de água. *Revista Brasileira de Sementes*, v.20, n.1, p. 184-188, 1998. http://www.abrates.org.br/revista/artigos/1998/v20n1/artigo30.pdf

BRASIL. Ministério da Agricultura, Pecuária e Abastecimento. *Regras para análise de sementes*. Ministério da Agricultura, Pecuária e Abastecimento. Secretaria de Defesa Agropecuária. Brasília: MAPA/ACS, 2009. 395 p. http://www.agricultura.gov.br/arq_editor/file/2946_regras_analise__sementes.pdf

CARVALHO, L.R.; SILVA, E.A.A.; DAVIDE, A.C. Classificação de sementes florestais quanto ao comportamento no armazenamento. *Revista Brasileira de Sementes*, v.28, n.2, p. 15-25, 2006. http://www.scielo.br/pdf/rbs/v28n2/a03v28n2.pdf

DELGADO, L.F.; BARBEDO, C.J. Tolerância à dessecação de sementes de espécies de *Eugenia*. *Pesquisa Agropecuária Brasileira*, v.42, n.2, p. 265-272, 2007. http://www.scielo.br/pdf/pab/v42n2/16.pdf

FERREIRA, A.G.; BORGHETTI, F. *Germinação*: do básico ao aplicado. Porto Alegre: Artmed, 2004. 323 p.

FERREIRA, R.A.; VIEIRA, M.G.G.C.; VON PINHO, E.V.R.; TONETTI, O.A.O. Morfologia da semente e de plântulas e avaliação da viabilidade da semente de sucupira-branca (*Pterodon pubescens* Benth. - Fabaceae) pelo teste de tetrazólio. *Revista Brasileira de Sementes*, v.23, n.1, p. 108-115, 2001. http://www.abrates.org.br/revista/artigos/2001/v23n1/artigo15.pdf

FOGAÇA, C.A.; MALAVASI, M.M.; ZUCARELI, C.; MALAVASI, U.C. Aplicação do teste de tetrazólio em sementes de *Gleditschia amorphoides* Taub. Caesalpinaceae. *Revista Brasileira de Sementes*, v.28, n.3, p. 101-107, 2006. http://www.scielo.br/pdf/rbs/v28n3/15.pdf

FOGAÇA, C.A.; KROHN, N.G.; SOUZA, M.A.; PAULA, R.C. Teste de tetrazólio em sementes de *Copaifera langsdorffi* e *Schizolobium parahyba*. *Floresta*, v.41, n.4, p. 895 - 904, 2011. http://ojs.c3sl.ufpr.br/ojs/index.php/floresta/article/viewArticle/25352

FONSECA, S.C.L.; FREIRE, H.B. Sementes recalcitrantes: problemas na pós-colheita. *Bragantia*, v.62, n.2, p.297-303, 2003. http://www.scielo.br/pdf/brag/v62n2/v62n2a16.pdf

FRANÇA-NETO, J.B.; KRZYZANOWSKI, F.C.; COSTA, N.P. *O teste de tetrazólio em sementes de soja*. Londrina: EMBRAPA-CNPSo, 1998, 72 p. (Documentos, 116).

GUEDES, R.B.; ALVES, E.U.; GONÇALVES, E.P.; VIANA, J.S.; FRANÇA, P.R.C.; SANTOS, S.S. Qualidade fisiológica de sementes armazenadas de *Amburana cearensis* (Allemão) A. C. Smith. *Semina: Ciências Agrárias*, v.31, n.2, p. 331-342, 2010. http://www.uel.br/revistas/uel/index.php/semagrarias/article/view/5296/4822

JUSTO, C.F.; ALVARENGA, A.A.; ALVES, E.; GUIMARÃES, R.M.; STRASSBURG, R.C. Efeito da secagem, do armazenamento e da germinação sobre a micromorfologia de sementes de *Eugenia pyriformis* Camb. *Revista Acta Botanica Brasilica*, v.21, n.3, p.539-551, 2007. http://www.scielo.br/pdf/abb/v21n3/a04v21n3.pdf

KERBAUY, G.B. *Fisiologia Vegetal*. Rio de Janeiro: Editora Guanabara Koogan S.A., 2008. 431p.

LABOURIAU, L.G. *A germinação de sementes*. Washington: Organização dos Estados Americanos, 1983. 174p.

LORENZI, H.; BACHER, L.; LACERDA, M.; SARTORI, S. *Frutas brasileiras e exóticas cultivadas*: de consumo *in natura*. Nova Odessa: Plantarum, 2006. p.198-199.

MALUF, A.M.; BILIA, D.A.C.; BARBEDO, C.J. Drying and storage of *Eugenia involucrata* DC. seeds. *Scientia Agricola*, v.60, n.3, p. 471-475, 2003. http://www.scielo.br/pdf/sa/v60n3/16400.pdf

MARCOS-FILHO, J. *Fisiologia de sementes de plantas cultivadas*. Piracicaba: FEALQ, 2005. 495p.

MARTINS, C.C.; MACHADO, C.G.; SANTANA, D.G.; ZUCARELI, C. Vermiculita como substrato para o teste de germinação de sementes de ipê-amarelo. *Semina: Ciências Agrárias*, v.33, n.2, p. 533-540, 2012. http://www.uel.br/revistas/uel/index.php/semagrarias/article/view/6370/10443

MASETTO, T.E.; DAVIDE, A.C.; FARIA, J.M.R.; SILVA, E.A.A.; REZENDE, R.K.S. Avaliação da qualidade de sementes de *Eugenia pleurantha* (MYRTACEAE) pelos testes de germinação e tetrazólio. *Agrarian*, v.2, n.5, p. 34-46, 2009. http://www.periodicos.ufgd.edu.br/index.php/agrarian/article/viewArticle/802

MENDES, A.M.S.; BASTOS, A.A.; MELO, M.G.G. Padronização do teste de tetrazólio em sementes de *Parkia velutina* Benoist (Leguminosae – Mimosoideae). *Acta Amazônica*, v.39, n.4, p. 823-828, 2009. http://www.scielo.br/pdf/aa/v39n4/v39n4a10.pdf

OLIVEIRA L.M.; CARVALHO, M.L.M.; DAVIDE, A.C. Teste de tetrazólio para avaliação da qualidade de sementes de *Peltophorum dubium* (Sprengel) Taubert – Leguminosae Caesalpinioideae. *Cerne*, v.11, n.2, p. 159-166, 2005. http://www.dcf.ufla.br/cerne/artigos/11-02-20097975v11_n2_artigo%2006.pdf

PINTO, T.L.F.; BRANCALION, P.H.S.; NOVEMBRE, A.D.L.C.; CICERO, S.M. Avaliação da viabilidade de sementes de coração-de-negro (*Poecilanthe parviflora* Benth. - Fabaceae-Faboideae) pelo teste de tetrazólio. *Revista Brasileira de Sementes*, v.30, n.1, p.208-214, 2008. http://www.scielo.br/pdf/rbs/v30n1/a26v30n1.pdf

SCALON, S.P.Q.; NEVES, E.M.S.; MASETTO, T.E.; PEREIRA, Z.V. Sensibilidade à dessecação e ao armazenamento em sementes de *Eugenia pyriformis* Cambess. (uvaia). *Revista Brasileira de Fruticultura*, v.34, n.1, p.269-276, 2012. http://www.scielo.br/pdf/rbf/v34n1/v34n1a36.pdf

SCALON, S.P.Q.; SCALON FILHO, H; RIGONI, M.R. Armazenamento e germinação de sementes de uvaia *Eugenia uvalha* Cambess. *Ciência e Agrotecnologia*, v.28, n.6, p.1228-1234, 2004. http://www.scielo.br/pdf/cagro/v28n6/a02v28n6.pdf

SENA, L.H.M.; MATOS, V.P.; FERREIRA, E.G.B.S.; SALES, A.G.F.A.; PACHECO, M.V. Qualidade fisiológica de sementes de pitangueira submetidas a diferentes procedimentos de secagem e substratos – Parte 1. *Revista Brasileira de Engenharia Agrícola e Ambiental*, v.14, n.4, p.405-411, 2010. http://www.scielo.br/pdf/rbeaa/v14n4/v14n04a09.pdf

Behavior of coffee seeds to desiccation tolerance and storage

Luciana Aparecida de Souza Abreu[1*], Adriano Delly Veiga[2],
Édila Vilela de Resende Von Pinho[1], Fiorita Faria
Monteiro[3], Sttela Dellyzette Veiga Franco da Rosa[4]

ABSTRACT – The technology developed by breeding programs is applied to coffee seeds; however, after processing and drying, they lose viability within a short period of time, thus making storage unsuitable. The objective of this research was to evaluate the quality of coffee seeds submitted to different drying methods and moisture contents during storage. The coffee seeds were submitted to conventional drying (slow shade drying) and fast drying in a static drier until they reached a moisture content of 40, 20, 12 and 5%. After this process, the seeds were stored in a cold chamber for 12 months, and seed quality was evaluated before and during storage by the germination test, electrophoretic patterns of heat resistant proteins, and the activity of isoenzyme systems. Conventional drying (slow shade drying) at 20% of moisture content maintains coffee seed quality until 12 months of storage.

Index terms: *Coffea arabica* L., longevity, desiccation tolerance, oxidative stress.

Comportamento de sementes de cafeeiro quanto à tolerância à dessecação e ao armazenamento

RESUMO – A semente do cafeeiro agrega toda a tecnologia desenvolvida pelos programas de melhoramento, porém, após o processamento e secagem, perdem a viabilidade em curto período de tempo, dificultando o seu armazenamento. Assim, objetivou-se com a pesquisa avaliar a qualidade de sementes de cafeeiro submetidas a diferentes métodos de secagem e ao armazenamento. Sementes de cafeeiro foram secadas até atingirem os teores de 40, 20, 12 e 5% de água por meio de dois métodos de secagem: lenta à sombra e rápida em secador mecânico. Após este processo, as sementes foram armazenadas em câmara fria por doze meses, sendo a qualidade, avaliada antes e ao final do armazenamento pelo teste de germinação, padrão eletroforético de proteínas resistentes ao calor e atividade das enzimas catalase, peroxidase, superóxido dismutase. A secagem lenta a sombra até 20% de teor de água propicia a conservação da qualidade em sementes de cafeeiro por até doze meses de armazenamento.

Termos para indexação: *Coffea arabica* L., longevidade, tolerância à dessecação, estresse oxidativo.

Introduction

Studies that shed light on the complex physiology of desiccation and deterioration of coffee seeds during storage are extremely important because coffee crops are economically and socially relevant for Brazil. These studies can be used as the theoretical basis for selecting adequate post-harvest processes which will benefit coffee seed producers.

The technology developed by breeding programs is fully applied to coffee seeds. However, after processing and drying, the seeds lose viability within a short period of time, and they do not maintain germination quality at satisfactory levels for longer than six months after harvest (Araújo et al., 2008).

Seed longevity is associated with desiccation tolerance of seeds. Coffee seeds are classified as having an intermediate behavior towards drying and storage (Ellis et al., 1991), based on the fact that they tolerate storage for up to twelve months at 15 °C after desiccation to approximately 10% water content. However, existing studies on the performance of coffee seeds after drying and storage are contradictory and inconclusive (Santos et al., 2013). Research conducted with the aim of determining techniques to extend the shelf life of coffee seeds have shown

[1]Departamento de Agricultura, UFLA, Caixa Postal, 3037, 37200 – Lavras, MG, Brasil.
[2]Embrapa Café, UFLA, Caixa Postal 3037, 37200-000,Lavras- MG, Brasil.
[3]Universidade Federal de Lavras, Caixa Postal, 3037, 37200-000, Lavras-MG, Brasil.

[4] Embrapa Café, Departamento de Agricultura, UFLA, Caixa Postal 3037, 37200-000 – Lavras, MG, Brasil.
*Corresponding author <luapsouza2003@yahoo.com.br>

conflicting results, especially for water content (Gentil, 2001).

Increased desiccation tolerance of coffee seeds is observed when slow drying is performed, probably due to the time allowed for induction and protection mechanisms. Thus, damage to membrane systems may occur during fast drying, preventing recovery processes. Thus, more time may be required for repairs during germination (Santos et al., 2013).

Several studies have shown that slow drying provides top quality coffee seeds (Veiga et al., 2007, Vieira et al., 2007), which makes it recommended and frequently used for seeds of this species. While slow drying is used as a method to simulate desiccation as it occurs in maturation, fast drying is the most appropriate way to evaluate the extent of desiccation tolerance at a particular stage of development. Vieira et al. (2007) found that slow drying results in seeds with better physiological quality, while Rosa et al. (2005) pointed out that slow drying resulted in poorer-quality seeds. Thus, the method of drying can greatly influence desiccation tolerance of coffee seeds.

Therefore, this study aimed to investigate the quality of coffee seeds subjected to different drying methods and storage.

Material and Methods

The research was conducted in the Central Seed Laboratory of the Federal University of Lavras (UFLA). Seeds of the specie *Coffea arabica* L., cultivar Catuaí Amarelo IAC 62, were used. They were collected in cropping fields of Procafé Foundation, in the city of Varginha, state of Minas Gerais.

The coffee fruits were picked selectively at the red ripe stage and pulped, and mucilage was removed mechanically before drying. Average seed water content was 52% at that point (control not subjected to drying). Ten kilograms of seeds were dried until they reached water contents of 40, 20, 12 and 5% through two drying methods: slow shade drying and fast drying in a small static dryer at 35 °C and air flow of about 20 $m^3.min^{-1}.t^{-1}$. Seed water content was determined by the oven method at 105 °C for 24 hours, using two replicates for each treatment (Brasil, 2009).

During the drying processes, samples consisted of 1 kg of seeds of each treatment with different water contents. The seeds were stored in plastic bags at 10 °C and with relative humidity of 50%. Before storage and at every four months of storage, the quality of coffee seeds was evaluated for physiological aspects by the following tests:

Germination test - performed with four replicates of 50 seeds per treatment. Paper towel rolls were used as substrate, moistened with a quantity of water equal to 2.5 times the dry weight of the substrate and maintained in germinators at 30 °C. Counts were made at fifteen and thirty days after sowing (Brasil, 2009), and the results were expressed as percentage

of normal seedlings. *Root protrusion* - conducted at fifteen days after the beginning of the germination test, by counting the seeds that had the taproot and at least two lateral roots. The results were expressed in percentage terms.

Isoenzyme analysis - samples of 100 grams of seeds were taken from the treatments before and after storage for electrophoretic analysis of enzymes. The samples were ground in a mortar on ice in the presence of PVP and liquid nitrogen and they were subsequently stored at -86 °C. For enzyme extraction, Tris HCl 0.2 M pH 8.0 + (0.1% β-mercaptoethanol) was used at a ratio of 250 μL per 100 mg of seeds. The material was homogenized by vortexing and kept overnight in a refrigerator, followed by centrifugation at 14,000 xg for 30 minutes at 4 °C. The electrophoretic runs were performed on polyacrylamide gels at 7.5% (separating gel) and 4.5% (concentrating gel). The gel/electrode system used was Tris-glycine pH 8.9. 50 μL of the supernatant from the sample was applied to the gels and the run was performed at 150 V for 4 hours. After the run, the gels were stained for enzymes catalase, esterase, peroxidase, and superoxide dismutase according to Alfenas (2006) and then analyzed visually.

For extraction of heat resistant proteins, (50 mM Tris-HCl pH 7,5; 500 mM NaCl; 5mM $MgCl_2$; 1 mM PMSF) were added to the buffer solution at a ratio of 1:10 (material weight: extraction buffer volume) and transferred to 1,500 μL microcentrifuge tubes. The homogenate was centrifuged for 45 minutes at 4 °C at 16,000 xg, and the supernatant was removed and incubated in a water bath at 85 °C for 15 minutes and then centrifuged again for 30 minutes. The supernatant was poured into microtubes, and the pellet was discarded. Before application to the gel, sample tubes containing 70 μL extract + 40 μL of sample buffer solution (2.5 mL glycerol, 0.46 g SDS, 20 mg of bromophenol blue, and the volume completed to 20 mL with extraction buffer Tris pH 7.5) were placed in a bath of boiling water for 5 minutes. 50 μL of this solution were applied to SDS-PAGE polyacrylamide gel at 12.5% in the separating gel and at 6% in the concentrating gel. Electrophoresis was performed at 150 V and the gel was stained with Coomassie Blue at 0.05%, according to Alfenas (2006), for 12 hours and bleached in a solution of 10% acetic acid.

Experimental design - completely randomized with four replications in a (2x4x4) + 1 factorial design, which corresponded to two drying methods (slow and fast), four times (0, 4, 8 and 12 months) and four water contents after drying (40, 20, 12, and 5%), plus the control (52% without drying). The analysis of data from the factorial scheme was performed using the statistical program SISVAR® (Ferreira, 2011), and regression analysis was performed for quantitative variation of storage time. The control (52%), within each period of storage, was compared with the treatments of the factorial by Dunnett's test using the GLM procedure of the software SAS® version 9.0.

Results and Discussion

By analyzing the germination of coffee seeds, all sources of variation were significant in the analysis of variance. Shade-dried seeds had higher values compared to artificial drying along the storage period evaluated (Figure 1). In the study of Veiga et al. (2007) on seeds harvested at the red ripe stage, there was better physiological quality when shade drying was used for all storage times evaluated. Thus, considering all water contents, only non stored, dried seeds (time zero) and shade-dried seeds dried with four months of storage had germination values above the marketable pattern, which is 70% (Carvalho et al., 2008).

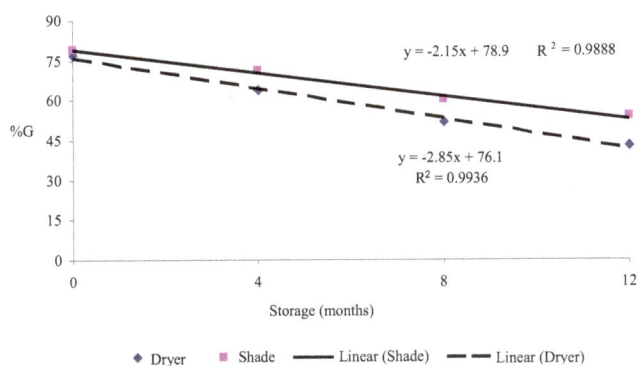

Figure 1. Estimated germination percentage values of coffee seeds subjected to slow drying (shade) and fast drying (dryer), stored for twelve months.

The significant triple interaction among factors method of drying, water content achieved after drying and storage time, showed that the best results for the twelve months were provided by shade drying to a water content of 20% (Figure 2). In such conditions, high quality seeds can be obtained throughout the storage period evaluated.

Figure 2 shows the drying methods separately for better visualization of germination estimates during storage. When fast drying was used, germination percentage was maintained up to eight months of storage, but only when drying was performed to 20% seed moisture. Araújo et al. (2008) found that coffee seeds are preserved better when stored at 18.5% moisture at a low temperature (7 °C) environment while maintaining their germinative ability for up to nine months.

For most tested treatments, there was a reduction in seed quality during storage, especially for seeds dried to 40% moisture. Seed drying to 5% water content was drastic to

physiological quality, regardless of drying method. Brandão Jr. et al. (2002), who evaluated desiccation tolerance of coffee seeds harvested at different stages of maturation, observed a higher level of desiccation tolerance in coffee seeds with increased development, and seeds dried to 15% moisture content maintained their physiological quality over nine months of storage. In contrast, non-dried seeds with 50% moisture content showed a linear decrease of germination during storage.

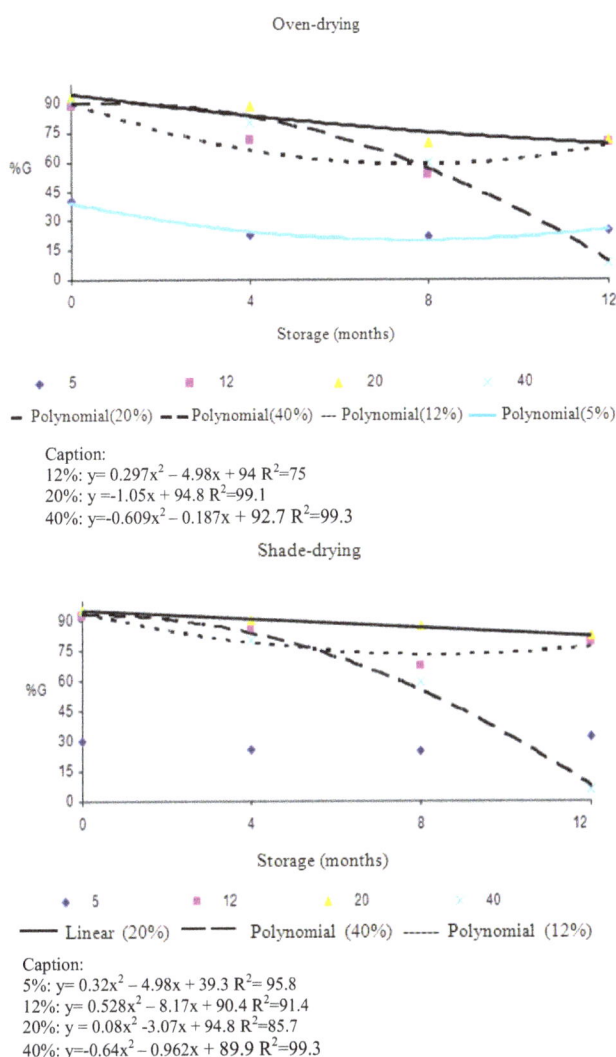

Figure 2. Estimated germination percentage values of coffee seeds subjected to slow drying (shade) and fast drying (dryer) to different water contents and stored for twelve months.

Similar results were found for root protrusion (Figure 3). The triple interaction between factors was significant, and indicated better results with shade drying up to water contents of 20% and 12% throughout the storage period. The results also highlighted the reduction of seed vigor during storage, especially when the seeds were subjected to water loss by 40% in both methods of drying, and the dried seeds, by 5% in the static dryer. When slow drying was used to water content of 5%, seed vigor was maintained during storage. According to Brandão Jr. et al. (2002), the reduction of vigor in seeds subjected to artificial drying causes intense changes such as crystal formation and disappearance of the vacuole and endomembranes, as shown in ultrastructural analyses.

By Dunnett's test, comparisons were made for germination and root protrusion between control (non dried seeds) and each treatment, within each storage time (Table 1).

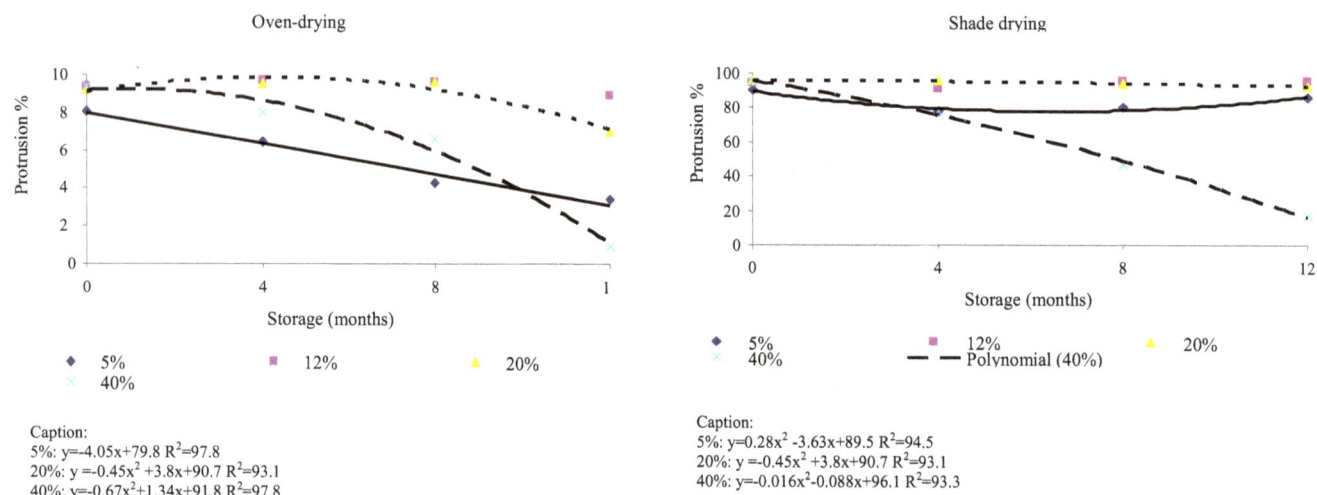

Oven-drying

Shade drying

Caption:
5%: $y=-4.05x+79.8$ $R^2=97.8$
20%: $y=-0.45x^2+3.8x+90.7$ $R^2=93.1$
40%: $y=-0.67x^2+1.34x+91.8$ $R^2=97.8$

Caption:
5%: $y=0.28x^2-3.63x+89.5$ $R^2=94.5$
20%: $y=-0.45x^2+3.8x+90.7$ $R^2=93.1$
40%: $y=-0.016x^2-0.088x+96.1$ $R^2=93.3$

Figure 3. Estimated root protrusion values of coffee seeds subjected to slow drying (shade) and fast drying (dryer), stored for twelve months.

Table 1. Comparison of germination and root protrusion (%) values of the control with coffee seeds subjected to slow drying (shade) and fast drying (dryer), stored for twelve months.

Treatment	Time (months)							
	0		4		8		12	
	G	RP	G	RP	G	RP	G	RP
5% oven	40[*]	81[*]	16[*]	63[ns]	22[**]	44[**]	24[**]	35[**]
5% shade	31[*]	91[ns]	26[*]	78[ns]	25[**]	81[**]	32[**]	86[**]
12% oven	87[ns]	93[ns]	75[**]	97[**]	54[**]	94[**]	71[**]	85[**]
12% shade	92[ns]	95[ns]	87[**]	91[**]	67[**]	97[**]	84[**]	94[**]
20% oven	93[ns]	92[ns]	90[**]	95[**]	71[**]	96[**]	72[**]	36[**]
20% shade	96[ns]	96[ns]	89[**]	96[**]	88[**]	95[**]	85[**]	93[**]
40% oven	91[ns]	96[ns]	80[**]	90[**]	62[**]	67[**]	7[ns]	8[ns]
40% shade	93[ns]	95[ns]	85[**]	88[**]	61[**]	67[**]	6[ns]	0[ns]
Control (52%)	92	97	59	64	0	0	0	0

[**]Significant and higher than the control by Dunnett's test at 5% probability;
[*]Significant and lower than the control by Dunnett's test at 5% probability;
[ns] Not significant by Dunnett's test at 5% probability.

For germination, a high value was observed before storage for the control; it was significantly different only from the treatments where drying reached up to 5%. For the second storage time, the control had considerable reduction. It was different from both higher values, observed for treatments where drying was performed to 40, 20 and 12%, and for lower values, observed when drying was performed to 5%.

For the subsequent months, significant differences were found for the treatments compared to the control because there was no seed germination in this treatment, except for the seeds dried to 40% after 12 months of storage. For high water contents, close to 52% without use of drying, the seeds should not be stored for more than four months. Veiga et al. (2007) found that after eight months of storage, the highest germination value was observed in seeds harvested at the yellowish-green stage and shade-dried, while the smallest was found in non dried seeds and seeds dried in the dryer.

For root protrusion, a significant difference before storage was observed only for seeds dried to 5% moisture in the dryer. After four months of storage, there was no root protrusion of seeds; the control had a considerable reduction in the amount of protrusion, and it was significantly equal to seeds dried to 5% moisture. In the eighth month of storage, there was no root protrusion in seeds of the control with 52% moisture, in contrast with all the treatments. This was also observed at twelve months, except for the seeds that were dried to 40% moisture, which were similar to the control after this period.

Isoenzyme analysis complements the evaluation of physiological seed quality. It is a rapid, sensitive and specific method for this purpose; it detects enzymes associated with seed metabolism, germination, mechanisms of seed protection during drying and storage and also enzymes linked to the deteriorating process. For this study, isoenzyme analyses were performed using only the treatments obtained before and after storage.

The reduction in the activity of free radical scavenger enzymes such as catalase is related to the loss of seed viability (Berjak, 2006), a fact confirmed in seeds dried to 5%, regardless of drying methods (Figure 4). For seeds with low water content dried artificially, this reduction was enhanced after twelve months of storage. For Berjak (2006), damage to membranes by desiccation can be caused by oxidation, which promotes phospholipid esterification or lipid peroxidation. This enzyme is involved in the removal of hydrogen peroxide (H_2O_2) from cells, and its increased activity may be associated with the decrease in mechanisms that prevent oxidative damage (Bailly et al., 2002).

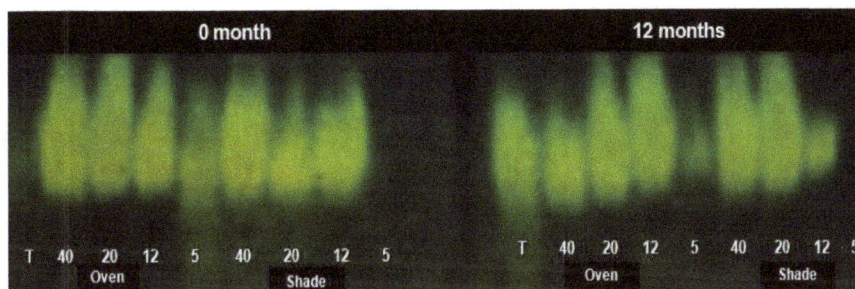

Figure 4. Enzymatic profiles of catalase (CAT) in seeds of *Coffea arabica* L., depending on drying methods, water content (%) and time of storage.

There is also an increase in the intensity of bands in the control after storage. This fact is related to increased activity of the enzyme in combating free radicals and antioxidative response under stress (Dussert et al., 2006). The results confirm that moister seeds substantially lose quality during storage, which can be seen in the results of the germination and vigor tests. Lower activity of catalase was observed in the seeds that were not dried, a finding similar to the one by Santos et al. (2014) in coffee seeds, which suggests that the loss of seed viability during the drying process is followed by an increase in the levels of reactive oxygen species (ROS) because of the physiological stress suffered. According to Takahashi et al. (2007), seeds have mechanisms to prevent the formation of ROS. They stated that the action of enzymes such as catalase removes hydrogen peroxide, a highly toxic compound to the seeds.

The expression of the esterase enzyme (Figure 5) has greater activity in seeds dried in a dryer before and after the storage period, especially coffee seeds that contain high water contents, which can be characterized as deteriorating seeds. Nakada et al. (2010) reported that this enzyme is very indicative of seed deterioration. However, for seeds from the treatment without drying (control), no bands were observed, especially at the end of storage. Brandão Júnior et al. (2002) did not identify bands of this enzyme in non dried coffee seeds (control), either.

Desiccation tolerance of seeds is achieved and maintained by means of several mechanisms, including the induction of heat-resistant proteins. The analysis of the electrophoretic profile of heat resistant proteins extracted from coffee seeds (Figure 6) shows a pattern of more significant bands in the seeds that were shade-dried before storage. The result seems to indicate more activity in seeds dried in the shade to water content of 20% before storage; such seeds had higher germination percentage and vigor.

This result shows the appearance or increase in intensity of bands with increasing water loss by seeds or higher content of heat-resistant proteins. This is indicative that drying induced the synthesis of this protein in seeds dried to low moisture contents. This fact is more clearly observed for the seeds quickly dried before storage and for seeds in both drying methods after storage.

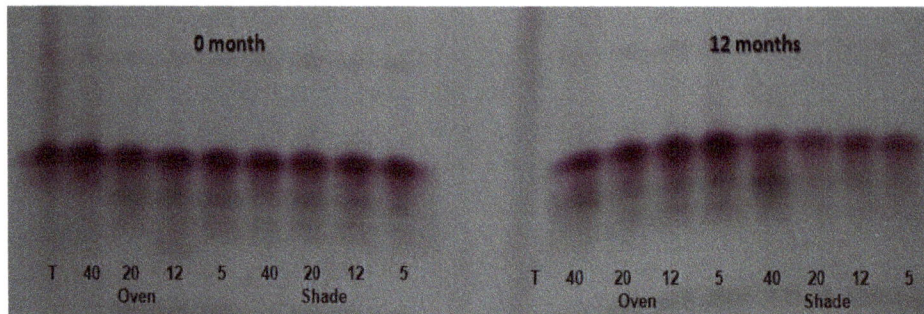

Figure 5.   Enzymatic profiles of esterase (EST) on seeds of *Coffea arabica* L., depending on drying methods, water content (%) and time of storage.

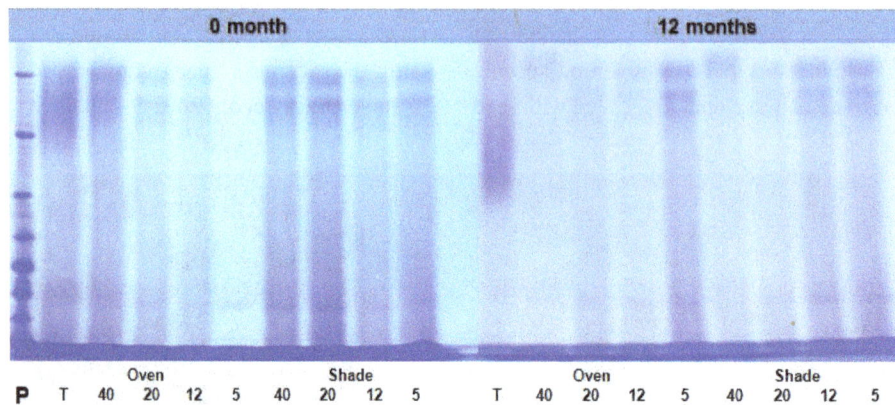

Figure 6.   Enzymatic profiles of heat-resistant proteins in seeds of *Coffea arabica* L., depending on drying methods, water content (%) and time of storage.

Seeds not subjected to drying showed no bands, which can relate this protein system to desiccation intolerance in these seeds, a fact that was also noted by Guimarães et al. (2002) in coffee seeds. Veiga et al. (2007) did not observe activity of heat resistant proteins in seeds that were not subjected to drying at all storage times evaluated.

These proteins are synthesized and accumulated at the later stages of seed development, before or during drying, and their stability, hydrophilicity and abundance in desiccation resistant organisms suggest a role associated with tolerance to drying because they protect seeds and are very important for preventing drying-induced damage (Vidigal et al., 2009). The drying process in seeds appears to induce the expression of alleles, thereby promoting the onset or exacerbation of bands. These results suggest low tolerance to drying and low longevity of stored coffee seeds, which can be correlated with the results obtained in the physiological tests.

Bands of the peroxidase enzyme tend to increase in intensity (activity) due to a decrease in the water content of the seeds subjected to fast drying (Figure 7). When slow drying

in the shade was used, the opposite effect was observed, and the intensity of the bands decreased as seeds lost water. There was also lower activity of bands in seeds dried in the shade compared with seeds dried in the dryer. This was more evident after storage. In moist seeds, that is, those that were not dried (control), the repair mechanisms that include free radical scavenger enzymes such as catalase and peroxidase were not triggered. This was confirmed by the absence of peroxidase activity for seeds not submitted to drying.

For the enzyme system of superoxide dismutase (Figure 8), activity was identified for seeds subjected to the two drying processes; however, no activity was observed for seeds harvested at 52% moisture (non dried) before and after storage. It should be noted that for seeds dried quickly to water content of 40%, the presence of bands was not identified, either. For seeds dried in the shade to 40% water content, slow drying may have led to the activation of this enzyme system. According to Berjak (2006), the absence or reduction in the activity of free radical scavenger enzymes increases the sensitivity of the seeds to oxidative stress.

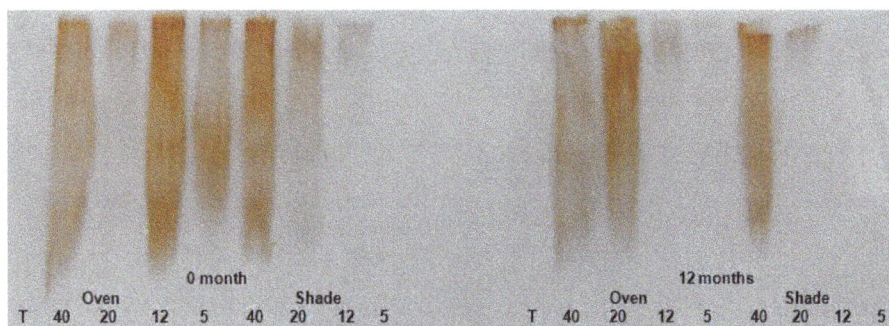

Figure 7. Enzymatic profiles of peroxidase (PO) in seeds of *Coffea arabica* L., depending on drying methods, water content (%) and time of storage.

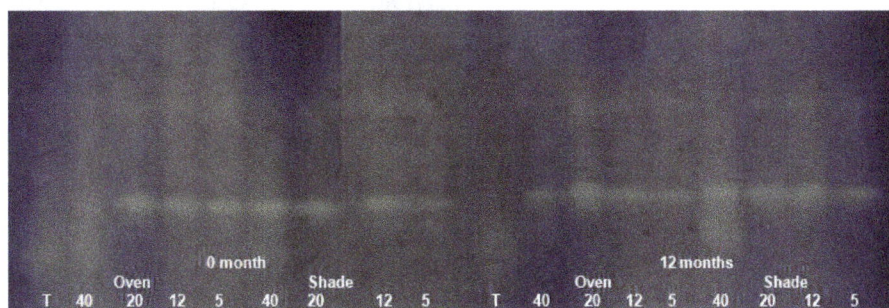

Figure 8. Enzymatic profiles of superoxide dismutase (SOD) in seeds of *Coffea arabica* L., depending on drying methods, water content (%) and time of storage.

Conclusions

Slow shade drying to 20% water content facilitates the maintenance of quality in coffee seeds for up to twelve months of storage.

Acknowledgements

The authors are thankful to the Coordination for the Improvement of Higher Education Personnel (CAPES) for granting a postdoctoral scholarship and financial support for the research.

References

ALFENAS, A. C. *Eletroforese e marcadores bioquímicos em plantas e microrganismos.* 2.ed. Viçosa: UFV, 2006. 627p.

ARAÚJO, R. F.; ARAUJO, E. F.; CECON, P.R; SOFIATTI, V. Conservação de sementes de café *(Coffea arabica L.)* despolpado e não despolpado. *Revista Brasileira de Sementes,* v.30, n.3, p.71-078, 2008. http://www.scielo.br/pdf/rbs/v30n3/10.pdf

BAILLY, C.; BOGATEK-LESZCZYNSKA, R.; CÔME, D.; CORBINEAU, F. Changes in activities of antioxidant enzymes and lipoxygenase during growth of sunflower seedlings from seeds of different vigour. *Seed Science Research*, v.12, n.1, p.47-55, 2002. http://journals.cambridge.org/action/displayabstract?frompage=online&aid=704312

BERJAK, P. Unifying perspectives of some mechanisms basic to desiccation tolerance across life forms. *Seed Science Research*, v.16, p.1-15, 2006. http://journals.cambridge.org/action/displayAbstract?fromPage=online&aid=705716&fileId=S0960258506000018

BRANDÃO JR, D. S.; VIEIRA, M. G. G. C.; HILHORST, H. W. M. Aquisição da tolerância à dessecação nos diferentes estádios de desenvolvimento de sementes de cafeeiro *(Coffea arabica* L.). *Ciência e Agrotecnologia*, v.26, n.4, p.673-681, 2002. http://www.scielo.br/scielo.php?pid=S0101-31222002000200004&script=sci_arttext

BRASIL. Ministério da Agricultura, Pecuária e Abastecimento. *Regras para Análises de Sementes.* Ministério da Agricultura, Pecuária e Abastecimento. Secretaria de Defesa Agropecuária. Brasília: MAPA/ACS, 2009. 395p. http://www.bs.cca.ufsc.br/publicacoes/regras%20analise%20sementes.pdf

CARVALHO, G. R., GUIMARÃES, P. T. G.; NOGUEIRA, A. M.; REZENDE, J. C. Normas e padrões para a comercialização de sementes e mudas de cafeeiros em Minas Gerais. *Informe Agropecuário*, Belo Horizonte: EPAMIG, v. 29, n. 247, p. 24-30, 2008.

DUSSERT, S.; DAVEY, M.W.; LAFFARGUE, A.; DOULBEAU,S.; SWENNEN, R,. ETIENNE, H. Oxidative stress, phospholipids loss and lipid hydrolysis during drying and storage of termediate seeds. *Physiologia Plantarum,* v.127, p.192-204, 2006. http://onlinelibrary.wiley.com/doi/10.1111/j.1399-3054.2006.00666.x/full

ELLIS, R. H.; HONG, T. D.; ROBERTS, E. H. An intermediate category of seed storage behavior? II. Effects of provenance, immaturity, and imbibition on desiccation-tolerance in coffee. *Journal of Experimental Botany*, v.42, n. 238, p. 653-657, 1991. http://jxb.oxfordjournals.org/content/42/5/653

FERREIRA, D. F. Sisvar: a computer statistical analysis system. *Revista Ciência e Agrotecnologia*, v.35, n.6, p.1039-1042, 2011. http://www. scielo. br/pdf/cagro/v35n6/a01v35n6.pdf

GENTIL, D. F. O. Conservação de sementes do cafeeiro: resultados discordantes ou complementares? *Bragantia*, v.60, n.3, p.149-154, 2001. http://www.scielo.br/pdf/brag/v60n3/a01v60n3.pdf

GUIMARÃES, R. M.; VIEIRA, M. G. G. C.; FRAGA, A. C.; VON PINHO, E. V .R.; FERRAZ, V.P. Tolerância à dessecação em sementes de cafeeiro (*Coffea arabica* L.). *Ciência e Agrotecnologia*, v.26, p.128-139, 2002. http://www. scielo.br/scielo.php?pid=S0101-31222002000200004&script=sci_arttext

NAKADA, P. G.; OLIVEIRA, J. A.; MELO, L. C.; GOMES, L. A. A.; VON PINHO, E. V. R. Desempenho durante o armazenamento de sementes de pepino submetidas a diferentes métodos de secagem. *Revista Brasileira de Sementes*, v.32, n.3, p.42-051, 2010. http://www.scielo.br/pdf/rbs/v33n1/13.pdf.

ROSA, S.D.V.F.; BRANDÃO JR, D. S.; VON PINHO, E. V. R.; VEIGA, A.D.; CASTRO, L.H. Effects of different drying rates on the physiological quality of *Coffea canephora* Pierre seeds. *Brazilian Journal of Plant Physiology*, v.17, n.2, p.199-205, 2005. http://www.scielo.br/scielo. php?script=sci_arttext&pid=S1677-04202005000200002&lng=en&tlng=en. 10.1590/S1677-04202005000200002.

SANTOS, G. C.; VON PINHO, E. V. R.; ROSA, S.D.V.F. Gene expression of coffee seed oxidation and germination processes during drying. *Genetics and Molecular Research*, v.12, n.4, p.6968-6982, 2013. http://www.funpecrp. com.br/gmr/year2013/vol12-4/pdf/gmr3263.pdf

SANTOS, F. C.; ROSA, S. D. V. F..; VON PINHO, E. V. R.; CIRILLO, M. A.; CLEMENTE, A. C. S. Desiccation sensitivity from different coffee seed phenological stages. *Journal of Seed Science*, v. 36, n.1, p. 25-31, 2014. http://www.scielo.br/scielo.php?script=sci_arttext&pid=S2317-15372014000100003&lng=en&tlng=en. 10.1590/S2317-15372014000100003.

TAKAHASHI, A.; OHTANI, N.; HARA, E. Irreversibility of cellular senescence: dual roles of p16INK4a/Rb-pathway in cell cycle control. *Cell Division*, v.2, n.10, p.1-5, 2007. http://www.celldiv.com/content/2/1/10

VEIGA, A.D.; GUIMARÃES, R.M.; ROSA, S.D.V.F.; VON PINHO, E.V.R.; SILVA, L.H.C.; VEIGA, A.D. Armazenabilidade de sementes de cafeeiro em diferentes estádios de maturação e submetidas a diferentes métodos de secagem. *Revista Brasileira de Sementes*, v.29, n.1, p.83-91, 2007. http://www.scielo.br/scielo.php?script=sci_arttext&pid=S0101-31222007000100012

VIDIGAL, D. S.; DIAS, D. C. F. S.; VON PINHO, E. V. R.; DIAS, L. A. S. Alterações fisiológicas e enzimáticas durante a maturação de sementes de pimenta (*Capsicum annuum* L.). *Revista Brasileira de Sementes*, v.31, n.2, p.129-136, 2009. http://www.scielo.br/pdf/rbs/v31n2/v31n2a15.pdf

VIEIRA, A. R.; OLIVEIRA, J.A.; GUIMARÃES, R.M.; PEREIRA, C.E.; CARVALHO, F.E. Armazenamento de sementes de cafeeiro: ambientes e métodos de secagem. *Revista Brasileira de Sementes*, v.29, n.1, p.76-82, 2007. http://www.scielo.br/scielo.php?pid=S0101-31222007000100011&script=sci_abstract&tlng=pt

Methodology of the tetrazolium test for assessing the viability of seeds of *Eugenia brasiliensis* Lam., *Eugenia uniflora* L. and Eugenia pyriformis Cambess

Edmir Vicente Lamarca[1*], Claudio José Barbedo[1]

ABSTRACT – *Eugenia brasiliensis* Lam. ("grumixameira"), *Eugenia uniflora* L. ("pitangueira") and *Eugenia pyriformis* Cambess. ("uvaieira") are forest and fruit species of pharmacological and gastronomic potential and have desiccation-sensitive seeds. The development of quick viability tests for the assessment of physiological quality of these seeds is needed. The tetrazolium test is an important method providing fast assessment of the seed physiological quality. Thus, this work aimed to develop a method for the tetrazolium test for determining viability of "grumixameira," "pitangueira" and "uvaieira" seeds. Initially the seeds of these species were soaked in water for 3 h at 25 ºC. Three concentrations of tetrazolium solutions were analyzed (0.100 %, 0.125 % and 0.250 %) for seed staining using three incubation periods (2, 3 and 6 h) at 35 ºC in the dark. After staining, seed viability was determined. Then, the seeds were subjected to different levels of controlled drying and were analyzed for their physiological quality by germination and electrical conductivity tests. The tetrazolium test is a suitable tool for determining viability after staining the seeds at 35 ºC using a 0.250 % concentration solution for 3 hours for "grumixameira" seeds, 0.125 % for 3 hours for "pitangueira" seeds and 0.100 % for 2 hours for "uvaieira" seeds.

Index terms: forest seeds, recalcitrant seeds, drying, viability test.

Metodologia do teste de tetrazólio para a avaliação da viabilidade de sementes de *Eugenia brasiliensis* Lam., *Eugenia uniflora* L. e *Eugenia pyriformis* Cambess

RESUMO – Grumixameira (*Eugenia brasiliensis* Lam.), pitangueira (*Eugenia uniflora* L.) e uvaieira (*Eugenia pyriformis* Cambess.) são espécies florestais e frutíferas de potencial farmacológico e gastronômico e apresentam sementes sensíveis à dessecação. O teste de tetrazólio é um importante componente para a rápida avaliação da viabilidade de sementes. Assim, o presente estudo teve como objetivo adequar a metodologia do teste de tetrazólio para a determinação da viabilidade de sementes de grumixameira, pitangueira e uvaieira. Sementes dessas três espécies foram pré-umedecidas em água por 3 h a 25 ºC. Após, foram coloridas com soluções de tetrazólio nas concentrações de 0,100%, 0,125% e 0,250% por 2, 3 e 6 horas, a 35 ºC no escuro. Após lavagem, determinou-se a viabilidade das sementes. As sementes foram também submetidas a diferentes níveis de secagem e foram analisadas quanto aos seus atributos fisiológicos pelos testes de germinação e condutividade elétrica. O teste de tetrazólio, com coloração das sementes a 35 ºC usando-se uma concentração da solução de 0,250% com incubação por 3 horas para sementes de grumixameira, a 0,125% por 3 horas para sementes de pitangueira e a 0,100% por 2 horas para sementes de uvaieira, mostra-se como uma eficiente ferramenta para a avaliação da viabilidade dessas sementes.

Termos para indexação: sementes florestais, sementes recalcitrantes, secagem, teste de viabilidade.

Introduction

'Grumixameira' (*Eugenia brasiliensis* Lam.), 'pitangueira' (*Eugenia uniflora* L.) and 'uvaieira' (*Eugenia pyriformis* Cambess.) are forest, tree and fruit species with great medicinal, industrial and gastronomic potential, besides being enjoyed for their own consumption *in natura* and widely used in agroforestry systems (Lamarca et al., 2013). These species belong to the Myrtaceae family in Brazil and occur in tropical and subtropical regions. They have large seeds with wide germination plasticity (Amador and Barbedo, 2011; Lamarca et al., 2011) and intolerant to desiccation (Delgado and Barbedo, 2007).

[1]Instituto de Botânica, Núcleo de Pesquisa em Sementes, Caixa Postal 68041, 04301012 – São Paulo, SP, Brasil.
*Corresponding author <edmirvicente18@gmail.com>

Seed quality is maximal at the time of physiological maturity. From that moment, they begin to deteriorate until they lose the ability to germinate. The assessment of quality of seeds by means of vigor tests can be understood as an important component for assessing the physiological quality, allowing to identify the actual state of deterioration and vigor, assisting in decision making regarding the use or disposal of seed lots (Marcos-Filho, 2005).

The main challenge for research on seed vigor tests is to identify the decay events that precede the loss of germination, such as those related to the membrane systems and the enzymatic activities. Thus, within this context, is highlighted the tetrazolium test, as already shown for seeds of brazilwood (*Caesalpinia echinata*) (Lamarca et al., 2009), sunflower (*Helianthus annuus*) (Silva et al., 2013), jabuticaba (*Plinia cauliflora*) (Hössel et al., 2013) and pineapple guava (*Acca sellowiana*) (Sarmento et al., 2013). The tetrazolium test is a rapid test for estimating the vigor and viability of seeds. It is based on the activity of dehydrogenase enzymes that reduce the 2,3,5 triphenyltetrazolium chloride in the living tissues, for where hydrogen ions are transferred. When the seeds are immersed in the solution of tetrazolium, this is propagated by the tissues causing in living cells the reduction reaction, resulting in the formation of a red non-diffusible compound, referred to as triphenilformazan indicating respiratory activity and that the tissue is viable. Non-viable tissues do not react with the solution, remaining in their natural color and damaged tissues exhibit intense red staining (França-Neto, 1994).

For seeds of various cultivated species, the tetrazolium test has already been extensively researched and has defined and applied its methodology. Regarding forest species, studies described in the literature demonstrate the advances in methodological adequacy test for seeds, but it is not possible to standardize the methodology for implementing the same, since each species needs its own processes, as seen by changes in the concentration optimal tetrazolium solution, for example, 0.050 % for Brazilian firetree or Brazilian fern tree (*Schizolobium parahyba*) (Ferreira et al., 2007) and brazilwood (Lamarca et al., 2009), 0.075% for angico-de-bezerro (*Piptadenia moniliformis* Benth.) (Azerêdo et al., 2011), 0.100% for pitanga-do-mato (*Myrcia crocea*) (Masetto et al., 2009) and jabuticabeira (Hössel et al., 2013) and 0.500% for macaúba (*Acrocomia aculeata* and *A. vinifera*) (Ribeiro et al., 2010) and pineapple guava (Sarmento et al., 2013).

However, in the forestry seeds there is little information in the literature about the appropriate methodology for the implementation of the tetrazolium test, especially when dealing with desiccation intolerant seeds, namely, the ones of short-term storage, such as, for example, seeds of *E.*

brasiliensis, E. uniflora and E. pyriformis. Thus, the present study aimed to suit the methodology of the tetrazolium test for seeds of grumixameira, pitangueira and uvaieira, to assess the viability of these seeds.

Material and Methods

Seed collection: the seeds of *E. brasiliensis, E. uniflora and E. pyriformis.* were obtained from ripe fruit freshly dispersed, in 2007, based on the information of maturation and collection of Delgado and Barbedo (2007). The fruits of grumixameira were from the municipality of São Paulo – SP (23°38'S, 46°37'W, 785 m) and of pitangueira and uvaieira from the municipality of Santo André – SP (23°40'S, 46°32'W, 791 m), Cwb regions, according to the climate classification of Köppen-Geiger (Peel et al., 2007). After collection, with the aid of running water and a sieve, the seeds were removed from fruits and stored in a cold chamber at 7 °C until the beginning of the experiments, not exceeding seven days after collection (Delgado and Barbedo, 2007).

Physical and physiological assessments: later, the seeds were characterized by water content, water potential, germination and electrical conductivity. The water content (WC) was gravimetrically determined by the oven method at 103 °C for 17 hours, and the results were presented in % of a wet basis (Brasil, 2009). The water potential (PΨ) was measured in seeds cut in the middle and analyzed in a potentiometer WP4 (Decagon Devices, Pullman, WA, USA), based on the temperature of the dew point of the air in equilibrium with the sample examined (Daws et al., 2004). The water potential values were checked by means of water sorption isotherms in solutions of polyethylene glycol 6000.

The germination test was conducted in germination chambers at 25 °C with constant light and 100% relative humidity, using the roll paper method (filter paper) with two leaves for the basis and one for coverage (Brasil, 2009). Four replications of 15 seeds each were used. Germination assessments were performed every 3 days for 70 days (Delgado and Barbedo, 2007), by registering the seeds that have issued a primary root, for the calculation of germinable seeds (GS) and those that produced normal seedlings, for the calculation of germination (G), both results presented in percentage. The first count (FC), by the germination test, was also recorded.

As a measure of seed vigor, the index of germination speed (IVG) was rated using the counts of the germination test itself, using the formula of Maguire (1962) and the germination average time (GT), according to Borghetti and Ferreira (2004). To determine the day of the first count, for each species was used as reference the germination test of the seeds without drying, determining the day when it presented with approximately 50% of germinable seeds, which occur at the 7th day of countdown for grumixameira, at the 14th day for pitangueira and the 22nd for uvaieira; a methodology adapted from Brasil (2009).

The electrical conductivity test (EC) was determined by means of digital conductivity meter bench MA150 (Marconi, Piracicaba), measured by means of a standard solution (146.7 µS.cm⁻¹). Thus, samples of intact seeds (15 per replicate) were placed in disposable plastic cups of 300 mL, containing 75 mL of deionized water (four cups, each being considered a replication), in a B.O.D. (Biochemical Oxigen Demand) type environmental chamber set to a constant temperature of 20 °C in the absence of light. The assessments were performed after 24 hours of conditioning (Vieira, 1994). The results obtained, in $\mu S\ cm^{-1}$, were divided by the equivalent dry weight of the sample (g), being presented by unit of dry weight in $\mu S.cm^{-1}.g^{-1}$. To obtain the equivalent in dry weight, seed samples were taken for determination of total dry weight in the same test for the determination of water content (Barbedo and Cicero, 1998; Martini Neto et al., 2014).

Concentration of tetrazolium solution and incubation period: different concentrations of salt 2,3,5 triphenyltetrazolium chloride for different periods of incubation were analyzed for seeds. Pre-tests showed the most appropriate method for the implementation of the tetrazolium test, namely, immersion of intact seeds for 3 hours at 25 °C in water, to determine the preconditioning, which was made to match the water content of the seeds with and without

drying and to soften the tissues and allow the development of a more obvious staining. After preconditioning, the longitudinal splitting of the seeds was performed, immediately immersing the two parts in the tetrazolium solution, so as to maximally avoid exposure of the cut parts to air, since the seeds of the species of the present study after cutting and in contact with air quickly oxidize, which would hinder the assessment. Such longitudinal sectioning was necessary because preliminary tests showed that there is no staining of intact seeds when incubated in tetrazolium (Figures 1A and B).

Therefore, the seeds of grumixameira, pitangueira and uvaieira (four replications of 15 seeds) were stained in the aqueous solution of tetrazolium in the concentrations of 0.100%, 0.125% and 0.250% for 2, 3 and 6 h, at 35 °C in the dark. At the end of staining, the seeds were washed in running water and left immersed in water until the beginning of the assessments, in which the surface of the embryo was assessed (Figure 1C). Embryos were assessed individually with the aid of a magnifying glass and tweezers, observing the presence and location of damage on their surface and were separated into categories of viable and non-viable, according to the different staining patterns, ranging from intense red until the milky white or absence of color (Figure 1) (Masetto et al., 2009). The color differentiation of tissues followed the criteria established by França-Neto (1994), namely, pinkish for a healthy tissue, intense red for a tissue in deterioration and color missing for a dead tissue.

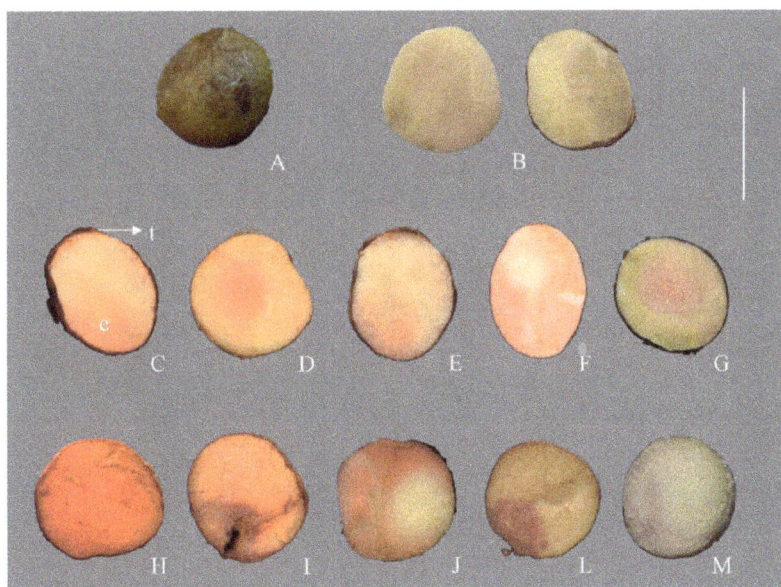

Figure 1.   Staining of seeds of *E. brasiliensis, E. uniflora and E. pyriformis*. by the tetrazolium test. A and B = intact seeds incubated, demonstrating that they do not acquire sufficient staining for analysis. C to M = classes for assessment of the viability of seeds. C – G (viable seeds), C = class I and D – G = class II. H – M (non viable seeds), H = class III, I – L = class IV and M = class V. On image C, E = surface of the embryo after the longitudinal section, and t = coat of the seeds. Scale of 1 cm.

Attributes of the physiological quality of seeds without or after drying: seeds of grumixameira, pitangueira and uvaieira, after removal of the control sample (without drying), were subjected to three levels of controlled drying in an oven. Drying was carried out intermittently, with 10 hours at 40 °C followed by 14 hours of standing at 20-25 °C, termed mild drying, intermediate drying and severe drying, trying to bring the seeds to target water content in 45%, 40% and 35% of a wet basis, namely, when the vigor and germination are affected (Delgado and Barbedo, 2007). To do so, periodically, based on the value of the dry weight of the seeds, samples were taken and assessed for water content, water potential, germination test, electrical conductivity test, as described above, and tetrazolium test.

The tetrazolium test was conducted based on the best staining results at 35 °C in the dark, obtained from the previous experiment, which were: a concentration of 0.250% of the tetrazolium solution for 3 hours of incubation for seeds of grumixameira; a concentration of 0.125% for 3 hours of incubation for seeds of pitangueira; and a concentration of 0.100% for 2 hours of incubation for seeds of uvaieira. The assessments were performed as previously described.

Classes for assessment of viability: after setting the best combination of concentration of tetrazolium solution, incubation time for staining and assessments of seeds with different drying levels, classes for viability assessment (Figure 1) were defined according to the tissue color of the embryo. For that, each seed was classified as viable (able to issue primary root and/or produce normal seedling) and non viable (without seed ability to issue root, but with living tissues and/or seeds without living tissues).

The classes provided were: classes I and II (viable seeds). Class I – embryos with uniform a rosy hue staining; Class II – areas with intense red or missing color (less than 50% of the total surface of the embryo) or embryo ends with missing color, but the central area in a rosy color. Classes III, IV and V (non viable seeds). Class III – embryos with intense red staining in its entirety; Class IV – wide areas in intense red or missing color (over 50% of the total surface of the embryo), including the most central area; Class V – completely discolored embryos.

Experimental design and statistical analyses: the experimental design was completely randomized in a 3 x 3 factorial diagram (solution concentration versus incubation period), with four replicates of 15 seeds. In the results of physiological quality the drying levels for each species were compared (with four replicates of 15 seeds). The data obtained were subjected to analysis of variance (F test) at 5% significance level. Where relevant, the averages were

compared by Tukey test, also at the level of 5%. Subsequently, the simple correlation coefficients between physical and physiological data were calculated for seeds of grumixameira, pitangueira and uvaieira, without or after three drying levels and the significance was determined by the t test at 5% probability (Santana and Ranal, 2004).

Results and Discussion

Concentration of tetrazolium solution and staining period: the analysis of variance showed a significant interaction between factors solution concentration and incubation period for the data of viable seeds obtained by the tetrazolium test (Table 1). It is noticed that the combinations that favored better analysis and visualization of damage to the surface of the embryo (Figure 1), as well as that neared the results of the germination test (Tables 1 and 2), were: 0.250% of the tetrazolium solution for 3 hours of incubation for seeds of grumixameira, 0.125% for 3 hours for seeds of pitangueira and 0.100% for 2 hours for seeds of uvaieira (Table 1), considered the ideal combinations for each species, all at a constant temperature of 35 °C and in the dark.

Table 1. Viable seeds (%) of grumixameira, pitangueira and uvaieira, analyzed by the tetrazolium test, due to the variation in staining period and the concentration of salt.

Species	Staining time	Concentration of the solution		
		0.100 %	0.125 %	0.250 %
Grumixameira	2 hours	50 aB*	53 aB	70 bA
	3 hours	35 bC	55 aB	85 aA
	6 hours	55 aA	65 aA	28 cB
C.V. (%)			15.05	
Pitangueira	2 hours	33 aB	68 aA	58 aA
	3 hours	45 aB	83 aA	15 bC
	6 hours	40 aA	10 bB	0 bB
C.V. (%)			26.65	
Uvaieira	2 hours	83 aA	45 aB	30 aB
	3 hours	53 bA	53 aA	0 bB
	6 hours	40 bA	5 bB	0 bB
C.V. (%)			27.74	

*Means followed by the same letter (lowercase in columns, uppercase in rows) do not differ by Tukey test at 5 %.

As the combinations (concentration and period) were distanced from the ideal, there was a decrease of seeds classified as viable (Table 1), since when the concentration and the period decreased, embryo development with weaker staining color or missing color increased (Figure 1M), and when the concentration and the period increased, embryo development with stronger coloration, namely, intense red, increased (Figure 1H). For example, for grumixameira, the combinations of 0.100% or 0.125% for 2 hours developed a very light rosy staining or missing color and the combinations of 0.100%, 0.125% or 0.250% for 6 hours developed an intense red staining, combinations that hampered assessments and classified a smaller amount of viable embryos (Table 1).

Although the species are of the same genus, it is observed that they differ among the ideal combinations for using the tetrazolium test. Study undertaken by Masetto et al. (2009), for "pitanga-do-mato" (*Eugenia pleurantha*), namely, a species of the same genus, demonstrated that the optimal combination of the tetrazolium test (0.100% for 4 hours) also differed from this study. The differences were even more pronounced for species of different genera but of the same family (Myrtaceae) from the ones of the present study, as seen by Hössel et al. (2013) for jabuticabeira (1.000% for 24 hours) and by Sarmento et al. (2013) for pineapple guava (0.500% for 4 hours). Other studies also revealed differences between the species, as seen by Lamarca et al. (2009) for seeds of brazilwood

(0.050% for 2 hours), by Ribeiro et al. (2010) for seeds of macaúba (0.500% for 4 hours) and by Carvalho et al. (2014) for seeds of sorghum (0.100% for 3 hours). Such variations emphasize that for the use of the tetrazolium test in seeds there is the need to adapt the methodology for each species.

Attributes of the physiological quality of seeds without or after drying: the results of the physical determinations and the attributes of the seed physiological quality without drying or after three levels of drying are shown in Table 2. It is noticed that the water content decreased as drying increased and water potential became more negative. It is observed that the decrease in water content influenced the seed physiological quality (Table 2), including significant correlations between water content and physiological attributes (Table 3), thus constituting three different drying levels (mild, intermediate and severe) and therefore different levels of physiological characteristics (Table 2). Studies with desiccation intolerant seeds demonstrate the efficiency of the use of controlled drying for obtaining seeds with different physiological quality, such as those conducted by Delgado and Barbedo (2007), Bonjovani and Barbedo (2008), Lamarca et al. (2011) and Ntuli et al. (2014).

Table 2. Water content (WC), water potential (PΨ), germinable seeds (GS), germination (G), index of germination speed (IVG), mean germination time (GT), first count (FC), electrical conductivity test (EC) and viable seed by the tetrazolium test (VSTZ) of seeds of grumixameira, pitangueira and uvaieira, without drying and after three levels of drying.

Level of drying	WC (%)	PΨ (-MPa)	GS (%)	G (%)	IVG	GT	FC (%)	EC (µS.cm⁻¹.g⁻¹)	VSTZ (%)
Grumixameira									
Without drying	46.16 a*	1.40 c	98 a	94 a	1.17 a	12.70 c	46 a	8.90 b	85 a
Mild	45.47 b	4.56 b	98 a	92 a	1.15 a	12.62 c	39 a	18.66 ab	77 ab
Intermediate	42.44 c	7.75 a	92 ab	88 a	0.69 b	20.48 b	17 b	20.17 ab	69 b
Severe	40.41 d	8.04 a	65 b	56 b	0.27 c	30.27 a	0 b	25.49 a	15 c
C.V. (%)	2.72	8.86	15.19	17.23	13.19	13.68	37.64	39.58	10.78
Pitangueira									
Without drying	59.62 a	0.86 c	100 a	100 a	0.76 b	15.60 c	66 a	4.56 b	83 a
Mild	56.98 a	1.53 c	97 a	90 a	1.04 a	17.86 c	75 a	7.15 b	75 a
Intermediate	48.95 b	4.45 b	95 a	92 a	0.63 c	26.18 b	18 b	5.29 b	50 b
Severe	37.45 c	11.32 a	58 b	37 b	0.20 d	44.91 a	0 c	20.84 a	30 c
C.V. (%)	2.89	20.98	8.99	9.73	8.43	6.01	18.99	47.31	11.28
Uvaieira									
Without drying	60.47 a	0.95 c	100 a	98 a	0.42 b	26.35 b	50 b	4.58 b	83 a
Mild	49.42 b	6.42 b	98 a	95 a	0.53 a	19.40 b	85 a	3.40 b	85 a
Intermediate	39.85 c	10.44 a	35 b	18 b	0.09 c	40.33 a	3 c	5.49 b	45 b
Severe	35.28 d	11.38 a	33 b	18 b	0.08 c	44.02 a	0 c	17.63 a	18 c
C.V. (%)	3.41	14.50	16.59	17.02	18.86	12.71	37.32	39.72	9.39

*Means followed by the same letter do not differ among themselves by the Tukey test at 5 %.

As for the physiological attributes, GS, G, IVG, FC and viable seed by the tetrazolium test (VSTZ) decreased with increasing drying and GT and EC increased (Table 2),

observing significant differences from the intermediate drying. When comparing the seeds with drying and the ones with intermediate drying, for grumixameira and pitangueira it is verified that GS and G did not significantly differ, but VSTZ did, as well as IVG, GT and FC. As for uvaieira, GS and G, as well as IVG, GT, FC and VSTZ differed between the seeds without drying and the ones with intermediate drying (Table 2). When comparing the seeds with intermediate drying and the ones with severe drying, it is verified that for grumixameira there was a difference for G, IVG, GT and VSTZ, for pitangueira there was a difference for GS, G, IVG, GT, EC and VSTZ and for uvaieira there was a difference for EC and VSTZ (Table 2). Damage by drying, observed by the tetrazolium test, were characteristic in seeds of grumixameira, pitangueira and uvaeira. It was observed that the tissues of the embryos will lose vitality inwardly. Figure 1G is a representation of damage by drying, in which the end of the embryo presents color missing (dead tissue), but the central region in a rosy color (alive and healthy tissue).

Table 3. Simple correlation coefficients (r) between physical and physiological seed data of grumixameira, pitangueira and uvaieira, without or after three levels of drying.

	PΨ	GS	G	IVG	GT	FC	EC	VSTZ
WC	-0.92*	0.78*	0.79*	0.55	-0.66*	0.82*	-0.71*	0.80*
PΨ		-0.85*	-0.86*	-0.75*	0.81*	-0.75*	0.55	-0.79*
GS			0.99*	0.80*	-0.89*	0.75*	-0.37	0.83*
G				0.78*	0.90*	0.75*	-0.39	0.83*
IVG					-0.91*	0.62*	-0.19	0.73*
GT						-0.74*	0.30	-0.83*
FC							-0.64*	0.85*
EC								-0.62*

WC = water content, PΨ = water potential, GS = germinable seeds, G = germination, IVG = index of germination speed, GT = mean germination time, FC = first count, EC = electrical conductivity test and VSTZ = viable seed by the tetrazolium test. (*) = r significant at 5 % of probability, without asterisk = non significant correlations.

The amount of physiological attributes that differ between the drying levels for each species must be associated with desiccation tolerance, since studies have reported differences in the degree of desiccation tolerance among seeds of grumixameira, pitangueira and uvaieira (Delgado and Barbedo, 2007). It is verified that the tetrazolium test proves to be consistent to analyze changes in the physiological quality of seeds of these three species. When changes in physiological quality were not identified by the final values of GS and G, but by IVG, GT and FC, the tetrazolium test (VSTZ) had also identified (Table 2). It is also observed that VSTZ showed a significant correlation with all the variables obtained by the germination test (GS, G, IVG, GT and FC) and also with the electrical conductivity test (EC), something that had occurred for the germination test only with the variable FC (Table 3).

The loss of seed physiological quality occurs gradually due to the deterioration process and may be of biochemical, physical, physiological or genetic origins, presented as early events in damage to the membrane systems, followed by a decrease of vigor until loss of ability to germinate (Marcos-Filho, 2005). Therefore, in the present study, the tetrazolium test could identify early events of loss of seed physiological quality, something which was not possible for the end result of the germination test. Also due to the positive correlation between the tetrazolium test and conductivity test (Table 3), this one used to measure the permeability damage to membrane systems (Barbedo and Cícero, 1998).

Several studies show the efficiency of the tetrazolium test to diagnose the state of physiological quality of seeds, as observed by Lamarca et al. (2009) for seeds of brazilwood, subjected to accelerated aging, by Rego et al. (2013) for seeds of guaçatonga (*Casearia sylvestris* Sw.) and "murta" (*Myrtus* L.), subjected to controlled drying, by Azerêdo et al. (2011) for seeds of "angico-de-bezerro", by Hössel et al. (2013) for seeds of jabuticabeira and by Sarmento et al. (2013) for seeds of pineapple guava, subjected to different storage conditions and times.

Finally, the results of this study showed that, since the tetrazolium solution concentration and staining the time are correctly ajusted, it is possible to establish the categories of seed viability for grumixameria, pitangueira and uveira. Thus, tetrazolium test shows up as an important tool for a quick and efficient diagnosis of the real state of physiological quality of these seeds, with or without controlled drying.

Conclusions

The tetrazolium test conducted on the solution concentration for the period of staining, at 0.250% for 3 hours for seeds of *E. brasiliensis*, at 0.125% for 3 hours for the seeds of *E. uniflora* and at 0.100% for 2 hours for the seeds of *E. pytiformis* at constant room temperature of 35 °C, proves efficient for assessing the physiological quality of these seeds.

Acknowledgments

The authors thank the Municipality of Santo André and the Institute of Botany for authorizing the collection of plant material; also CAPES (Coordenação de Aperfeiçoamento de Pessoal de Nível Superior) for the postgraduate scholarship awarded to E.V. Lamarca and CNPq (Conselho Nacional de Desenvolvimento Científico e Tecnológico – National Counsel of Technological and Scientific Development) the fellowship awarded to C.J. Barbedo.

References

AMADOR, T.S.; BARBEDO, C.J. Potencial de inibição da regeneração de raízes e plântulas em sementes germinantes de *Eugenia pyriformis*. *Pesquisa Agropecuária Brasileira*, v.46, n.8, p.814-821, 2011. http://www.scielo.br/pdf/pab/v46n8/05.pdf

AZERÊDO, G.A.; PAULA, R.C.; VALERI, S.V. Viabilidade de sementes de *Piptadenia moniliformis* Benth. pelo teste de tetrazólio. *Revista Brasileira de Sementes*, v.33, n.1 p.061-068, 2011. http://www.scielo.br/pdf/rbs/v33n1/07.pdf

BARBEDO, C.J.; CICERO, S.M. Utilização do teste de condutividade elétrica para previsão do potencial germinativo de sementes de ingá. *Scientia Agricola*, v.55, n.2, p.249-259, 1998. http://www.scielo.br/scielo.php?script=sci_arttext&pid=S0103-90161998000200013

BONJOVANI, M.R.; BARBEDO, C.J. Sementes recalcitrantes: intolerantes a baixas temperaturas? embriões recalcitrantes de *Inga vera* Willd. subsp. *affinis* (DC.) T. D. Penn. toleram temperatura sub-zero. *Revista Brasileira de Botânica*, v.31, n.2, p.345-356, 2008. http://www.scielo.br/pdf/rbb/v31n2/v31n2a17.pdf

BORGHETTI, F.; FERREIRA, A.G. Interpretação de resultados de germinação. In: FERREIRA, A.G.; BORGHETTI, F. (Eds.). *Germinação: do básico ao aplicado*. Porto Alegre: Artmed, 2004. p.209-222.

BRASIL. Ministério da Agricultura, Pecuária e Abastecimento. *Regras para análise de sementes*. Ministério da Agricultura, Pecuária e Abastecimento. Secretaria de Defesa Agropecuária. Brasília: MAPA/ACS, 2009. 395 p. http://www.agricultura.gov.br/arq_editor/file/2946_regras_analise__sementes.pdf.

CARVALHO, T.C.; GRZYBOWSKI, C.R.S.; OHLSON, O.C.; PANOBIANCO, M. Adaptation of the tetrazolium test method for estimating the viability of sorghum seeds. *Journal of Seed Science*, v.36, n.1, p.246-252, 2014.http://www.scielo.br/pdf/jss/v36n2/v36n2a14.pdf

DAWS, M.I.; LYDALL, E.; CHMIELARZ, P.; LEPRINCE, O.; MATTHEWS, S.; THANOS, C.A.; PRITCHARD, H.W. Developmental heat sum influences recalcitrant seed traits in *Aesculus hippocastanum* across Europe. *New Phytologist*, v.162, n.1, p.157-166, 2004. http://onlinelibrary.wiley.com/doi/10.1111/j.1469-8137.2004.01012.x/pdf

DELGADO, L.F.; BARBEDO, C.J. Tolerância à dessecação de sementes de espécies de *Eugenia*. *Pesquisa Agropecuária Brasileira*, v.42, n.2, p.265-272, 2007. http://www.scielo.br/pdf/pab/v42n2/16.pdf

FERREIRA, R.A.; OLIVEIRA, L.M.; TONETTI, O.A.O.; DAVIDE, A.C. Comparação da viabilidade de sementes de *Schizolobiumparahyba*(Vell.) Blake – Leguminosae Caesalpinioideae, pelos testes de germinação e tetrazólio. *Revista Brasileira de Sementes*, v.29, n.3, p.73-79, 2007. http://www.scielo.br/pdf/rbs/v29n3/a11v29n3.pdf

FRANÇA-NETO, J.B. O teste de tetrazólio em sementes de soja. In: VIEIRA, R.D.; CARVALHO, N.M. (Ed.). *Testes de vigor em sementes*. Jaboticabal: FUNEP, 1994. p.87-102.

HÖSSEL, C.; OLIVEIRA, J.S.M.A.; FABIANE, K.C.; WAGNER JÚNIOR, A.; CITADIN, I. Conservação e teste de tetrazólio em sementes de jabuticabeira. *Revista Brasileira de Fruticultura*, v.35, n.1, p.255-261, 2013. http://www.scielo.br/pdf/rbf/v35n1/29.pdf

LAMARCA, E.V.; LEDUC, S.N.M.; BARBEDO, C.J. Viabilidade e vigor de sementes de *Caesalpinia echinata* Lam. (pau-brasil – *Leguminosae*) pelo teste de tetrazólio. *Revista Brasileira de Botânica*, v.32, n.4, p. 793-803, 2009. http://www.scielo.br/pdf/rbb/v32n4/a17v32n4.pdf

LAMARCA, E.V.; SILVA, C.V.; BARBEDO, C.J. Limites térmicos para a germinação em função da origem de sementes de espécies de *Eugenia* (Myrtaceae) nativas do Brasil. *Acta Botanica Brasilica*, v.25, n.2, p. 293-300, 2011. http://www.scielo.br/pdf/abb/v25n2/a05v25n2.pdf

LAMARCA, E.V.; BAPTISTA, W.; RODRIGUES, D.S.; OLIVEIRA JÚNIOR, C.J.F. Contribuições do conhecimento local sobre o uso de *Eugenia* spp. em sistemas de policultivos e agroflorestas. *Revista Brasileira de Agroecologia*, v.8, n.3, p.119-130, 2013. http://www.aba-agroecologia.org.br/revistas/index.php/rbagroecologia/article/view/13256/9905.

MAGUIRE, J. D. Speed of germination-aid in selection and evaluation for seedling emergence and vigor. *Crop Science*, v.2, n.2, p. 176-177, 1962.

MARCOS-FILHO, J. *Fisiologia de sementes de plantas cultivadas*. Piracicaba: FEALQ. 2005. 495p.

MARTINI NETO, N.; LAMARCA, E.V.; BARBEDO, C.J. Kinetics of solute leachate from imbibing *Caesalpinia echinata* Lam. (Brazilwood) seeds. *Revista Ceres*, v.61, n.1, p.90-97, 2014. http://www.scielo.br/pdf/rceres/v61n1/v61n1a12.pdf

MASETTO, T.E.; DAVIDE, A.C.; FARIA, J.M.R.; SILVA, E.A.A.; REZENDE, R.K.S. Avaliação da qualidade de sementes de *Eugenia pleurantha* (MYRTACEAE) pelos testes de germinação e tetrazólio. *Agrarian*, v.2, n.5, p.33-46, 2009. http://www.periodicos.ufgd.edu.br/index.php/agrarian/article/view/802/482

NTULI, T.M.; BERJAK, P.; PAMMENTER, N.W. Tissue diversity in respiratory metabolism and free radical processes in embryonic axes on the white mangrove (*Avicennia marina* L.) during drying and wet storage. *African Journal of Biotechnology*, v.13, n.17, p.1813-1823, 2014. http://www.academicjournals.org/article/article1398764350_Ntuli%20et%20al.pdf

PEEL, M.C.; FINLAYSON, B.L.; MCMAHON, T.A. Updated world map of the Köppen-Geiger climate classification. *Hydrology and Earth System Sciences*, v.11, p.1633-1644, 2007. http://www.hydrol-earth-syst-sci.net/11/1633/2007/hess-11-1633-2007.pdf

REGO, S.S.; NOGUEIRA, A.C.; MEDEIROS, A.C.S.; PETKOWICZ, C.L.O.; SANTOS, A.F. Physiological behaviour of *Blepharocalyx salicifolius* and *Casearia decandra* seeds on the tolerance to dehydration. *Journal of Seed Science*, v.35, n.3, p.323-330, 2013.http://www.scielo.br/pdf/jss/v35n3/08.pdf

RIBEIRO, L.M.; GARCIA, Q.S.; OLIVEIRA, D.M.T.; NEVES, S.C. Critérios para o teste de tetrazólio na estimativa do potencial germinativo em macaúba. *Pesquisa Agropecuária Brasileira,* v.45, n.4, p.361-368, 2010. http://www.scielo.br/pdf/pab/v45n4/a03v45n4

SANTANA, D.G.; RANAL, M.A. *Análise da germinação: um enfoque estatístico.* Universidade de Brasília: Brasília, 2004. 248p.

SARMENTO, M.B.; SILVA, A.C.S.; VILLELA, F.A.; SANTOS, K.L.; MATTOS, L.C.P. Teste de tetrazólio para avaliação da qualidade fisiológica em sementes de goiabeira-serrana (*Acca sellowiana* O. Berg Burret). *Revista Brasileira de Fruticultura,* v.35, n.1, p.270-276, 2013. http://www.scielo.br/pdf/rbf/v35n1/31.pdf

SILVA, R.C.; GRZYBOWSKI, C.R.S.; FRANÇA-NETO, J.B.; PANOBIANCO, M. Adaptação do teste de tetrazólio para avaliação da viabilidade e vigor em sementes de girassol. *Pesquisa Agropecuária Brasileira,* v.48, n.1, p.105-113, 2013. http://www.scielo.br/pdf/pab/v48n1/14.pdf

VIEIRA, R.D. Teste de condutividade elétrica. In: VIEIRA, R.D.; CARVALHO, N.M. (Eds.). *Testes de vigor emsementes.*Jaboticabal: FUNEP, 1994. p.103-125.

Physiological maturation of cowpea seeds

Narjara Walessa Nogueira[1]*, Rômulo Magno Oliveira de Freitas[1], Salvador Barros Torres[1], Caio César Pereira Leal[1]

ABSTRACT - The seed maturation process is genetically controlled and involves an arranged sequence of morphological and physiological changes extending from fertilization to its total independence from the mother-plant. These changes also include a set of preparatory phases for the germination process, which are characterized for the synthesis and accumulation of nutrient reserves. Thereby, this study was developed aiming at assessing development and physiological quality of cowpea seeds during maturation process. To this, the cowpea pods of cultivar BRS-Guariba were harvested from the tenth day after anthesis (DAA) until the twenty sixth DAA, with four days intervals. Immediately after each harvest, seeds were manually extracted from the pods and then subjected to the following determinations: moisture content, first count of germination, final germination percentage, length of shoots and roots, hypocotyl diameter, and seedling dry mass. The experiment was conducted in a completely randomized design, with five treatments (DAA), and four replications to each treatment. Results have shown that cowpea seeds have fairly fast physiological maturation, and that seeds harvested between 14 and 18 DAA have better vigor as well as higher germination rates; thus the harvest performed during this period does not cause damages to seeds.

Index terms: *Vigna unguiculata* L., physiological quality, vigor.

Maturação fisiológica de sementes de feijão-caupi

RESUMO - A maturação da semente é controlada geneticamente, envolvendo uma sequência ordenada de alterações verificadas a partir da fecundação até que se tornem independentes da planta-mãe. Essas alterações compreendem um conjunto de etapas que preparam para o sucesso da futura germinação, caracterizada pela síntese e acúmulo de reservas. Dessa forma, este estudo teve por objetivo avaliar o desenvolvimento e qualidade fisiológica de sementes de feijão-caupi durante o processo de maturação. Para isso, vagens da cultivar BRS-Guariba de feijão-caupi foram colhidas a partir do décimo dia após a antese (DAA) até o vigésimo sexto DAA, com intervalos de quatro dias. Após cada coleta, as sementes foram manualmente extraídas das vagens e submetidas às seguintes determinações: grau de umidade, primeira contagem de germinação, porcentagem final de germinação, comprimento de parte aérea e raízes, diâmetro do hipocótilo e massa seca das plântulas. O experimento foi conduzido em delineamento experimental inteiramente casualizado, com cinco tratamentos (DAA) e quatro repetições cada. As sementes de feijão-caupi apresentam maturação fisiológica bastante rápida e aquelas colhidas entre 14 e 18 DAA têm melhor vigor e maior porcentagem de germinação; assim, colheita deve ser realizada durante esse período, pois não causa danos às sementes.

Termos para indexação: *Vigna unguiculata* L., qualidade fisiológica, vigor.

Introduction

The Northeast Brazil is a tropical region, which has about 1.6 million square kilometers of areas with semi-arid climate, where more than 30 million people live, and where there is predominance of rainfed agriculture using subsistence crops as the cowpea [*Vigna unguiculata* (L.) Walp.]. This crop has a significant socioeconomic importance to the population of the Northeast Region, since it is used both as food supplement as well as for the establishment of hand labor in the field, and is also the main component of the agricultural production of the northeastern farmers (Dutra et al., 2007; Bezerra et al., 2008; Rocha et al., 2009; Silva et al., 2010).

Although such crop has a broad adaptation to different growing conditions, the fields cropped with cowpea still have low productivity, with a world mean productivity of 370 kg.ha^{-1} (FAO, 2009). Nevertheless, Brazil annually produces about 480 thousand metric tons of cowpea, grown on 1.3 million

[1]Laboratório de Análise de Sementes, Departamento de Fitotecnia, UFERSA, 59625-900 - Mossoró, RN, Brasil.
*Corresponding author < narjarawalessa@yahoo.com.br>

hectares, with an approximate mean productivity of only 360 kg.ha^{-1}. Such low productivity is mainly due to the low technological level of producers of this leguminous and the use of seeds with low physiological quality in their cowpea crops (Freire-Filho et al., 2005; Cardoso and Ribeiro, 2006).

The seed is the main input in production systems of the major crops; and the high physiological quality of these seeds is the main factor responsible by the initial stand and by vigor of plants in the field, and consequently by the high crop productivity. However, it is essential that seed has reached the full physiological maturity before being harvested to be able to express all its germination potential. Hence, obtaining high physiological quality seeds will depend on the ideal harvest moment; once such moment should correspond to that in which the physiological maturity is reached and the seed has the maximum dry mass accumulation, high vigor, and high potential for germination (Carvalho and Nakagawa, 2000).

The seed maturation process is genetically controlled and involves an arranged sequence of morphological and physiological changes extending from fertilization to its total independence from the mother-plant. These changes also include a set of preparatory phases for the germination process, which are characterized for the synthesis and accumulation of nutrient reserves (Marcos-Filho, 2005).

The literature on this subject presents several studies on the influence of the physiological maturity, as well as seed quality on productivity of several important crops. As examples, it can be cited the studies performed by David et al. (2003), Faria et al. (2005), and Araújo et al. (2006) with seeds of maize (*Zea mays L.*); Botelho et al. (2010) and Bolina (2012) with seeds of common bean (*Phaseolus vulgaris* L.) and Eskandari (2012) with seeds of southern pea (*Vigna sinensis* (L.). However, information on physiological maturity and ideal harvest time of cowpea seeds are still scarce, especially for the cv. BRS-Guariba, which is widely grown in the crop fields of Northeast Region of Brazil.

Thereby, this study was developed aiming at assessing development and physiological quality of cowpea seeds during maturation process.

Material and Methods

Seeds of cv. BRS-Guariba of cowpea used in this study were produced in a seed production field installed in the didactic vegetable garden of the Department of Plant Sciences, Federal Rural University of the Semi-Arid, located in the municipality of Mossoró, state of Rio Grande do Norte, Northeastern Region of Brazil, and collected between September and December, 2011. Municipality of Mossoró

is located between the geographic coordinates 5°11'S and 37°20'W, at an altitude of 18 m.

The Figure 1 shows the climatological data (air temperature and relative humidity) observed during the experiment. During the research there was only one rain of 5 mm at 31 days after sowing (DAS).

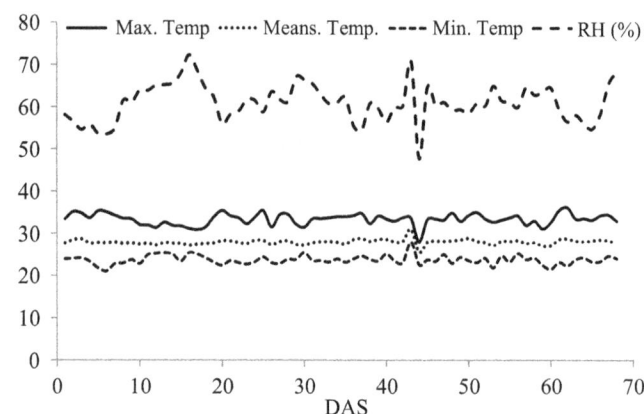

Figure 1. Temperatures minimal, means and maximal (°C) and relative humidity (RH) recorded at the Meteorological.

The soil samples collected within the experimental area were subjected to chemical analysis, which revealed the following fertility level of soil area: pH (water) = 6.1; organic matter content = 10.1 g.kg^{-1}; P = 220.4 mg.dm^{-3}; K^{+} = 157.3 mg.dm^{-3}; Ca^{2+} = 3.65 cmolc.dm^{-3}; Mg^{2+} = 1.0 cmolc.dm^{-3}; and Al^{3+} = 0.15 cmolc.dm^{-3}.

The soil of experimental area was tilled with a single plowing followed by two disking; and the sowing was performed with a spacing of 0.3 m x 0.5 m, with aid of a manually operated seeder for the simultaneous procedures of seeding and fertilizer application, set to sow from two to four seeds in each of the planting pits together with an adequate fertilizer amount. On the tenth day after sowing a thinning was made on the plants emerged in each of the planting pits; thereby keeping only one plant per planting pit.

Water requirement was daily monitored by using a set of soil tensiometers installed at 20 cm deep; and whenever needed irrigation was applied by a dripping system, with flow regulated to 1.7 L .h^{-1}. According to the need, all other cultural practices were also performed.

During flowering, which has approximately started 35 DAS, and throughout the crop cycle, always in the morning period, the blooming flowers were marked with colored ribbons to identify the anthesis date. Harvesting of pods was performed according to the harvesting periods pre-established to the treatments (10,

14, 18, 22 and 26 DAA); and soon after harvesting, such pods were taken to the Seed Testing Laboratory, where seeds were immediately extracted from the pods.

The seed quality was assessed by the following tests: *Seed moisture content* - this test was performed with two seed subsamples, containing 10 grams each, to each treatment. The moisture content of the seeds was determined by the oven method, at 105±3 °C, for 24 h (Brasil, 2009), and results were expressed as percentage (wet basis); *Germination* - was conducted with four replications, containing 50 seeds each, to each treatment. To this, the seeds were evenly distributed onto two towel paper sheets, moistened with distilled water, in a volume equivalent to 2.5 times the weight of the dry substrate, and covered with another sheet of same paper likewise moistened. Thereupon, the set (seed + towel paper) was made into rolls, which were then placed in a vertical position into a seed germinator, set at 30 °C constant temperature, and 8/14 h (L/D) photoperiod. Assessments were performed at the eighth DAS by counting the number of normal emerged seedlings; and results were expressed on percentage germination (Brasil, 2009); *First count of germination* - the test was performed during the standard germination test; but, percentage of normal emerged seedlings was determined at the fifth DAS (Brasil, 2009); *Germination speed* - this test was also conducted during germination test; however, the normal emerged seedlings were daily counted until the eighth DAS; and the mean number of days for germination was computed by the equation proposed by Edmond and Drapala (1958).

To the normal emerged seedlings were also carried out the following determinations: *Root length* - for this determination, the normal emerged seedlings were measured from the base of hypocotyl until the tip of main root, with aid of a ruler graduated in centimeters. Such determination was performed on all normal emerged seedlings, in all replications of all treatments; and results were expressed as cm.seedling^{-1}; *Shoot length* - on this determination the normal emerged seedlings were measured from the hypocotyl base to the apex of the seedling, also with aid of a ruler graduated in centimeters. This determination was also performed on all normal seedlings, in all replications of all treatments; and results were also expressed on cm.seedling^{-1}; *Hypocotyl diameter* - these data were obtained by measuring the hypocotyl basis of the seedling at soil line, with aid of a digital caliper, and carried out on all normal emerged seedlings, in all replicates of each treatment. Results were expressed as mm.seedling^{-1}; and *Seedling dry mass* - for this determination, all the normal emerged seedlings at the experimental unit were placed into a forced hot air circulating drying oven, set at 65 °C, until reach constant weight and immediately weighed on an analytical balance with 0.001 g precision. Results were expressed as g.seedling^{-1}.

The experiment was carried out on a completely randomized design, with five treatments (10, 14, 18, 22 and 26 DAA), and four replicates per treatment. Data were previously subjected to exploratory analysis to assess the assumptions of normality of residuals, homogeneity of variance of treatments, and additivity of the model, before applying the ANOVA. The variables analyzed were then subjected to regression analysis, and adjustments of the regression polynomial curves were computed as function of the age of pods.

Results and Discussion

Results obtained in this experiment showed that, depending on the harvesting date of the cowpea pods there were significant differences (p <0.01) between the means achieved for germination, first count of germination, and germination speed as well as for length of shoots and roots, hypocotyl diameter, and seedling dry mass.

The mean values obtained for seed moisture content presented a decreasing behavior over time; (Figure 1A); from the first harvest date (10 DAA - 87.2%) there was a sharp drop on mean values until the second (14 DAA - 53.6%) and third (18 DAA - 14%) harvest dates, and after that date the values slowly decreased until last date assessed (26 DAA - 12 %). Thus, the higher moisture content, and hence the greater accumulation of dry seeds, occurred in the last harvest, at 26 DAA.

Gradual reduction on moisture content throughout seed maturation process has also been observed in mayze (David et al., 2003; Faria et al., 2005; Araújo et al., 2006), common bean (Botelho et al., 2010; Bolina, 2012) and Southern pea (Eskandari, 2012).

The polynomial linear regression curve presented on Figure 2B shows that the cowpea seeds harvested at 10 DAA did not germinate and germination percentage of seeds harvested at 14 DAA reached 96%. It also shows that, from the third harvest (18 DAA - 92%) until the last harvest, the germination percentage remained high and stable with mean percentages of 95% and 93%, respectively for the seeds harvested at 22 DAA and 26 DAA.

The highest percentage of germination was observed between the second and third collection (14 DAA and 18 DAA) in the same period, also were observed the highest values for the first count of germination (Figure 2C), root length (Figure 3A), shoot length (Figure 3B) and seedling dry mass (Figure 3D).

The higher vigor, indicated by the higher percentage and first count of germination, greater roots and shoots lengths and increased seedling dry matter obtained between 14 and 18 DAA coincides with the point of highest physiological seed quality, the point of physiological maturity.

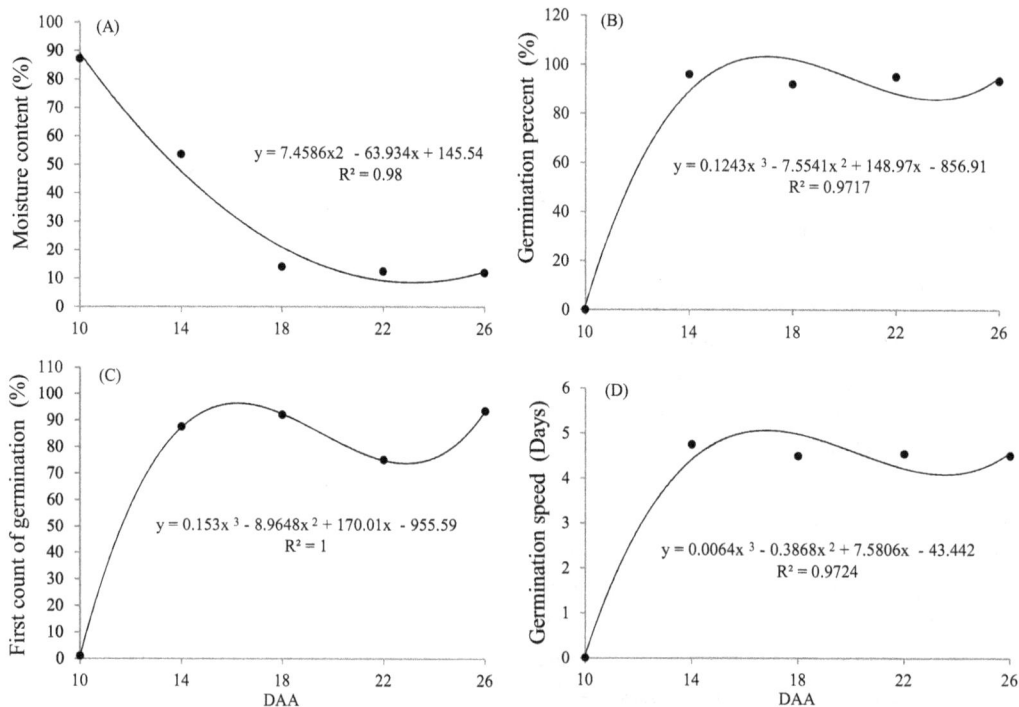

Figure 2. Graphical representation of the polynomial linear regression curves plotted to mean values obtained for moisture content (A), germination percentage (B), first count of germination (C), germination speed (D) of cowpea seeds, cv. BRS-Guariba, harvested at different times (days after anthesis-DAA).

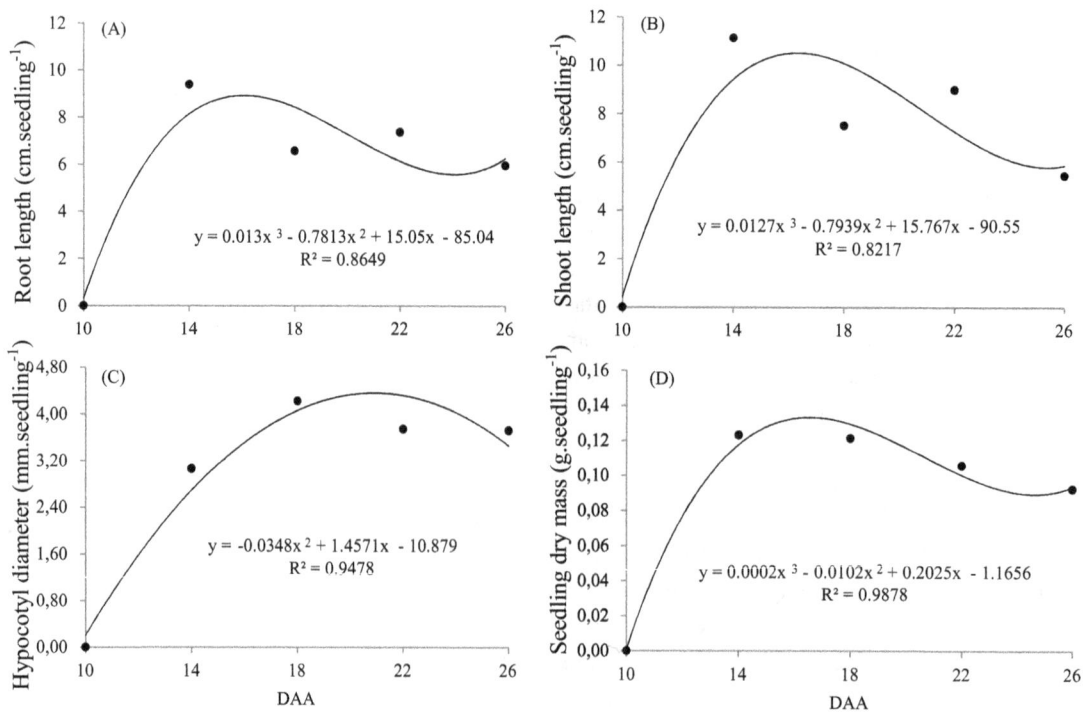

Figure 3. Graphical representation of the polynomial linear regression curves plotted to mean values obtained for root length (A), shoot length (B), hypocotyl diameter (C) and dry mass (D) of cowpea seedlings, cv. BRS-Guariba, as function of days after anthesis (DAA) in which seeds were harvested.

The seeds harvested at 10 DAA did not germinate probably due to presence of inhibitors, such as abscisic acid, which according Kermode (2005), when in high concentrations in the seed may prevent its early germination; thus favoring both the reserves accumulation as well as the embryo maturation. Similarly, Braga-Júnior (2009) in his work with castor bean seeds has verified that, to the seeds harvested at 10 DAA the germination was also nil, starting only after 21 DAA. Moreover, this author also observed that seed germination ranged between 86% and 91%, when seeds were harvested at 28 DAA, reaching a maximum value of 91.2% at 56 DAA.

Stabilization of germination percentage observed in this study on cowpea seeds has been also observed by David et al. (2003) with popcorn seeds [*Zea mays* L. var. *everta* (Sturtev.) L. H. Bailey], obtaining 77% germination for the seeds harvested at 30 DAA, and 99% germination for the seeds harvested at 44 DAA.

From the maturity point the germination percentage remained stable until the end of the studied period (93 DAA). Likewise, Araújo et al. (2006), in a study on physiological maturity of seeds of sweet corn [*Zea mays* L. var. *saccharata* (Sturtev.) L. H. Bailey] found that the seeds harvested at 27 DAA had germination of 66% reaching 93% only when harvested at 34 DAA; from this maturation phase the seed germination remained stable.

The values obtained for first count of the germination test (65%) and of the standard germination test (80%) of the seeds harvested between 14 DAA and 26 DAA (Figures 2B and 2C, respectively) were higher than those values obtained by Teixeira et al. (2010) to this same cowpea cultivar.

The polynomial linear regression curve plotted with mean of the values obtained for hypocotyl diameter shows that there was a sharp increase until 18 DAA (4.22 cm.seedling⁻¹); from this harvesting date, there was a slowly decreased (3.74 cm.seedling⁻¹ at 22 DAA), and afterwards remained stable until the last assessment, which was performed at 26 DAA (3.72 cm.seedling⁻¹) (Figure 3C).

Conclusions

Seeds of cowpea cv. BRS-Guariba only reach physiological maturity at 14 days after anthesis (DAA).

The ideal point of physiological maturity of the seeds of cv. BRS-Guariba of cowpea is situated between 14 and 18 DAA.

The seeds of cowpea of cv. BRS Guariba can be harvested between 14 and 18 DAA without damage to their vigor, for it is during this period that physiological maturation of the seed reaches their maximum and hence the maximum germination potential.

References

ARAUJO, E. F.; ARAUJO, R. F.; SOFIATTI, V.; SILVA, R. F. Maturação de sementes de milho-doce - Grupo super doce. *Revista Brasileira de Sementes*, v.28, n.2, p. 69-76, 2006. http://www.scielo.br/pdf/rbs/v28n2/a09v28n2

BEZERRA, A. A. C.; TÁVORA, F. J. A. F.; FREIRE FILHO, F. R.; RIBEIRO, V. Q. Morfologia e produção de grãos em linhagens modernas de feijão-caupi submetidas a diferentes densidades populacionais. *Revista de Biologia e Ciências da Terra*, v.8, n.1, p.85-93, 2008. http://www.redalyc.org/articulo.oa?id=50080109

BOLINA, C. C. Maturação fisiológica da semente e determinação da época adequada de colheita do feijão (*Phaseolus vulgaris* L.). *Revista Científica Linkania Master*, v.2, n.3, 2012. http://linkania.org/master/article/view/78/50

BOTELHO, F. J. E.; GUIMARÃES, R. M., OLIVEIRA, J. A.; EVANGELISTA, J. R. E.; ELOI, T. A.; BALIZA, D. P. Desempenho fisiológico de sementes de feijão colhidas em diferentes períodos do desenvolvimento. *Ciência e Agrotecnologia*, v.34, n.4, p.900-907, 2010. http://www.scielo.br/scielo.php?script=sci_arttext&pid=S1413-70542010000400015

BRAGA-JUNIOR, J. M. Maturação, qualidade fisiológica e testes de vigor em sementes de mamona. UFP, Areias. 2009. 62p. http://www.livrosgratis.com.br/arquivos_livros/cp085203.pdf

BRASIL. Ministério da Agricultura, Pecuária e Abastecimento. *Regras para Análise de Sementes*. Ministério da Agricultura, Pecuária e Abastecimento. Secretaria de Defesa Agropecuária. Brasília: MAPA/ACS, 2009. 395p. http://www.agricultura.gov.br/arq_editor/file/2946_regras_analise__sementes.pdf

CARDOSO, M. J.; RIBEIRO, V. Q. Desempenho agronômico do feijão-caupi, cv. Rouxinol, em função de espaçamentos entre linhas e densidades de plantas sob regime de sequeiro. *Revista Ciência Agronômica*, v.37, n.1, p.102-105, 2006. http://www.redalyc.org/articulo.oa?id=195317425018

CARVALHO, N. M.; NAKAGAWA, J. *Sementes*: ciência, tecnologia e produção. 4 ed. Jaboticabal: FUNEP, 2000. 588 p.

DAVID, A. M. S. S.; ARAÚJO, E. F.; MIRANDA, G. V.; DIAS, D. C. F. S.; GALVÃO, J.C. C.; CARNEIRO, V. Maturação de sementes de milho-pipoca. *Revista Brasileira de Milho e Sorgo*, v.2, n.3, p. 121-131, 2003. http://rbms.cnpms.embrapa.br/index.php/ojs/article/view/78/79

DUTRA, A. S.; TEÓFILO, E. M.; MEDEIROS, FILHO, S.; DIAS, F. T. C. Qualidade fisiológica de sementes de feijão caupi em quatro regiões do estado do Ceará. *Revista Brasileira de Sementes*, v.29, n.2, p.111-116, 2007. http://www.scielo.br/scielo.php?pid=S0101-31222007000200015&script=sci_arttext

EDMOND, J. B.; DRAPALA, W. J. The effects of temperature, sand, soil, and acetone on germination of okra seeds. *Proceedings of the American Society for Horticultural Science*, v.71, p.428-434, 1958.

ESKANDARI, H. Seed quality changes in cowpea (*Vigna sinensis*) during seed development and maturation. *Seed Science and Technology*, v.40, n.1, p.108-112, 2012. http://www.ingentaconnect.com/content/ista/sst/2012/00000040/00000001/art00012

FARIA, M. A. V. R.; PINHO, R. G. V.; PINHO, E. V. R. V.; GUIMARÃES, R. M.; FREITAS, F. E. O. Qualidade de sementes de milho colhidas em diferentes estádios de maturação em duas épocas de produção. *Revista Ceres*, v.52, n.300, p.293-304, 2005. http://www.ceres.ufv.br/ceres/revistas/V52N300P02305.pdf

FAO - Food And Agriculture Organization Of The United Nations. 2009. *Crops*: Cow peas, dry. Fao Stat. Avaliable at: http://faostat.fao.org/site/567/ DesktopDefault.aspx?PageID=567#ancor Acessed on: Feb.10th, 2014.

FREIRE-FILHO, F. R.; RIBEIRO, V. Q.; BARRETO, P. D.; SANTOS, A. A. *Melhoramento Genético*. In: FREIRE FILHO, F. R.; LIMA, J. A. A.; RIBEIRO, V. Q. (Eds). *Feijão-caupi*: avanços tecnológicos. Brasília: EMBRAPA, 2005. p. 487-497.

KERMODE, A. R. Role of abscisic acid in seed dormancy. *Journal of Plant Growth Regulation*, v.24, p. 319-344, 2005. http://link.springer.com/ article/10.1007%2Fs00344-005-0110-2

MARCOS-FILHO, J. *Fisiologia de sementes de plantas cultivadas*. Piracicaba: FEALQ, 2005. 495p.

ROCHA, M. M.; CARVALHO, K. J. M.; FREIRE FILHO, F. R.; LOPES, A. C. A.; GOMES, R. L. F.; SOUSA, I. S. Controle genético do comprimento do pedúnculo em feijão-caupi. *Pesquisa Agropecuária Brasileira*, v.44, n.3, p.270-275, 2009. http://www.scielo.br/scielo.php?script=sci_ arttext&pid=S0100-204X2009000300008

SILVA, V. P. R.; AZEVEDO, P. V.; BRITO, R. S.; CAMPOS, J. H. B. C. Evaluating the urban climate of a typically tropical city of northeastern Brazil. *Environmental Monitoring and Assessment*, v.161, n.1-4, p. 45-59, 2010. http://www.ncbi.nlm.nih.gov/pubmed/19184489

TEIXEIRA, I. R.; SILVA, G. C.; OLIVEIRA, J. P. R.; SILVA, A. G.; PELÁ, A. Desempenho agronômico e qualidade de sementes de cultivares de feijão-caupi na região do cerrado. *Revista Ciência Agronômica*, v.41, n.2, p.300-307, 2010. http://www.scielo.br/scielo.php?script=sci_ arttext&pid=S1806-66902010000200019

Methods for overcoming seed dormancy in *Ormosia arborea* seeds, characterization and harvest time

Aparecida Leonir da Silva[1*], Denise Cunha Fernandes dos Santos Dias[2],
Liana Baptista de Lima[3], Glaucia Almeida de Morais[4]

ABSTRACT - *Ormosia arborea*, a Leguminosae, presents seeds with tegumentary dormancy. The purpose of this study was to evaluate the efficiency of dormancy breaking methods, characterize seeds obtained from different mother plants, and to determine the best period to collect *Ormosia arborea* seeds. The seeds were harvested from mother plants in two different periods (June and August/2011). The seeds were then subjected to biometrical analyses, determination of moisture content and germination tests. Determination of the soaking curve and evaluation of the dormancy breaking methods were performed using the seeds collected in the second period. The soaking curve confirmed the tegumentary dormancy, and the chemical scarification for 15 minutes was the more adequate procedure to overcome this dormancy. The biometry revealed average values higher than those on literature, and there was difference between both harvesting periods. The mass correlates with the other evaluated parameters, and can be indicated for selecting seeds for seedling production. The two harvesting periods of *Ormosia arborea* seeds were considered appropriated for seed supplying, due to the high germination potential. Nevertheless, the best period for harvesting is when fruits are already opened, mature, and with low moisture content (no additional drying time needed), what hinders the occurrence of fungi.

Index terms: soaking curve, chemical scarification, mechanical scarification.

Métodos para superação da dormência em sementes de *Ormosia arborea*, caracterização e época de colheita

RESUMO - *Ormosia arborea* é uma Leguminosae que apresenta sementes com dormência tegumentar. Objetivou-se avaliar a eficiência de métodos para superação da dormência desta espécie, caracterizar lotes de sementes coletados em diferentes matrizes e determinar a melhor época de coleta. As sementes coletadas de matrizes, em duas épocas (junho e agosto/2011), foram submetidas às análises biométricas, determinação do teor de água e testes de germinação. Com sementes da segunda coleta, determinou-se a curva de embebição e avaliou-se métodos para superação da dormência. A curva de embebição confirmou a dormência tegumentar e a escarificação química por 15 minutos foi o procedimento mais adequado para a superação desta dormência. A biometria revelou valores médios maiores do que os relatados na literatura e houve diferença entre as épocas de coleta. A massa correlacionou-se com os demais parâmetros avaliados, podendo ser indicada para a seleção de sementes na produção de mudas. As duas épocas de coletas de sementes de *Ormosia arborea* utilizadas foram consideradas adequadas para o fornecimento de sementes, em função do elevado potencial germinativo obtido. Entretanto, a melhor época para a coleta das sementes é quando os frutos já estão abertos, maduros, com baixo teor de água, sem necessidade de secagem adicional, o que dificulta a ocorrência de fungos.

Termos para indexação: curva de embebição, escarificação química, escarificação mecânica.

Introduction

The need to restore the natural environment has set forth a rise in demand for seedlings pertaining to native forest species. Most of these species are dispersed by seeds and the success in seedling formation depends on the knowledge about the germination process of each seed as well as the quality of the utilized seed (Rego et al., 2009).

Ormosia arborea (Vell.) Harms belongs to the Fabaceae family and is popularly known as *olho-de-cabra* (goat eye) (Lorenzi, 2000). These seeds exhibit primary dormancy, which is characterized as the expression of dormancy as

[1]Departamento de Biologia Vegetal, Universidade Federal de Viçosa, 36570-000-Viçosa, MG, Brasil.
[2]Departamento de Fitotecnia, Universidade Federal de Viçosa, 36570-000-Viçosa, MG, Brasil.

[3]Universidade Federal de Mato Grosso do Sul, Centro de Ciências Biológicas e da Saúde, Laboratório de Botânica, 79070-900-Campo Grande, MS, Brasil.
[4]Universidade Estadual de Mato Grosso do Sul, 79740-000- Ivinhema, MS, Brasil.
*Corresponding author:< cida_leonir@hotmail.com>

soon as its development is complete. The dormancy must be overcome by an efficient method that raises germination rates and enables a greater availability of seedlings, enhancing the use of the species for environmental recovery. Silva and Morais (2012) noted a greater efficacy in seedling emergence of *Ormosia arborea* when the seeds were scarified in sulfuric acid for 20 minutes. With the intention to produce quality seedlings, a thorough selection of the mother plant is an essential step of the process. The information obtained from monitoring the mother plant provides an indication of the maturation progress and aids in the decision of the best period to begin collecting seeds (Fowler and Martins, 2001).

Until seeds and fruits reach maturity, they undergo several modifications during the maturation process including alteration of their size and shape. Knowledge of the biometry of fruits and seeds is a basic requirement to conserve and explore a plant species. It is also an important tool to detect genetic variability within a population of the same species. The biometry of fruits and seeds can also be used to study the relationship between the genetic variability and environmental factors (Carvalho et al., 2003). Likewise, basic information on germination, cultivation and the potentiality of native species are often demanded for several purposes (Gusmão et al., 2006). This information is essential for comprehending the eco-physiology of the species, its adaptations in the field and to support reforestation programs. However, studies of this nature are scarce and, because of the lack of basic information, the seedling production is often times inefficient.

Seeking to contribute to the seedling production of *Ormosia arborea*, this study is aimed to evaluate the efficiency of methods to overcome seed dormancy, characterize lots of seeds collected in two distinct periods and to determine the best period for collecting seeds.

Material and Methods

Seeds of *Ormosia arborea* were collected from nine mother plants in two different periods (06/16 and 08/06/2011). The mother plants were located at the left riverside of the Ivinhema river, in the city of Nova Andradina, in the state of Mato Grosso do Sul (MS). During the first seed harvest, the fruits were initiating the opening of the pods, but most of them were still closed. In order to fully open the pods of the first lot and release their seeds, the fruits were placed to dry in the sun. In the second harvest, the fruits were already open and the seeds were exposed, remaining sustained by the funiculi.

Soaking curves were built in order to verify if the seed coat was impermeable and how long would it take for the seeds to reach phase II of the germination process. The seeds utilized

to build the soaking curves were from the second harvest (which contained more available seeds) and were divided into three groups. The first group contained non-scarified seeds (control group). The second group contained seeds subjected to mechanical scarification and, lastly, the third group contained seeds subjected to chemical scarification (immersion in sulfuric acid for 15 minutes according to Marques et al., 2004). One hundred seeds were utilized for each procedure, which were divided in four replicates of 25 seeds. The soaking curves data were subjected to Kruskal-Wallis test (followed by Dunn test at 5% of probability). Seeds were immersed in distilled water and kept in a B.O.D. incubator at 25 °C. The mass gain was determined during the entire soaking process. For the first six hours, the gain was determined hourly, then at the 12 hour mark and 24 hour mark. After 24 hours had passed, the determination of mass gain was made every 24 hours until the end of the experiment. Seeds were dried in absorbent paper every time before weighing. Soaking was considered as an increase in mass with respect to the initial mass, and the curve was terminated when the first seed emitted the primary root.

Seeds from nine mother plants of the second seed harvest were utilized to study methods for overcoming seed dormancy. The following treatments were tested: treatments 1 and 2 – mechanical scarification of the seed, using sandpaper number 100 on the opposite side of the hilum, followed by immersion in distilled water for 3 and 5 days, respectively, with daily water renewal. Treatments 3, 4 and 5 – chemical scarification of the seeds via immersion into concentrated sulfuric acid (H_2SO_4) for 15, 30 and 45 minutes, respectively, followed by rinsing with running water for 5 minutes. Intact seeds were used as the control group.

Germination experiments were conducted with five replicates of 25 seeds placed inside plastic boxes of the gerbox type. The boxes were filled with ¾ of vermiculite humidified with 80 mL of distilled water, and kept in a B.O.D. incubator at 25 °C for 30 days (Fowler and Martins, 2001) under constant light, adding water when necessary. Seeds that produced a primary root of at least a 2 mm of length were considered germinated. The germination was assessed daily (Fowler and Martins, 2001). The results were expressed in germination percentage. Using daily data average time, average speed and relative frequency and synchronization index. The average time and average speed were calculated according to Labouriau and Valadares (1976) as follows:

Average time of germination: $t = \dfrac{\left[\sum_{i=1}^{k} ni.ti\right]}{\sum_{i=1}^{k} ni}$

In which, t: average time of incubation; ni = number of germinated seeds per day; ti= time of incubation (days).

Average speed: $V = \dfrac{1}{t}$

In which: V= average speed of germination; t= average time of germination.

The relative frequency and the synchronization index were calculated according to Labouriau and Pacheco (1978):

Relative frequency: $Fr \dfrac{ni}{\sum_{i=1}^{k} ni}$

In which, Fr = Relative frequency of germination; ni = number of germinated seeds per day; Σni= total number of germinated seeds.

Synchronization index or informational entropy of germination: $E = \sum_{i=1}^{k} fi . log2 . fi$

In which, E= synchronization index; fi: relative frequency of germination; $log_{2\ =}$ base 2 logarithm.

Treatments were analyzed by Analysis of Variance followed by Tukey test at 5% of probability. The method considered more efficient was the one that provided a greater germination percentage in less time.

Seeds obtained from both harvests were subjected to the biometric evaluation, moisture content and germination capacity, for the characterization of the lot and determination of the best period for harvesting. With the aid of a digital caliper ruler (0.01 mm) and precision scale (0.0001 g), 100 seeds from the mother plant were utilized to determine length, thickness, width and mass.

Frequency distribution of seeds in mass classes, on the first harvest and second harvest, were made with 100 seeds from each mother plant. Length was considered as the longest axis of the seed; width and thickness were obtained at the largest portion of the seed. The biometric data were subjected to Kruskal-Wallis test (followed by Dunn test at 5% of probability) in order to verify if there was any difference between mother plants from the same harvest.

A t Test at 5% probability was used to analyze the biometric parameters of seeds from both harvests. The D'Agostino normality test was applied to the distribution of the amplitude of the seeds. The data was evaluated through Spearman's Correlation Coefficient for the existence of correlation between parameters. All tests were performed using the BioEstat 5.3 software.

As described in the *Regras para Análise de Sementes* (Brasil, 2009), the seeds were incubated at 105 °C for 24 hours in order to determine the moisture content of the seeds.

After the testing was concluded, the best method to break dormancy was used to scarify seeds from two lots. Each lot was harvested in a distinct period and nine replicates of 50 seeds were used from each harvest. The seeds were placed to

germinate according to method previously described here. The average time data was subjected to t Test for two independent samples and to Analysis of Variance.

Results and Discussion

Non-scarified seeds (control group) had a very low water uptake, with a total gain of 0.13 g of water after 144 hours. For the mechanical and chemical treatments, the average water uptake was 23.26 g and 22.79 g, respectively (Figure 1).

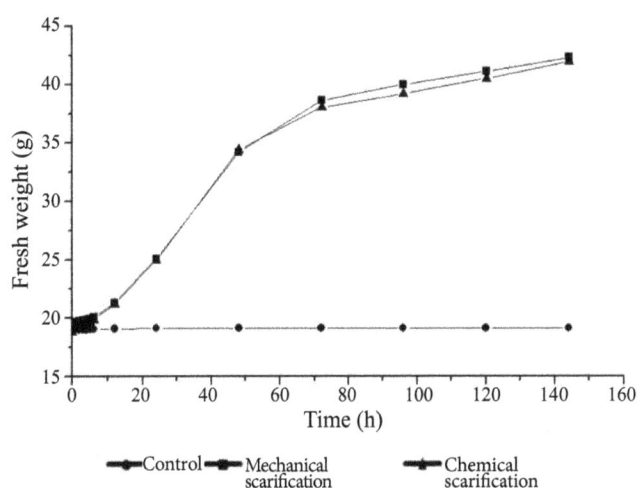

Figure 1. Soaking curves for *Ormosia arborea* seeds. Scarified and non-scarified seeds.

The water uptake values did not differ significantly between the scarification methods, however both methods statistically differed from the control group (H = 21.1938; $p < 0.0001$). Both scarification methods provided an increase of moisture content within 72 hours, and from this point, the protrusion of the primary root was verified for scarified seeds. According to Carvalho and Nakagawa (2000), when a dry seed is exposed to water, a rapid increase in fresh weigh occurs (due to the matric potential). This rapid increase in fresh weight is followed by a slow water absorption, when the main metabolic events of the germination process occur.

Lopes et al. (2004) verified the same imbibition period of 72 hours, at 30 °C, for seeds of *Ormosia arborea* subjected to mechanical scarification. They also verified the absence of imbibition for intact seeds. Differently, Basqueira et al. (2011) verified that *Ormosia arborea* seeds subjected to mechanical scarification, showed progressive absorption between one and six hours at 25 °C, stabilizing after that period.

Non-scarified seeds showed an unchanged soaking curve during the entire weighing process, which indicates the presence of an impermeable seed coat. Conversely, scarification eliminated the seed coat impermeability, allowing water to be absorbed and the primary root to be emitted after 72 hours, at the end of phase II. According to Perez (2004), members of the Fabaceae family present seeds with an impermeable seed coat which is responsible for preventing water to be absorbed. Bewley et al. (2013) highlighted that in the germination process the imbibition capacity of the seed is fundamental, since a deficiency in water absorption may compromise the function of physiological activities. Thus, for a successful germination it is important to define the best method

to break this impermeable seed coat.

Chemical scarification treatments were more effective in breaking dormancy when compared to mechanical scarification. This resulted in high germination percentage with a significant difference in percentage between seeds scarified for 15 and 60 minutes. Mechanical scarification resulted in lower germination percentages, with no significant difference between the two soaking times (Table 1). Non-scarified seeds from the control group did not germinate during the evaluated period.

Table 1.   Average time, average speed and synchronization index for the germination of *Ormosia arborea* seeds subjected to different scarification treatments: T1 - mechanical + water immersion for 3 days; T2 – mechanical + water immersion for 5 days; T3, T4 e T5 - chemical (15, 30 and 60 minutes, respectively). Control – intact seeds.

Treatments	Germination (%)	Average time (days)	Average speed (days⁻¹)	Synchronization index (bits)
T1	55 d	9.656 a	0.105	-3.2651
T2	57 cd	8.239 a	0.121	-3.1079
T3	91 a	14.833 a	0.067	-4.0088
T4	82 ab	13.215 a	0.076	-3.8582
T5	71 bc	14.326 a	0.070	-3.8411
Control	0	0 b	0	0
CV%	23.4			

Averages followed by the same letter, did not differ statistically from each other by Tukey test at 5% probability.

In a study conducted by Lopes et al. (2004), chemical scarification also provided high germination percentages in seeds of *Ormosia arborea*, whereas mechanical scarification did not differ from control. Teixeira et al. (2009) and Gonçalves et al. (2011) also verified the greater efficiency of chemical scarification, however chemical scarification did not significantly differ from mechanical scarification using sandpaper in the latter study. Marques et al. (2004) recommended both procedures for breaking dormancy in *Ormosia arborea* seeds. Conversely, Lorenzi (2000) only indicated the mechanical scarification method to promote seed germination in this species. He reported a germination rate of over 50%, with emergence occurring in 20-50 days. Fowler and Martins (2001) also recommended the mechanical scarification method using sandpaper.

In this study, approximately 37% of the mechanically scarified seeds were contaminated by fungi and did not germinate, whereas for chemically scarified seeds this loss was only 7%. Similar results were found by Lopes et al. (2004) where fungi deteriorated about 48% of the seeds that were mechanically scarified and pre-imbibed for 24 hours.

The average germination time varied from 8.239 day (T2) to 14.833 days (T3). These treatments exhibited the highest and lowest synchrony, respectively. The frequency distribution graphs (Figure 2) revealed a tendency of displacement to the left in T2 and T1, with the lowest values of average time (although without significant difference from

other treatments) and highest values of synchrony (Table 1).

In the first and second harvests, the respective average values of length (12.9 and 12.4 mm), thickness (8.5 and 8.3 mm), width (11.3 and 11.1 mm) and mass (0.827 and 0.776 g) were greater than those reported by Gurski et al. (2012) for *Ormosia arborea* seeds (11.5 mm; 7.5 mm and 9.2 mm) (Table 2).

A significant difference ($p < 0.0001$) in seed biometry collected from different mother plants was found for both the first and the second for all parameters, considering general average seeds from the second harvest had statistically lower values when compared to those from the first. Mass proved to be the most heterogeneous measurement with a higher coefficient of variation, while length proved to be the most stable measurement (Table 2). This variation in seed biometry occurred between individuals from the same species, and frequently in seeds originated from the same individual (Vaughton and Ramsey, 1998).

Analyzing the mass frequency distribution of the seeds, the most frequent groups were those containing seeds of 0.651- 0.75 g and 0.751- 0.85 g (Figure 3). The difference between harvests was caused by the decrease in sample variation however the second harvest showed higher uniformity of seed mass and a normal distribution (Figure 3). The minimum and maximum values obtained from each mother plant show that the amplitude variation can be high for seeds within the same mother plant and therefore none of the mother plants had exclusively large or small seeds (Table 2 and Figure 3).

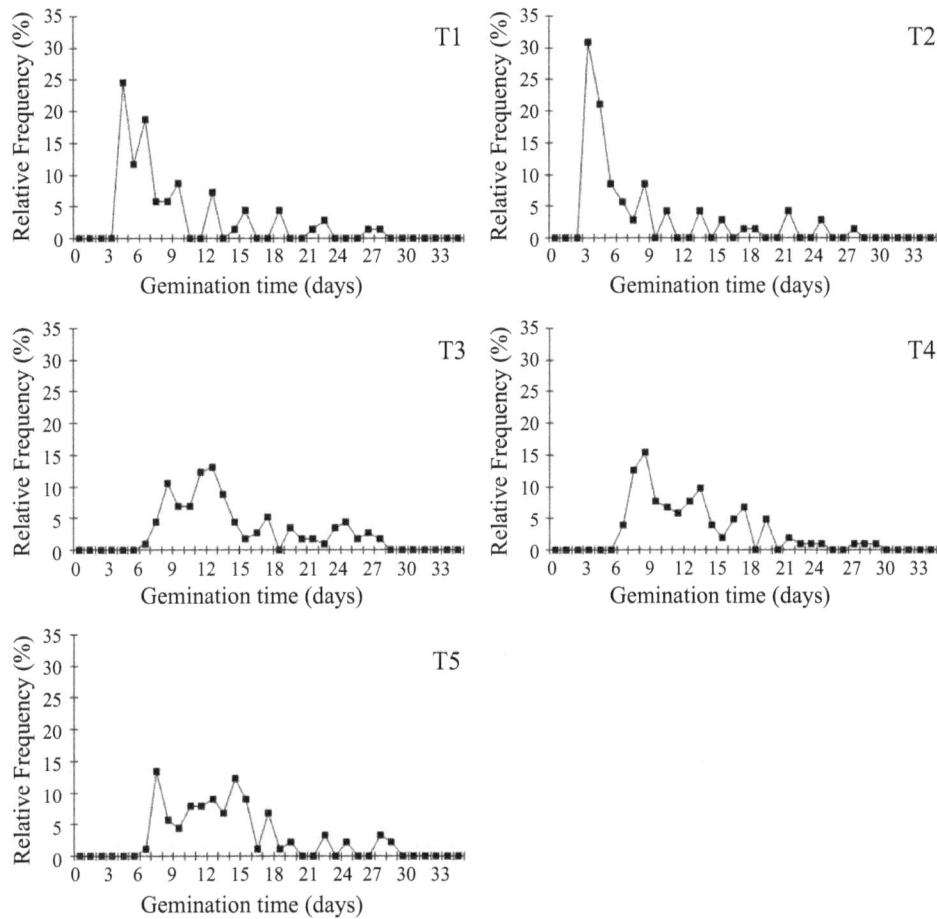

Figure 2. Relative frequency polygons for the germination of *Ormosia arborea* seeds subjected to different scarification treatments: T1 - mechanical + immersion in water for 3 days; T2 – mechanical + immersion in water for 5 days; T3, T4 e T5 - chemical (15, 30 and 60 minutes, respectively).

Table 2. Biometrical data of *Ormosia arborea* seeds: length, thickness and width in millimeters and mass in grams.

Mother plant	Length		Thickness		Width		Mass	
	1ª Harvest	2ª Harvest	1ª Harvest	2ª Harvest	1ª Harvest	2ª Harvest	1ª Harvest	2ª Harvest
1	11.96 e	11.85 e	8.32 bc	8.24 cde	10.30 e	10.23 d	0.70 bc	0.69 b
2	13.21 c	12.65 b	9.14 a	9.06 a	11.48 c	11.42 b	0.92 a	0.88 a
3	11.93 e	11.87 de	8.41b	8.26 cde	10.74 d	10.80 c	0.71b	0.70 b
4	14.11 a	13.27 a	8.88 a	8.47 bc	12.03 b	11.45 b	0.99 a	0.89 a
5	13.75 ab	12.68 b	9.17 a	8.63 b	13.06 a	12.41 a	1.05 a	0.91 a
6	13.39 bc	12.50 bc	8.96 a	8.38 bd	12.75 ab	12.15 a	1.01 a	0.87 a
7	11.97 e	12.00 de	8.35 bc	8.23 de	10.64 de	10.91 c	0.70 bc	0.72 b
8	13.16 c	12.70 b	7.13 d	7.08 f	10.39 de	10.14 d	0.64 c	0.61 c
9	12.42 d	12.20 cd	8.07 c	8.10 e	10.64 de	10.65 c	0.72 b	0.72 b
General Average	12.88 A	12.41 B	8.49 A	8.27 B	11.34 A	11.13 B	0.83 A	0.78 B
Maximum	16.18	14.56	11.55	10.76	15.50	14.10	1.67	1.34
Minimum	9.43	10.21	4.59	5.66	8.76	8.31	0.33	0.40
Coefficient of variation	7.97%	6.39%	9.56%	8.48%	10.47%	8.64%	22.45%	17.77%
Standard Deviation	1.026	0.793	0.812	0.701	1.187	0.961	0.186	0.138
H=	557.483	302.379	510.718	387.439	608.519	556.932	634.766	558.854

Mean values followed by the same letter did not differ statistically from each other at 5% probability by test *t*: lowercase between column, and uppercase between row.

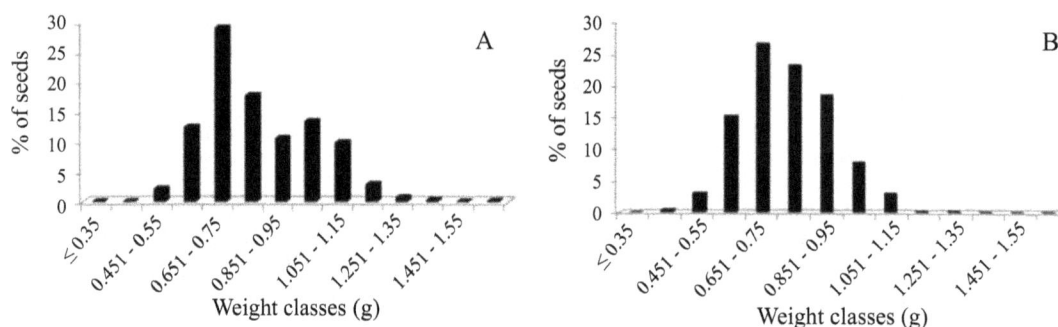

Figure 3. Frequency distribution of *Ormosia arborea* seeds in mass classes: A. First harvest; B. Second harvest.

This is an important aspect for mother plant selection. According to Fontenele et al. (2007), the difference between minimum and maximum values allows us to explore the genetic variability of features. This exploration focus on the maintenance of areas with representative populations of this species and its genetic variability

According to the correlation analysis (Table 3), a significant relation between mass and the other measurements such as length, thickness and width, was found. The following significant correlations, in that order, had decreasing values in both harvests: width x mass; thickness x mass; length x mass; and thickness x width. There was no significant correlation between length and thickness of seeds.

Table 3. Spearman's correlation (r_s) for biometric variables of *Ormosia arborea* seeds from the first and second harvest.

	1ª Harvest	2ª Harvest
	r_s	r_s
length x thickness	0.3658	0.1548
length x width	0.5971**	0.3522
length x mass	0.6928**	0.5718**
thickness x width	0.6529**	0.4744**
thickness x mass	0.8218**	0.7476**
width x mass	0.8654**	0.8099**

**Highly significant considering p value.

The minimum moisture content observed, considering all mother plants, was 6.80% and the maximum was 27.40%. The mean value for the first harvest was 13.05% and for the second harvest 11.64%. There was a slight reduction of moisture content in seeds collected from open fruits. Due to this reduction a difference in the average values of seed dimension could be observed between the different harvests (Table 2), in which the values were lower for the second harvest. However, other environmental factors cannot be discarded as well as

climatic conditions and physiological aspects. This water loss represents a pre-programmed phase of drying (or desiccation) in seed maturation, also a feature of orthodox seeds (Castro et al., 2004).

Based on the germination test (Table 1), the most efficient method to break dormancy was the chemical scarification in sulfuric acid for 15 minutes. This method was utilized to evaluate the germination capacity of lots collected in different periods. The germination percentages of *Ormosia arborea* were 89% for the first harvest and 98% for the second harvest (Table 4). These values were higher than those found by Lopes et al. (2004) (61%) for the same species subjected to the same immersion period of 15 minutes in acid, but at 30 °C. This difference can be caused by genetic and environmental factors related to the mother plant, which results in different responses. Aspects such as, lot origin, period of harvest and distinct conditions of experimentation can also affect the final results (Perez, 2004).

Thus, with the intention to reduce fungi growth, and consequently seed loss, as well as to facilitate seed storage, the best period observed to harvest seeds from this species was at end of the desiccation phase, which is characterized by spontaneous fruit opening.

When the germination data (Table 4) is compared with the average biometry of the seeds by harvest period (Table 2), it was observed that seeds with larger dimensions did not necessarily present higher percentage of germination. According to Carvalho and Nakagawa (2000), the germination is not influenced by seed size, which would affect the seedling vigor, but it is dependent on other factor such as viability, climatic conditions, etc. A plant originated from a small seed would have its growth retarded in the beginning, whereas after this initial growth the plant would recover reaching its normal growth (Carvalho and Nakagawa, 2000).

The data related to seed biometry, specially the mass, could help the selection of seeds for an efficient seedling establishment and would determine the success in the growth of the plants. The fast emergence of seedlings from heavier seeds was verified by Parker et al. (2006) for *Pinus strobus* L. (Pinaceae), although under laboratorial conditions the difference in seed mass did not influence the germination.

Table 4.   Average time, average speed and synchronization index for the germination of *Ormosia arborea* seeds from the first and second harvest.

	Germination[ns] (%)	Average time[ns] (days)	Average speed (days-^{1})	Synchronization Index (bits)
1ª Harvest	89	10.995	0.090	- 3.5143
2ª Harvest	98	14.786	0.067	- 4.2459

*significant difference by test *t* at 5% probability; [ns] no significant difference by test *t* at 5% probability.

The average germination time was 10.995 days for the first harvest and 14.786 days for the second harvest, with no significant difference between both harvest periods (Table 4).

Conversely from what was found by Oliveira et al. (2008) for another Leguminosae species, the germination percentage of *Ormosia arborea* seeds was high (independent from harvest period) and there was no significant difference in the average percentage.

Conclusions

The chemical scarification by immersion in sulfuric acid for 15 minutes is efficient for dormancy breaking of *Ormosia arborea* seeds.

Both harvesting periods utilized for this study were considered adequate for seed supplying, due to the high germination potential achieved under laboratorial conditions. Nevertheless, the best period for harvesting is when the fruits are already opened and the seeds exposed. Hence the seeds will not need any additional drying time and will be less susceptible to fungi attack due to the lower moisture content of the seed.

Acknowledgements

We thank Prof. Dr. Douglas de Araujo for revising the text.

References

BASQUEIRA, R.A.; PESSA, H.; SOUZA-LEAL, T.; PEDROSO-DE-MORAES, C. Superação de dormência em *Ormosia arborea* (Fabaceae: Papilionoideae) pela utilização de dois métodos de escarificação mecânica em diferentes pontos do tegumento. *Revista em Agronegócios e Meio Ambiente*, v.4, n.3, p.547-561, 2011. http://www.cesumar.br/pesquisa/periodicos/index.php/rama/article/view/1876/1301

BEWLEY, J.D.; BRADFORD, K.J., HILHORST, H.W.M.; NONOGAKI, H. *Seeds: Physiology of Development, Germination and Dormancy*. New York: Plenum, 2013. 392p.

BRASIL. Ministério da Agricultura, Pecuária e Abastecimento. *Regras para análise de sementes*. Ministério da Agricultura, Pecuária e Abastecimento. Secretaria de Defesa Agropecuária. Brasília: MAPA/ACS, 2009. 395p. http://www.bs.cca.ufsc.br/publicacoes/regras%20analise%20sementes.pdf

CARVALHO, N.M; NAKAGAWA, J. *Sementes*: ciência, tecnologia e produção. 4 ed. Campinas: Fundação Cargill, 2000. 588p.

CARVALHO, J.E.U.; NAZARÉ, R.F.R.; OLIVEIRA, W.M. Características físicas ou físico-químicas de um tipo de bacuri (*Platonia insignis* Mart.) com rendimento industrial superior. *Revista Brasileira de Fruticultura*, v.25, n.2, p.326-328, 2003. http://www.scielo.br/scielo.php?script=sci_arttext&pid=S0100-29452003000200036

CASTRO, R.D.; BRADFORD, K.J.; HILHORST, H.W.M. Desenvolvimento de sementes e conteúdo de água. In: FERREIRA, A.G.; BORGHETTI, F. (Orgs.). *Germinação*: do básico ao aplicado. Porto Alegre: Artmed, 2004. p. 51-67.

FONTENELE, A.C.F.; ARAGÃO, W.M.; RANGEL, J.H.A. Biometria de frutos e sementes de *Desmanthus virgatus* (L) Willd nativas de Sergipe. *Revista Brasileira de Biociências*, v.5, n.1, p.252-254, 2007. www.ufrgs.br/seerbio/ojs/index.php/rbb/article/download/275/244

FOWLER, A.P.; MARTINS, E.G. *Manejo de sementes de espécies florestais*. Colombo: Embrapa Florestas, 2001. 71p. (Embrapa Florestas. Documentos, 58). www.infoteca.cnptia.embrapa.br/bitstream/doc/289390/1/doc58.pdf

GONÇALVES, E.P.; SOARES, F.S.J.; SILVA, S.S.; TAVARES, D.S.; VIANA, J.S.; CARDOSO, B.C.C. Dormancy Breaking in *Ormosia arborea* Seeds. *International Journal of Agronomy*, v.2011, 5p., 2011. http://www.hindawi.com/journals/ija/2011/524709/

GURSKI, C.; DIAS, E.S.; MATTOS, E.A. Caracteres das sementes, plântulas e plantas jovens de *Ormosia arborea* (Vell.) Harms e *Ormosia fastigiata* Tul. (Leg-papilionoideae). *Revista Árvore*, v.36, n.1, p.37-48, 2012. http://www.scielo.br/scielo.php?pid=S0100-67622012000100005&script=sci_arttext

GUSMÃO, E.; FILHO, S.M.; VIEIRA, F.A.; JUNIOR, E.M.F. Biometria de frutos e endocarpos de murici (*Byrsonima verbascifolia* Rich.ex A. Juss). *Cerne*, v.12, n.1, p.84-91, 2006. http://redalyc.uaemex.mx/src/inicio/ArtPdfRed.jsp?iCve=74412110

LABOURIAU, L.G.; PACHECO, A. On the frequency of isothermal germination in seeds of *Dolichos biflorus* L. *Plant and Cell Physiology*, v.21, p.507-512, 1978. http://pcp.oxfordjournals.org/content/19/3/507.abstract

LABOURIAU, L.G.; VALADARES, M.E.B. On the germination of seeds of *Calotropis procera* (Ait.) Ait. F. *Anais da Academia Brasileira de Ciências*, v.48, p. 263-284, 1976.

LOPES, J.C.; DIAS, P.C.; MACEDO, C.M.P. Tratamentos para superar a dormência de sementes de *Ormosia arborea* (Vell.) Harms. *Brasil Florestal*, n.80, p.25-35, 2004. http://www.researchgate.net/profile/Paulo_Dias15/publication/263182712_Tratamentos_para_superar_a_dormncia_de_sementes_de_Ormosia_arborea_%28Vell.%29_Harms/links/0f31753a1aefa8bc4a000000

LORENZI, H. *Árvores Brasileiras:* Manual de Identificação e Cultivo de Plantas Arbóreas Nativas do Brasil. 3. ed. Nova Odesssa/SP: Instituto Plantarum, 2000. 352p.

MARQUES, M.A.; RODRIGUES, T.J.D.; PAULA, R.C. Germinação de sementes de *Ormosia arborea* (Vell.) Harms submetidas a diferentes tratamentos pré-germinativos. *Científica*, v.32, n.2, p.141-146, 2004. www.cientifica.org.br/index.php/cientifica/article/download/79/62

OLIVEIRA, D.A.; NUNES, Y.R.F.; ROCHA, E.A.; BRAGA, R.F.; PIMENTA, M.A.S.; VELOSO, M.D.M. Potencial germinativo de sementes de fava-d'anta (*Dimorphandra mollis* Benth. – FABACEAE: MIMOSOIDEAE) sob diferentes procedências, datas de coleta e tratamentos de escarificação. *Revista Árvore*, v.32, n.6, p.1001-1009, 2008. http://www.scielo.br/scielo.php?script=sci_arttext&pid=S0100-67622008000600005

PARKER, W.C.; NOLAND, T.L. MORNEAULT, A.E. The effects of seed mass on germination, seedling emergence, and early seedling growth of eastern white pine (*Pinus strobus* L.). *New Forests*, v.32, n.1, p.33-49, 2006. http://link.springer.com/article/10.1007%2Fs11056-005-3391-1#page-1

PEREZ, S.C.J.G.A. Envoltórios. In: FERREIRA, A.G.; BORGHETTI, F. (Orgs.). *Germinação*: do básico ao aplicado. Porto Alegre: Artmed, 2004. p.125-134.

REGO, S.S.; NOGUEIRA, A.C.; KUNIYOSHI, Y.S.; SANTOS, A.F. Germinação de sementes de *Blepharocalyx salicifolius* (H.B.K.) Berg. em diferentes substratos e condições de temperaturas, luz e umidade. *Revista Brasileira de Sementes*, v.31, n.2, p.212-220, 2009. http://www.scielo.br/scielo.php?pid=S0101-31222009000200025&script=sci_arttext

SILVA, A.L.; MORAIS, G.A. Biometry and dormancy breaking of *Ormosia arborea* seeds. *Communications in Plant Sciences*. Lages, v.2, p.3-4, 2012. http://complantsci.files.wordpress.com/2012/12/complantsci_2_2_22.pdf

TEIXEIRA, W.F.; RODRIGUES, E.A.; AMARAL, A.F. Estudo de superação de dormência de *Ormosia arborea* sob diferentes testes, para produção de mudas para reflorestamento de áreas degradadas no município de Patos de Minas, MG. *Perquirere - Revista do Núcleo Interdisciplinar de Pesquisa e Extensão do UNIPAM*, n.6, p.26-30, 2009. http://www.unipam.edu.br/perquirere/file/file/2009/Estudo_de_superacao_de_dormencia.pdf

VAUGHTON, G.; RAMSEY, M. Sources and consequences of seed mass variation in *Banksia marginata* (Proteaceae). *Journal of Ecology*, v.86, p.63-573, 1998. http://www.jstor.org/discover/10.2307/2648421?uid=3737664&uid=2129&uid=2&uid=70&uid=4&sid=21101639546833

Morphological aspects of fruits, seeds, seedlings and *in vivo* and *in vitro* germination of species of the genus *Cleome*

Tatiana Carvalho de Castro[1*], Claudia Simões-Gurgel[1], Ivan Gonçalves Ribeiro[1], Marsen Garcia Pinto Coelho[2], Norma Albarello[1]

ABSTRACT - The genus *Cleome* is widely distributed in drier areas of the tropics and subtropics. *Cleome dendroides* and *C. rosea* are Brazilian native species that occur mainly in Atlantic Forest and sandy coastal plains, respectively ecosystems negatively affected by human impacts. *Cleome spinosa* is frequently found in urban areas. Many *Cleome* species have been used in traditional medicine, as *C. spinosa*. In the present work, was investigated *C. dendroides*, *C. rosea* and *C. spinosa* germinative behavior under *in vivo* conditions, as well as was established suitable conditions to *in vitro* germination and seedling development. The *in vivo* germination was performed evaluating the influence of temperature, substrate and light. It was observed that only *C. spinosa* seeds presents physiological dormancy, which was overcome by using alternate temperatures. The substrate influenced significantly the germination of *C. rosea* and the seeds of *C. dendroides* showed the highest germination percentages in the different conditions evaluated. The post-seminal development stages under *in vivo* and *in vitro* conditions were defined. It was observed that the development was faster under *in vitro* than *in vivo* conditions. An effective methodology for *in vitro* germination, enabling the providing of material to experiment on plant tissue culture was established to *C. dendroides* and *C. spinosa*.

Index terms: *Cleome dendroides*, *Cleome rosea*, *Cleome spinosa*, post-seminal development.

Aspectos morfológicos de frutos, sementes, plântulas e germinação *in vivo* e *in vitro* de espécies do gênero *Cleome*

RESUMO - O gênero *Cleome* encontra-se distribuído em áreas tropicais e subtropicais do mundo. *Cleome dendroides* é uma espécie endêmica de Mata Atlântica do estado do Rio de Janeiro, enquanto *C. rosea* é uma espécie nativa que ocorre principalmente em restingas, ambos ecossistemas que sofrem constante impacto antrópico. *Cleome spinosa* é frequentemente encontrada em áreas urbanas. Muitas espécies de *Cleome* têm sido utilizadas na medicina tradicional, como *C. spinosa*. No presente trabalho, avaliou-se a influência do substrato, da temperatura e da luz sobre a germinação *in vivo* das três espécies, bem como determinaram-se as condições para germinação *in vitro* e as etapas do desenvolvimento pós-seminal. Observaram-se que apenas sementes de *C. spinosa* apresentam dormência fisiológica, a qual é superada utilizando-se temperaturas alternadas. O substrato influenciou significativamente a germinação de *C. rosea* e as sementes de *C. dendroides* apresentaram alta porcentagem de germinação nas diferentes condições avaliadas. As etapas do desenvolvimento pós-seminal sob condições *in vivo* e *in vitro* foram definidas. Observou-se que o desenvolvimento sob condições *in vitro* foi mais acelerado que *in vivo*. Uma eficiente metodologia para germinação *in vitro* foi estabelecida para *C. dendroides* e *C. spinosa*, fornecendo material juvenil em condições assépticas para futuros experimentos de cultura de tecidos vegetais.

Termos para indexação: *Cleome dendroides*, *Cleome rosea*, *Cleome spinosa*, desenvolvimento pós-seminal.

Introduction

Cleome is the largest genus from Cleomaceae family, with over 200 species distributed in drier areas of the tropics and subtropics (Iltis, 1960). It consists mainly of annual or perennial herbaceous plants and, rarely, shrubs. Several species of *Cleome* are used in traditional medicine and many of them have been subject of pharmacological and phytochemical studies (Aparadh et al., 2012). In Brazil, some of these species have been used in traditional medicine, such

[1]Departamento de Biologia Vegetal, Universidade do Estado do Rio de Janeiro, Instituto de Biologia Roberto Alcântara Gomes, 20550-013 - Rio de Janeiro, RJ, Brasil.

[2]Departamento de Bioquímica , Universidade do Estado do Rio de Janeiro, Instituto de Biologia Roberto Alcântara Gomes, 20550-170 - Rio de Janeiro, RJ, Brasil.
*Corresponding author <tatianaccastro17@gmail.com>

as *C. spinosa*, used in the treatment of diseases related to respiratory tract, and several inflammatory disorders (Agra et al., 2007). Several species may be employed as ornamental due to their attractive inflorescences.

Biotechnological studies have been conducted with *Cleome* species in order to develop efficient protocols for mass propagation and to enhance the production of bioactive molecules under *in vitro* conditions. Such studies have been performed with *C. rosea* Vahl, Brazilian native species mainly found in coastal sandy plains, ecosystems negatively affected by human impacts and *C. spinosa* Jacq., found in urban areas and highly vulnerable to human activities and insect infestations. *In vitro* protocols were established for these species (Simões et al., 2004; Albarello et al., 2006; Albarello et al., 2007; Simões et al., 2009a; Simões et al., 2010a; Simões-Gurgel et al., 2012), and the medicinal potential of *in vivo* and *in vitro* materials was also evaluated (Simões et al., 2006; Simões et al., 2009b; Simões et al., 2010b; Simões-Gurgel et al., 2012; Albarello et al., 2013).

Germination is a critical stage in plants the life cycle and tends to be highly unpredictable over space and time (Baskin and Baskin, 2001). Successful establishment of plants largely depends on successful germination. The germination process may be affected by environmental and internal seed factors and frequently, ideal conditions are species-specific and need to be determined through experimentation.

There is little information available in literature about seed germination of *Cleome*. These studies report low and non-uniform germination, presence of seed dormancy and a significantly variation in germination rates depending on the seed lot used (Ochuodho and Modi, 2007; Raboteaux and Anderson, 2010; Tlig et al., 2012; K'Opondo, 2011). As seeds, in general, are more resistant to disinfection treatments required to obtain aseptic material, when compared to tissues of a developed plant, physiological knowledge of *in vitro* germination allows the establishment of efficient methods that ensure the supply of stock plants in excellent phytosanitary conditions to be used as source of explants for plant tissue culture. There are no studies related to *in vitro* germination of the three species selected.

The purpose of this research was to investigate *C. dendroides*, *C. rosea* and *C. spinosa* germinative behavior under *in vivo* conditions, as well as to establish suitable conditions to *in vitro* germination and seedling development.

Materials and Methods

Seeds of the three species selected were collected in Rio de Janeiro state, Brazil. Fruits were collected from populations located at 22°56'60" S and 43°10'0" W (*C. dendroides*), at 22°58'01" S and 42°58'36" W (*C. rosea*), and at 22°54'17" S and 43°15'52" W (*C. spinosa*). The authenticity of the specimens was previously confirmed and vouchers specimen were deposited in the Herbarium of the Rio the Janeiro State University, Rio de Janeiro, Brazil (*C. dendroides* / HRJ11.104, *C. rosea* / HRJ7185, *C. spinosa* / HRJ7639).

The variables evaluated for fruits were fruit length, thickness, shape and color, whereas for seeds, were seed length, thickness, shape, color, fresh and dry weight, moisture content, viability, coat permeability and mean number of seeds per fruit.

Analysis followed the Rules for Seed Testing (Brasil, 2009). Ten seed lots with one (*C. spinosa* and *C. rosea*) or five (*C. dendroides*) seeds each were weighed to determine the fresh weight (FW) and then subjected to drying in an oven at 105±3 °C for 24 h to determine the dry weight (DW). The moisture content (MC) was defined as: MC = (FW-DW)/FW x 100.

Seed viability was determined through the triphenyl tetrazolium chloride (TTC) test using five lots of 20 seeds. Seeds were longitudinally sectioned and soaked in 2% TTC solution at 30 °C for 24h in the dark. The viability was assessed using a stereomicroscope (Olympus SD30).

To evaluate seed-coat permeability, eleven lots of five seeds were immersed in distilled water for 24 h at 26 ± 1 °C. During successive periods the fresh weight of each lot was measured to establish the imbibition curve.

Seeds were washed with detergent Fisoex® (10%), rinsed in tap water, and placed in sodium hypochlorite (NaOCl) at different concentrations (1.0; 2.0 or 2.5% w/v) containing Tween 80 (0.05% w/v) for different periods (5; 10 or 20 min). Seeds were rinsed thrice with sterile distilled water and were germinated in transparent plastic germination boxes (Gerbox) (11 x 11 x 3.5 cm). Different substrates were tested: vermiculite (average grain diameter 1.4 to 4.0 mm), sand and towel paper. Substrates were autoclaved at 121 °C for 45 min and moistened with sterile distilled water. The effect of constant (20 and 25 °C) and alternate (15-25 and 20-30 °C) temperatures was evaluated. The experiments were conducted in germination chamber (Eletrolab EL202), under a 16 h photoperiod (20 mmol m^{-2}s^{-1}) and relative humidity of 80%. Another set of experiments was conducted in the absence of light using the best conditions of temperature and substrate established for each species in the previous experiment.

Based on the method proposed by Labouriau (1983), germination rate (GR), mean germination time (GT), coefficient of velocity of germination (GV) and relative frequency of germination (GF) were measured and defined as:

Germination rate: GR = N/A x 100

where: N = total number of germinated seeds; A = total number of seeds;

Mean germination time: GT = Σ $n_i t_i$ / Σ n_i

where: n_i = total number of germinated seeds per day; t_i = incubation time (days); Σ n_i = total number of germinated seeds.

Coefficient of velocity of germination: GV = GT^{-1}

Relative frequency of germination related to the incubation time: GF (%) = n_i.100 / Σ n_i

Evaluations were performed every two days during 50 days. When needed substrates were moistened to keep suitable conditions for germination and seedling development. Twelve treatments (three substrates x four temperatures) were considered with three replications of 20 seeds. Experiments were repeated twice.

Seeds were washed with detergent Fisoex® (10%), rinsed in tap water, and disinfected under aseptic conditions in laminar flow hood. Two protocols of decontamination were evaluated: exposure to atmosphere saturated with formaldehyde (paraformaldehyde 80%) for 1, 2 and 3 h, and immersion in a solution of NaOCl (1.0; 1.5; 2.0 and 2.5% w/v) containing Tween 80 (0.05% w/v) for 10, 15, 20 and 30 min. After that, the seeds were rinsed thrice with sterile distilled water and inoculated into flasks containing 30 mL of MS medium (Murashige and Skoog, 1962) added with 30 g.L^{-1} sucrose and solidified with 8 g.L^{-1} agar. The pH of media was adjusted to 5.8 prior to autoclaving (121 °C - 15 min).

Flasks were closed with polypropylene caps and maintained in a growth chamber at 26 ± 2 °C, under a 16 h photoperiod provided by cool white fluorescent tubes (45 mmol m^{-2}s^{-1}) or in germination chambers (Eletrolab EL202) under alternate temperatures (20-30 °C), a 16 h photoperiod (20 mmol m^{-2}s^{-1}) and relative humidity of 80%. Four seeds were inoculated in each flask in a total of 30 flasks per treatment. Experiments were repeated twice.

The germination was carried out on MS medium with reduced salt concentrations (MS1/2, MS1/3 and MS1/4). Seeds of C. rosea were also inoculated on MS medium containing glucose (30 g.L^{-1}) as source of carbohydrate and phytagel (2 g.L^{-1}) as solidified agent. The influence of aeration was evaluated using aerated caps (Sigma B-3031) to close the flasks.

Morphological aspects were recorded since root protrusion to primary leaves expansion (Brasil, 2009). Root, hypocotyl and epicotyl length as well as the number of leaves, were measured. Seedlings were considered normal when essential structures were perfectly developed (root system well developed, hypocotyl, two cotyledons, epicotyl with apical bud and primary leaves expanded), according to Rules for Seed Testing (Brasil, 2009). The external morphology of seeds, fruits and seedlings were described and illustrated.

Graphic records were obtained from fresh material using a stereomicroscope (Olympus SD30) and photographs were taken using Sony H9 camera.

Data were subjected to analysis of variance (ANOVA). Data from germination percentage, when necessary, were transformed into arcsine prior to analysis to normalize their distribution. Means were compared by Tukey test at 5% probability (p $\leqslant$ 0.05). The analyses were performed using the GraphPad Prism version 5.0 software.

Results and Discussion

Although the three species under study presented morphological similarities, some traits showed to be species-specific (Table 1). Fruits are siliqua-shaped capsules, dry and dehiscent, and in C. dendroides they are also inflated. In C. rosea the capsules are greenish (Figure 1A), while in C. dendroides (Figure 1B), they are greenish and turn pale brown when riped and in C. spinosa capsules are pale yellow turning brown when riped (Figure 1C). The fruit dehiscence occurs from base to apex separating the two valves, to release the seeds. Some seeds may remain attached to the replum, a strip holding the valves edges (Figure 1D).

Seeds of three species are small and present particular morphological traits in each species. In C. spinosa the seeds are pale yellow to dark brown with kidney-like to cochlear shape with exarillate seed coat, having open terminations and crests on the surface (Figure 1E). Cleome rosea seeds are also pale yellow to dark brown and cochlear. They present vesicular aril and tegument with crusts and transversal ribs (Figure 1F). Cleome dendroides has brown orbicular seeds. They are applanate, with papilose and muricate surface with finger-shape extensions (Figure 1G). Figure 1H shows C. dendroides seed internal structure. Embryos are cotyledon-type with flat, folded and overlapping cotyledons. They present slightly larger thickness than the radicle-hypocotyl axis.

All species had viability higher than 80% and moisture content ranged from 6% to 15% (Table 1). Imbibition curves showed a fast imbibition phase up to the first two hours, with seeds water content reaching around 55%-60% (data not shown), demonstrating high seed-coat permeability.

The substrate and the use of constant or alternate temperatures had no effect on GR in C. dendroides (Table 2). Lowest GT values and highest GV values were reached on sand and paper. Cleome rosea germination was significantly influenced by the different substrates. Highest values of GR were obtained in vermiculite, although no statistical differences were observed on GT and GV. The use of alternate temperatures was an essential condition to seed germination in *C. spinosa*. As observed in *C. dendroides*, lowest GT values and highest GV values were reached in sand and paper. In the three species, the germination was observed both in the presence and absence of light.

When considering all physical conditions evaluated, it was found that *C. dendroides* and *C. spinosa* seeds had the highest germination capacity, showing GR values above 90%, while GR values for *C. rosea* stayed below 80%. In addition, in all species, seedling development was superior in vermiculite (Figures 2A-C). No abnormal seedlings were observed in all tested conditions.

Table 1. Characterization of fruits and seeds in different species of the genus *Cleome*.

Plant Material	Variables	Species		
		C. dendroides	*C. rosea*	*C. spinosa*
Fruits	Length (cm)	6.5 ± 2.1	9.0 ± 2.5	17.0 ± 2.0
	Thickness (cm)	1.5 ± 0.5	0.8 ± 0.4	1.35 ± 0.2
	Shape	Siliqua-shaped capsules	Siliqua-shaped capsules	Siliqua-shaped capsules
	Color	Greenish to pale brown	Greenish	Pale yellow to brown
Seeds	Length (mm)	2.04 ± 0.07	1.49 ± 0.32**	2.34 ± 0.24**
	Thickness (mm)	1.19 ± 0.09	1.09 ± 0.30**	1.35 ± 0.22**
	Shape	Orbicularis	Cochlear	Kidney-like to cochlear
	Color	Brown	Pale yellow to dark brown	Pale yellow to dark brown
	Fresh weight (mg)	2.73 ± 0.15*	1.00 ± 0.20**	2.00 ± 0.04**
	Dry weight (mg)	2.34 ± 0.21*	0.94 ± 0.15**	1.70 ± 0.03**
	Moisture (%)	12.17 ± 4.20	6.00 ± 2.40**	15.00 ± 1.30**
	Viability (%)	90 - 100	80 - 90**	90 - 100**
	Coat permeability	High	High	High
	Number per fruit	380 ± 55	90 ± 15	139 ± 20

*average weight of five seeds.
** dark brown seeds.

Figure 1. Morphological aspects of fruits and seeds of the genus *Cleome*: A - fruit of *C. rosea*; B - fruit of *C. dendroides*; C - fruit of *C. spinosa*; D - seeds of *C. rosea* still attached to replum; E - seed of *C. spinosa*; F - seed of *C. rosea*; G - seed of *C. dendroides*; H - the internal structure of *C. dendroides* seed (a - extensions of testa; b – testa; c - endosperm; d - hilum; e - hypocotyl-radicle axis). Scale bars: 1 cm.

Table 2. Effect of temperature and substrate on germination rate (GR), mean germination time (GT), and coefficient of velocity of germination (GV) in species of the genus *Cleome*.

Parameters	Substratum	*C. dendroides*			*C. rosea*				*C. spinosa*			
		Temperature (°C)										
		20	25	20-30	20	25	15-25	20-30	20	25	15-25	20-30
GR (%)	Paper	96.7a	96.7a	96.7a	60.0b	33.3bc	66.6b	26.6c	0b	0b	100a	96.7a
	Sand	93.3a	100a	93.3a	66.6b	40.0bc	46.6bc	53.3b	0b	0b	96.7a	96.7a
	Vermiculite	96.7a	100a	96.7a	80.0a	33.3bc	73.3a	80.0a	6.6b	0b	93.3a	90.0a
GT (days)	Paper	9.8b	7.3c	7.4c	13.1b	16.4b	15.7b	23.8a	-	-	2.9c	3.1c
	Sand	9.3b	6.8c	7.3c	18.2b	17.2b	27.6a	20.2a	-	-	2.5c	2.2c
	Vermiculite	12.9b	10.2b	15.2a	14.7b	13.6b	22.9a	28.8a	11.5a	-	5.7b	4.0b
GV (days)	Paper	0.10b	0.14a	0.13a	0.08a	0.06a	0.06a	0.04b	-	-	0.35a	0.32a
	Sand	0.11b	0.15a	0.14a	0.06a	0.06a	0.04b	0.05b	-	-	0.40a	0.45a
	Vermiculite	0.08b	0.10b	0.06c	0.07a	0.07a	0.04b	0.04b	0.10c	-	0.18b	0.25b

Means of each parameter evaluated for each species followed by the same letter do not differ by Tukey test (5%).
- no germination.

Figure 2. *In vivo* and *in vitro* germination of *Cleome*. *In vivo* germination of *C. dendroides*, at 25 °C, on different substrates: A - paper; B - sand; C - vermiculite. *In vitro* germination: D - *C. dendroides*; E - *C. spinosa*; F - *C. rosea*. Scale bars: 1 cm.

The relative frequency polygons for germination, which determine the proportion of seeds germinated daily (Figure 3), showed that *C. spinosa* seeds presented the fastest germination when kept in alternate temperatures (Figure 3A). In this species, radicle protrusion took place at 1 - 3 days, on paper and sand. Maintenance of *C. dendroides* seeds in alternate temperatures of 20-30 °C or in constant temperature of 25 °C, accelerated radicle protrusion, which took place around the seventh day, on paper and sand (Figure 3B). These results showed that both *C. spinosa* and *C. dendroides* have a synchronic germination process. *Cleome rosea* showed a delayed and intermittent germination, in some cases, the

beginning of the process occurred after 40 days (Figure 3C).

Germination under *in vitro* conditions was successfully established to *C. dendroides* and *C. spinosa*, but it was not efficient to *C. rosea*. The exposure of seeds to formaldehyde saturated atmosphere was more efficient protocol of decontamination when compared to the immersion in sodium hypochlorite solution. To *C. dendroides*, three hours exposure to formaldehyde resulted in an efficient decontamination level (100%) and in GR of 100%. The use of sodium hypochlorite (2.5% for 20 min) allowed low contamination (10%), seeds showed lower GR (90-95%) and delayed the start of germination process. Similar results were observed in *C. spinosa*. Decontamination in formaldehyde atmosphere for 1 h resulted in GR above 80%, while the best results obtained with sodium hypochlorite (1.5%, 20 min) were 48%. To *C. rosea* the highest GR (85%) was reached with sodium hypochlorite at 1% for 10 min.

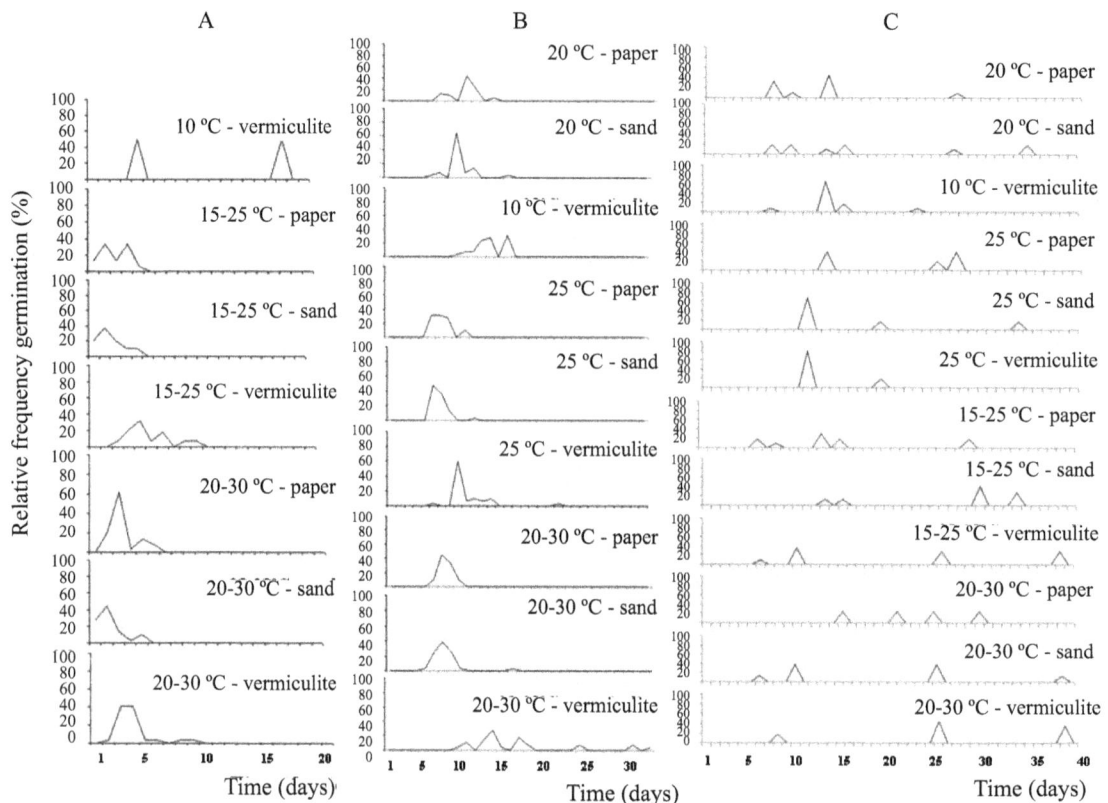

Figure 3. Relative frequency polygon from the germination of *C. spinosa* (A), *C. dendroides* (B) and *C. rosea* (C) under different temperatures and substrates.

After decontamination, seeds where inoculated on MS medium in regular or with reduced salt concentration (MS1/2 and MS1/4). *Cleome dendroides* seedlings showed hyperhydricity on MS medium, but developed normally on MS1/4 (Figure 2D). Germination occurred both in constant (26 ± 2 °C) and alternate (20-30 °C) temperatures. As for *C. spinosa,* salt concentration had no influence on seedling development (Figure 2E) and alternate temperatures were required for germination.

In vitro germination of *C. rosea* resulted in seedlings presenting an intense callusing in folioles, cotyledons and stem (Figure 2F). To establish healthy seedlings new assays were performed with modifications on culture conditions, however the new conditions did not prevent callus formation and, in addition, glucose caused a decrease in GR (60%) and seedlings size.

Cleome dendroides in vivo germination started after around seven days, with radicle protrusion, which breaks the tegument in the hilum (Figure 4B). After nine days, primary root was white, cylindrical, thin, and reached 8 mm (Figure

4C). On the day 11, hypocotyl growth was faster than the root, showing cylindrical shape and bright green color. In this stage, the cotyledons are carried above the soil surface by the elongating hypocotyl (Figure 4D) characterizing the epigeal germination. Around day 12, cotyledons were free from the tegument, which is typical of phanerocotyledonous germination. The cotyledons were orbicularis (diameter 0.5 cm), opposite, glabrous, dark green on the adaxial surface and bright green on the abaxial surface, entire margin and 1 cm petiole. Cotyledon expansion was observed on the day 13 (Figure 4E). At this stage, hypocotyl elongation ceased (1.5 - 2.0 cm) and it was observed the development of a root system of the axial type, with cylindrical, thin, hairy and white secondary roots (Figure 4F). Epicotyl was recorded around day 25. It was cylindrical and bright green. Around day 30, the pair of primary leaves was formed. Cotyledons abscission occurred after about 35 days. The young plant, after 40 (Figure 4G), presented the pair of opposite primary leaves, trifoliate, membranous, and oblong-elliptical leaflets with attenuated base, obtusely acuminate at apex, smooth edges, pubescent adaxially and slightly pubescent on the abaxial surface with printed rib on the adaxial and prominent on the abaxial surface, dark green on the adaxial and bright green on the abaxial surface, but more clear than the cotyledons. Central leaflets were larger than lateral ones. The cylindrical petioles had about 0.8 cm, leaflets were sessile and the axial root system become fully established.

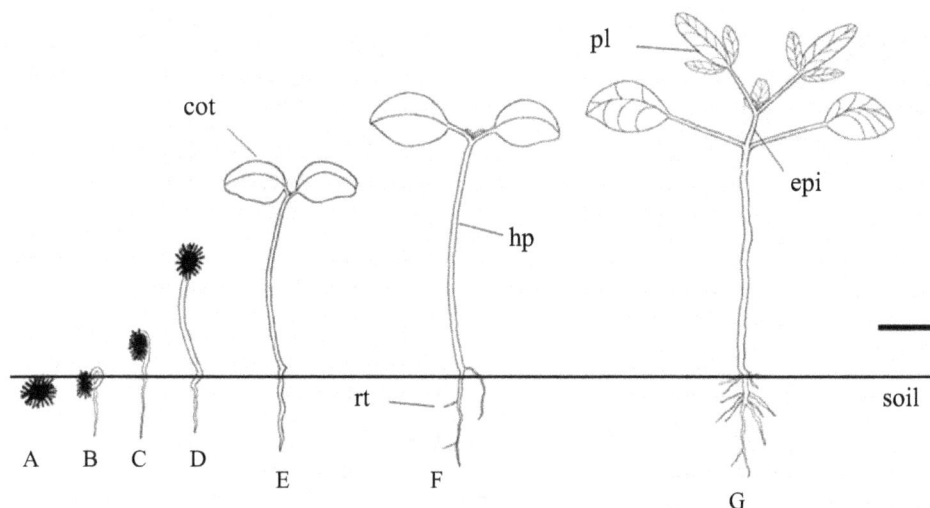

Figure 4. Morphological aspects of *in vivo* germination of *C. dendroides*, in sand, at 25 °C: A - seed; B - 7 days after primary root protrusion; C - 9 days; D -11 days; E - 13 days; F - 20 days; G - 40 days. cot - cotyledon; hp - hypocotyl; rt - root; pl - primary leaves; epi-epicotyl. Scale bars: 4 mm.

Cleome spinosa in vivo germination took place around day 3. On day 5, primary root reached 0.5 cm and hypocotyl with the same dimensions. Hypocotyl was bright green and cylindrical. During its development, elevated foliar cotyledons already released the tegument, characterizing the epigeal germination. Around the day 15, the epicotyl (0.2 cm) was formed. On day 20, the first pair of membranous and trifoliate leaves was established and after 30 days the second pair of leaves with alternate phyllotaxy was formed. The compound leaves presented the central leaflets larger than the lateral ones.

Cleome rosea in vivo germination was observed 7 - 9 days after sowing. From day 10 hypocotyl entered a period of active growth, stretching and elevating the cotyledons above the substrate characterizing the epigeal germination. On day 15 the early development of the epicotyl and first pair of leaves were observed. Between 18 and 20 day, first pair of trifoliate leaves was formed and after 40 days they possessed a bright-green, glabrous stem axis consisting of the hypocotyl (3.5 cm) and epicotyl (3.8 cm) and two pairs of trifoliate leaves. At day 60, plants had an axial root system, well developed, with many secondary roots and three pairs of compound leaves, arranged in alternate phyllotaxy, with membranous and hairy leaflets, with entire margin and bright green color. Central leaflets were larger than the laterals.

Cleome dendroides in vitro germination took place after 7-8 days of culture, by the rupture of the tegument in the hilum and root protrusion. After 8 days, hypocotyl differentiation and primary root elongation occurred. After 9 days, the tegument remained adhered to the cotyledons and hypocotyl elongation occurred. The emergence and expansion of cotyledons was observed after 9 - 10 days. The pair of primary leaves emerged around day 17 and epicotyl and secondary roots around the day 20. After 25 days, the seedling had all the essential structures: roots, hypocotyls, cotyledons and epicotyl with the pair of primary leaves expanded.

The start of *C. spinosa in vitro* germination took place after 1 - 2 days after inoculation. Hypocotyl and primary root developed after three days. The elevation of foliaceous cotyledons occurred after 5 days and epicotyl was formed after 7 days. The first pair of membranous and trifoliolate leaves was observed after 12 days and the visualization of the second pair of leaves occurred after 18 days.

Mature fruits of both *C. rosea* and *C. spinosa* contained seeds of varying color and size, prevailing dark brown seeds with larger dimensions, higher dry weight, and lower moisture. Considering *C. dendroides* seeds they are only brown. The analyzed lots showed moisture content ranging from 6 to 15%. Tegument color, along with these other characteristics, may provide data on the quality of the seed lot (Ochuodho and Modi, 2010). Physiological maturity of seeds is associated with low moisture content and at this stage, seeds reach maximum germination (Black et al., 2006). Thus, dark brown seeds of all species evaluated were considered fully mature.

The topographical tetrazolium test enables fast and efficient determination of the percentage of viable seeds, especially those that present dormancy, the recalcitrant and species that require long time to germinate in laboratory (Brasil, 2009). This test is routinely used in quality control programs for various species (Carvalho et al., 2013). The TTC test results showed seed viability ranging from 80 to 100% and the high viability of the lots was confirmed in germination tests.

Considering the tegument permeability, the results indicated that *Cleome* seeds do not exhibit dormancy in the tegument (physical dormancy) since it presented no resistance to water entry. Physical dormant seeds restrict imbibition by different mechanisms, such as the presence of one or more layers of lignified palisade cells on the tegument, making it impermeable (Baskin and Baskin, 2001).

The influence of substrate and temperature on the germination process showed that the optimum conditions varies among the species evaluated. The substrate influenced most significantly the germination in *C. rosea*, while for *C. spinosa* the alternating temperature was an essential condition. On the other hand, *C. dendroides* reached high germination rates in all conditions tested. Some species of the genus *Cleome* present no dormancy and germinate without the need of pretreatment. *Cleome amblyocarpa* germinates in the absence of light, and 25 °C was considered the ideal temperature, with germination percentage of 81% (Tlig et al., 2012). However, several species of *Cleome* exhibit physiological dormancy. Ekpong (2009) studying *C. gynandra* observed germination rate of 17.3% in freshly harvested seeds, at alternate temperatures (20 - 30 °C), without any pre-germination treatment. According to the author, the species requires a post-maturation period of three months, at environment temperature in order to break dormancy, after what germination rates were at 90%. Physiological dormancy was also reported for *C. lutea* (Cane, 2008), *C. serrulata* (Cane, 2008) and *C. hassleriana* (Raboteaux and Anderson, 2010).

In the present study the germination process was not influenced by light, characterizing the *Cleome* seeds as neutral photoblastic. However, experiments carried out to evaluate the influence of temperature and light on germination of *C. gynandra* from Africa (Ochuodho and Modi, 2007) concluded that the species showed photoinhibition. The authors recommend that germination should be performed in

the absence of light and/or alternate temperature of 20-30 °C.

Presented results show that *C. spinosa* needs the breaking of physiological dormancy in order to germinate, and this can be achieved with the use of alternate temperatures. This is in contrast to what was observed in *C. dendroides* and *C. rosea* that showed no dormancy. Alternate temperatures may favor dormancy breaking, by modulating hormone synthesis, altering endogenous levels and influencing seed germination (Black et al., 2006).

Germination was synchronic in *C. spinosa* and *C. dendroides* and asynchronous in *C. rosea*. Asynchronous germination leads to a wider distribution in seed germination time, which enhances the species' chances of survival under natural conditions by favoring the formation of seed banks (Baskin and Baskin, 2001). Among the studied species, *C. rosea* is the only that occurs in restinga, coastal ecosystems characterized by elevated temperatures and saline sandy soils. These abiotic features may require longer germination periods to ensure the species' establishment success. Asynchrony may be attributed to environmental oscillations during seed formation, as well as between individual intrinsic genetic variability within populations. Zamith and Scarano (2004), analyzing seedling production in species from Rio de Janeiro's restingas, observed large amplitude in germination time requirement.

In vitro conditions promote continuous physiological stress, which sometimes leads to hyperhydricity. Hyperhydric plants or buds present turgid, thick, wrinkled, twisted, translucent, with glass appearance, pale green stems and leaves, rigid and easily breakable. Shoots with shortened internodes and this morphological appearance suggest that water is in excess. The type of sealing may reduce or control the water on the atmosphere of culture flasks (Vasconcelos et al., 2012). In this present study, hyperhydricity was observed in plants of *C. dendroides* kept on MS medium, and the type of sealing showed no influence on its prevention. *In vitro* germination of *C. dendroides* was achieved on MS¼ medium with high GR yields. Besides the relative humidity, the concentration and type of gelling agent, and growth regulators, must be controlled to avoid hyperhydricity (Vasconcelos et al., 2012). The decrease in salt concentration, in particular of ammonium ions, may also be effective in preventing hyperhydricity (Guerra and Nodari, 2006), as seen in *C. dendroides*. The germination process under *in vitro* conditions of *C. rosea* seeds was not efficient. Despite the high percentage (85%) of seed germination in MS medium, plants presented low development of stem axis when compared to plants germinated *in vivo*. Callus formation was observed over the entire surface of the leaves.

The use of glucose as carbon source reduced germination

efficiency in comparison to sucrose. Carbohydrates play an important role in maintaining proper osmolarity on culture medium, as well as serving as nutrient source. The low germination percentage of *C. rosea* seeds inoculated in medium supplemented with glucose could be the result of an inadequate osmolarity to allow proper water exchange between the medium and seeds.

Calli observed in *C. rosea* leaves may be related to the presence of ethylene in the culture flasks. Under *in vitro* conditions ethylene may be produced after the buckling as a byproduct of combustion accumulating in vials. George (2008) reported the formation of callus-like structures on leaves and stems of intact plants kept on *in vitro* conditions, when they were maintained in ethylene enriched atmosphere for a prolonged time. The stimulation of cell division observed in *C. rosea* leaves and consequent callus formation may also have been due to auxin accumulation in their synthesis sites or in close regions, caused by the antagonistic action of ethylene on polar auxin transport (George, 2008). No reduction in leaf callogenesis was observed when flasks sealed with lids that allow greater gas exchange. According to Armstrong et al. (1997), only a forced ventilation system was able to significantly reduce ethylene levels inside culture flasks. Considering the possibility that *C. rosea* has a high sensitivity to ethylene, new methods to decrease its concentration in culture flasks must be employed in order to develop an effective protocol for germination the species under *in vitro* conditions.

During post-seminal development of *C. dendroides* and *C. spinosa*, it was observed faster development *in vitro* in comparison to *in vivo* conditions, which may be due to carbon supply (sucrose) and other nutrients present in culture medium, in addition to natural seed reserves.

Conclusions

The present study shows that optimal conditions for the germination are different for each of the species. *Cleome spinosa* has physiological dormancy, while *C. dendroides* and *C. rosea* do not exhibit dormancy under *in vivo* conditions. The different stages of *in vivo* post-seminal development of the three species are defined. In addition, an effective methodology for *in vitro* germination, enabling the providing of material to experiment on plant tissue culture is established to *C. dendroides* and *C. spinosa*.

Acknowledgements

The authors thank the technical support of Maria Francisca Santoro de Assunção (PROATEC/UERJ) and Adriana Maria Lanziotti (Qualitec/UERJ). The authors also thank Fundação de Amparo à Pesquisa do Estado do Rio de Janeiro (FAPERJ) and Conselho Nacional de Desenvolvimento Científico e Tecnológico (CNPq) for providing financial support.

References

AGRA, M. F.; FREITAS, P. F.; BARBOSA-FILHO, J. M. Synopsis of the plants known as medicinal and poisonous in Northeast of Brazil. *Brazilian Journal of Pharmacognosy*, v.17, p.114-140, 2007. http://www.scielo.br/pdf/rbfar/v17n1/a21v17n1.pdf

ALBARELLO, N.; SIMÕES, C.; ROSAS, P.F.G.; CASTRO, T.C.; GIANFALDONI, M.G.; CALLADO, C.H.; MANSUR, E. *In vitro* propagation of *Cleome spinosa* (Capparaceae) using explants from nursery-grown seedlings and axenic plants. *In Vitro Cellular and Developmental Biology – Plant*, v.42, p.601-606, 2006. http://link.springer.com/article/10.1079%2FIVP2006828

ALBARELLO, N.; RIBEIRO I.G.; SIMÕES, C.; CASTRO, T.C.; GIANFALDONI, M.G.; CALLADO, C.H.; KUSTER, R.M.; COELHO, M.G.P.; MANSUR, E. Histological analysis of calluses from *in vitro* propagated plants of *Cleome spinosa* Jacq. *Brazilian Journal of Biosciences*, v.5, p.699-701, 2007. http:// www.ufrgs.br/ seerbio/ojs/index.php/rbb/article/viewFile/599/504

ALBARELLO, N.; SIMÕES-GURGEL, C.; CASTRO, T. C.; GAYER, C.R.M.; COELHO, M.G.P.; MOURA, R.S.; MANSUR, E. Anti-inflammatory and antinociceptive activity of field growth plants and tissue culture of *Cleome spinosa* (Jacq.) in mice. *Journal of Medicinal Plants Research*, v.7, p.1043-1049, 2013. http://www.academicjournals.org/article/article1380718349_Albarello%20et%20al.pdf

APARADH, V.T.; MAHAMUNI, R.J.; KARADGE, B.A. Taxonomy and physiological studies in spider flower (*Cleome species*): a critical review. *Plant Sciences Feed*, v.2, p. 25-46, 2012. http://www.academia.edu/2054890/taxonomy_and_physiological_studies_in_spider_flower_cleome_species_a_critical_review

ARMSTRONG, J.; LEMOS, E.E.P.; ZOBAYED, S.M.A.; JUSTIN, S.; ARMSTRONG, W. A humidity-induced convective through flow ventilation sytem benefits *Annona squamosa* L. explants and coconut calloid. *Annals of Botany*, v.79, p.31-40, 1997. http://aob.oxfordjournals.org/content/ 79/1/31.full.pdf+html

BASKIN, C.C.; BASKIN, J.M. *Seeds. Ecology, biogeography, and evolution of dormancy and germination*. 1.ed. San Diego: Academic Press, 2001, 666p.

BLACK, M. J.; BEWLEY, J. D.; HALMER, P. *The Encyclopedia of Seeds:* Science, Technology and Uses, 2006. 828p.

BRASIL. Ministério da Agricultura, Pecuária e Abastecimento. *Regras para Análise de Sementes*. Ministério da Agricultura, Pecuária e Abastecimento. Secretaria de Defesa Agropecuária. Brasília: MAPA, 2009. 395p. http://www.bs.cca.ufsc.br/ publicacoes/regras%20analise%20sementes.pdf

CANE, J.H. Breeding biologies, seed production and species-rich bee guilds of *Cleome lutea* and *Cleome serrulata* (Cleomaceae). *Plant Species Biology*, v.23, p.152-158, 2008. http:// onlinelibrary.wiley.com/ doi/10.1111/j.1442-1984.2008.00224.x/pdf

CARVALHO, T.C.; KRZYZANOWSKI, F.C.; OHLSON, O.C.; PANOBIANCO, M. Tetrazolium test adjustment for wheat seeds. *Journal of Seed Science*, v.35, p.361-367, 2013. http://dx.doi.org/10.1590/S2317-15372013000300013

EKPONG, B. Effects of seed maturity, seed storage and pregermination treatments on seed germination of *Cleome* (*Cleome gynandra* L.). *Scientia Horticulturae*, v.119, p.236-240, 2009. http://ac.els-cdn.com/S0304423808003324/1-s2.0-S0304423808003324-main.pdf?_tid=f2a3b83e-1d5b-11e4-863c-00000aab0f01&acdnat=1407324327_d06ff8bd17e82982ca36aac8c709ac38

GEORGE, E.F. Plant Tissue Culture Procedure - Background. In: GEORGE, E.F., HALL, M.A., CLERK, G.J. *Plant propagation by tissue culture*. 3ªed. v.1 The background Springer: The Netherlands, 2008. 479p.

GUERRA, M.P.; NODARI, R.O. *Apostila de Biotecnologia* – LFDGV/CCA/UFSC. Florianópolis: Edição da Steinmacher, 2006, 41pp. http://www.lfdgv.ufsc.br/Apostila%20Biotecnologia.pdf

ILTIS, H.H. Studies in the Capparidaceae. VII. Old World Cleomes adventive in the New World. *Brittonia*, v.12, p.279-294, 1960. http://www.jstor.org/discover/10.2307/2805120?uid=3737664&uid=2&uid=4&sid=21104035229631

K'OPONDO, F.B. Morphological characterization of selected spider plant (*Cleome gynandra* L.) types from western Kenya. *Annals of Biological Research*, v.2, p.54-64, 2011. http:// scholarsresearchlibrary.com/ABR-vol2-iss2/ABR-2011-2-2-54-64.pdf

LABOURIAU, L.G. *A germinação das sementes*. Washington: Secretaria da OEA, 1983. 173p.

MURASHIGE, T.; SKOOG, F. A revised medium for rapid growth and bioassays with tobacco tissue cultures. *Physiologia Plantarum*, v.15, p.473-497, 1962.

OCHUODHO, J.O.; MODI, A.T. Light-induced transient dormancy in *Cleome gynandra* L. seeds. *African Journal of Agricultural Research*, v.2, p.587-591, 2007. http://www. academicjournals.org/article/ article1380898274_Ochuodho%20and%20Modi.pdf

OCHUODHO, J.O.; MODI, A.T. Association of seed coat colour with germination of three wild mustard species with agronomic potential *in Second RUFORUM Biennial Meeting Entebbe*, 229-232p, 2010. http://news.mak.ac.ug/documents/ RUFORUM/ Ochuodho.pdf

RABOTEAUX, N.N.G.; ANDERSON, N.O. Germination of *Cleome hassleriana* and *Polanisia dodecandra* seed lots in response to light, temperature and stratification. *Research Journal of Seed Science*, v.3, p.1-17, 2010. http://scialert.net/fulltext/?doi=rjss.2010.1.17&org=10

SIMÕES, C.; SANTOS, A.S.; ALBARELLO, N.; FIGUEIREDO, S.F.L. Shoot organogenesis and plantlet regeneration from stem explants of *Cleome rosea* Vahl (Capparaceae). *Journal Plant Biotechnology*, v.6, p.199-204, 2004. http://210.101.116.28/W_files/kiss2/05801997_pv.pdf

SIMÕES, C.; MATTOS, J.C.P.; SABINO, K.C.C.; CALDEIRA-DE-ARAÚJO, A.; COELHO, M.G.P.; ALBARELLO, N.; FIGUEIREDO, S.F.L. Medicinal potencial from *in vivo* and acclimatized plants of *Cleome rosea* Vahl ex DC. (Capparaceae). *Fitoterapia*, v.77, p.94-99, 2006. http://www.researchgate.net/publication/ 7393807_Medicinal_potential_from_in_vivo_and_acclimatized_plants_of_Cleome_rosea

SIMÕES, C.; ALBARELLO, N.; CALLADO, C.H.; CASTRO, T.C.; MANSUR, E. New approaches for shoot production and establishment of *in vitro* root cultures of *Cleome rosea* Vahl. *Plant Cell Tissue and Organ Culture*, v.98, p.79-86, 2009a.http://link.springer.com/article/10.1007%2Fs11240-0099540-z

SIMÕES, C.; BIZARRI, C,H,B.; CORDEIRO, L,S.; CASTRO,T, C; COUTADA,L,C,M.; SILVA,A,R,J.;ALBARELLO, N.;MANSUR,E. Anthocyanin production in callus cultures of Cleome rosea: Modulation by culture conditions and characterization of pigments by means of HPLC-DAD/ESIMS. *Plant Physiology and Biochemistry*, v.47, p.895-903, 2009b. http://www.sciencedirect.com/science/article/pii/S0981942809001466

SIMÕES, C.; CASTRO, T.C.; CORDEIRO, L.S.; ALBARELLO, N.; MANSUR, E.; ROMANOS, M.T.V. Antiviral activity of *Cleome rosea* extracts from field-grown plants and tissue culture-derived materials against acyclovir-resistant *Herpes simplex* viruses type 1 (ACVr-HSV-1) and type 2 (ACVr-HSV-2). *World Journal of Microbiology & Biotechnology*, v.26, p.93-99, 2010b. http://link.springer.com/article/ 10.1007%2Fs11274-009-0147-7#page-1

SIMÕES, C.; ALBARELLO, N.; CALLADO, C.H.; CASTRO, T.C.; MANSUR, E.A. Somatic embryogenesis and plant regeneration from callus cultures of *Cleome rosea* Vahl. *Brazilian Archives of Biology and Technology*, v.53, p.679-686, 2010a. http://www.scielo.br/scielo.php?script=sci_arttext&pid=S1516-89132010000300024

SIMÕES-GURGEL, C.; ROCHA, A.S.; CORDEIRO, L.S.; GAYER, C.R.M.; CASTRO, T.C.; COELHO, M.G.P.; ALBARELLO, N.; MANSUR, E.; ROSA, A.C.P. Antibacterial activity of field-grown plants, *in vitro* propagated plants, callus and cell suspension cultures of *Cleome rosea* Vahl. *Journal of Pharmacy Research*, v.5, p.3304-3308, 2012. http://connection.ebscohost.com/c/articles/84424609/antibacterial-activity-field-grown-plants-vitro-propagated-plants-callus-cell-suspension-cultures-cleome-rosea-vahl

TLIG, T.; GORAI, M.; NEFFAT, M. Factors influencing seed germination of *Cleome amblyocarpa* Barr. & Murb. (Capparidaceae) occurring in southern Tunisia. *Revue d Ecologie – La Terre et la Vie*, v.67, p.305-312, 2012. http://www.researchgate.net/publication/235220501_ Factors_influencing_seed_germination_of_Cleome_amblyocarpa_Barr.__Murb._(Capparidaceae)_occurring_in_southern_Tunisia

VASCONCELOS, A.G.V.; TOMAS, L.F.; CAMARA, T.R.; WILLADINO, L. Hiperidricidade: uma desordem metabólica. *Ciência Rural*, v.42, p.837-844, 2012. http://www.scielo.br/scielo.php?pid=S0103-84782012000500013&script=sci_arttext

ZAMITH, L.R.; SCARANO, F.R. Produção de mudas de espécies das Restingas do município do Rio de Janeiro, RJ, Brasil. *Acta Botanica Brasilica*, v.18, n.1, p.161-176, 2004. http://www.scielo.br/pdf/abb/v18n1/v18n1a14.pdf

Standard germination test in physic nut (*Jatropha curcas* L.) seeds

Glauter Lima Oliveira[1], Denise Cunha Fernandes dos Santos Dias[1*],

Paulo Cesar Hilst[1], Laércio Junio da Silva[1], Luiz Antônio dos Santos Dias[1]

ABSTRACT - Defining adequate methods to assess seed germination is important to control the quality of commercial lots, especially for species that are not yet included in the Rules for Seed Testing. This study aimed to establish an adequate procedure for germination test in physic nut seeds (*J. curcas* L.). Three seed lots, in eight replications of 25 seeds each, were sown on paper towel rolls moistened with a water volume equivalent to 2.7 the weight of the dry paper and in sterilized sand moistened up to 60% of its water-holding capacity. The seeds of each treatment were maintained on germination chambers at temperatures of 20, 25, 30 and 20-30 ºC. Daily counts were made to define the ideal date to perform the first and the last count test. Criteria for classifying seedlings as normal and abnormal were also established. The experiment was conducted in a completely randomized design in a split plot arrangement, and the means were compared by Tukey's test (P<0.05). For maximum germination potential of physic nut seeds, the germination test should be conducted at 25 ºC and 30 ºC, using sand or paper towel as a substrate, with the counts at 7 and 12 days after sowing.

Index terms: viability, methodology, seeds, physic nut.

Teste de germinação em sementes de pinhão manso (*Jatropha curcas* L.)

RESUMO - Estabelecer procedimento adequado para a condução do teste de germinação com espécies que ainda não constam nas Regras para Análise de Sementes é de grande importância para viabilizar a certificação e comercialização de lotes. Objetivou-se estabelecer metodologia para a condução do teste de germinação em sementes de pinhão manso (*J. curcas* L.). Sementes de três lotes, em oito repetições de 25, foram semeadas nos seguintes substratos: rolo de papel toalha umedecido com volume de água equivalente a 2,7 vezes o peso do papel seco, confeccionando-se rolos; areia esterilizada e umedecida até 60% da sua capacidade de retenção (entre areia). As sementes foram mantidas em germinadores nas temperaturas de 20, 25, 30 ºC e 20-30 ºC, realizando-se contagens diárias para definir a data ideal para as avaliações do teste. Foram estabelecidos também critérios para a classificação das plântulas como normais e anormais. O experimento foi conduzido em DIC, em esquema de parcela subdividida. As médias dos tratamentos foram comparadas pelo teste de Tukey (P<0,05). Para a obtenção do potencial máximo de germinação das sementes de pinhão manso, o teste de germinação deve ser conduzido nas temperaturas de 25 ºC ou 30 ºC, utilizando-se semeadura entre areia ou em rolo de papel, com contagens aos 7 e 12 dias após a semeadura.

Termos para indexação: viabilidade, metodologia, sementes, pinhão manso.

Introduction

Physic nut (*Jatropha curcas* L.) is an oilseed crop that belongs to the Euphorbiaceae family, and it has properties that are suitable for biodiesel production (Tiwari et al., 2007; Dias, 2011). The crop is mainly propagated by seeds collected from mother plants selected by producers, but they are marketed without a strict quality control, because there is still not an organized system in Brazil for producing and marketing these seeds (Silva et al., 2012).

Considering the demand for plant propagation material for the establishment of physic nut crops, the Ministry of Agriculture (MAPA), through the normative instruction number 4, of 01/14/2008, authorized the registration of the species *Jatropha curcas* L. on the National Register of Cultivars (RNC) without the requirement for a minimum supply of propagation material to be made available by a natural person or entity. Moreover, the seeds can be marketed on condition that the Statement of Commitment and Responsibility has been signed between seed producers and farmers. This statement will be required until standards

[1]Departamento de Fitotecnia, Universidade Federal de Viçosa, 36570-000 - Viçosa, MG, Brasil.
*Corresponding author <dcdias@ufv.br>

for identification and quality of the propagation material have been established (Brasil, 2008). Currently, there is no official standard for marketing physic nut seeds according to Normative Instruction number 45 of 09/17/2013 (Brasil, 2013), which sets standards for producing and marketing seeds of various crops. In addition, there is no information in the Rules for Seed Testing (Brasil, 2009) on conducting germination tests with seeds of physic nut.

In this context, standardized methods should be established for seed quality assessment to enable quality control of marketed lots (Silva et al., 2008; Pinto et al., 2009; Pascuali et al., 2012). The germination test is one of these methods. It must be carried out under controlled conditions so as to ensure standardization and reproducibility of results, which are optimum for the species. In this test, factors such as temperature, humidity, substrate and seeding method greatly influence the results, which should express the maximum potential germination of the lot (Marcos-Filho, 2005).

Information on the ideal conditions for the germination of physic nut seeds are not conclusive. Fogaça et al. (2007) found higher values for germination under the following conditions: sowing in sand under alternating temperature of 20-30 °C, with an 8-hour photoperiod or sowing in vermiculite or on paper roll, at 30 °C. This temperature was also ideal for seed germination both for seeding in sand or on paper roll. There was no germination below 20 °C, or at 45 °C or above (Dias et al., 2007), while for Martins et al. (2008), the germination test for physic nut seeds should be performed in sand and under alternating temperatures of 20-30 °C. In contrast, Neves et al. (2009) observed higher germination at 25 and 30 °C compared with 20 and 20-30 °C. Temperatures of 25 °C or 20-30 °C were the most suitable for seed germination on paper substrate with final counting at 12 days after sowing (Vanzolini et al., 2010). More recently, Pascuali et al. (2012), when evaluating different temperatures and substrates for germination, found higher germination at 30 °C in sand substrate and in the absence of light. However, Nobre et al. (2007) found higher germination at 30 °C or at 20-30 °C under light.

Defining criteria for seedling assessment is also important when it aims to establish an appropriate methodology for the germination test. In this sense, there are few studies on the characterization of post-seminal development of physic nut.

The purpose of this study was to establish appropriate methodology for conducting the standard germination test of physic nut seeds.

Material and Methods

The research was conducted in the Seed Laboratory in the Department of Plant Science, Federal University of Viçosa,

and used three lots of physic nut seeds with initial moisture content of approximately 8.5%, as determined by the oven method at 105 ± 3 °C, for 24 hours (Brasil, 2009).

The seeds of each lot were submitted to the germination test using eight replicates of 25 seeds, with the substrates and sowing methods described below:

Germination test on paper roll: seeds were distributed on two sheets of paper towel, sterilized in an oven at 120 °C for an hour, and moistened with water equivalent to 2.7 times the weight of the dry substrate, according to the results obtained in preliminary tests. Then, the seeds were covered with a sheet of paper and rolled up. *Germination test in sand:* 1500 mL polypropylene boxes were used; they contained sand that was washed, sterilized and moistened until 60% water-holding capacity, according to Brasil (2009). Sowing was done about 1.5 cm deep.

After sowing, each treatment described above was kept in germinators at temperatures of 20, 25, 30 °C and 20-30 °C (16 h at 20 °C and 8 h at 30 °C, every 24 h) under light for eight hours a day. Daily counts were performed of the number of normal seedlings obtained each day to determine the accumulated germination curves, so as to identify the best times to perform the first and second counts. Seedlings were also illustrated and criteria were established for classifying them as normal or abnormal. The following evaluations were made:

Germination percentage: consisted of the number of seedlings characterized as normal obtained at 10 and 12 days after sowing; values are expressed in percentage terms. *First count of germination test:* represented by the number of normal seedlings obtained at 5 and 7 days after start of the germination test, and results were expressed in percentage terms. Dates for counts were based on data obtained for the curves of accumulated germination. *Germination rate (GR) and Germination Rate Index (GRI):* were conducted together with the germination test; daily counts were made of the number of normal seedlings until stabilization of counts. GR (days) and GRI (seedlings.day^{-1}) were calculated according to Nakagawa (1999). *Seedling dry matter (DM):* normal seedlings obtained after each count of the germination test were dried in a forced air circulation oven, at 60 ± 2 °C, until constant weight. The results were expressed in g.seedling^{-1}.

The experiment was conducted in a completely randomized design with four replications. The data underwent the Shapiro-Wilk normality test. Then, they underwent analysis of variance in a split plot design: the plots contained the temperatures (20, 25, 30 and 20-30 °C) and the substrates (paper towel and sand), and the subplots were comprised of the assessment times (5, 7, 10 and 12 days) of the germination test. The means for the temperature factor were compared by Tukey's test ($p<0.05\%$), and the mean of factors substrate and assessment times, by the F-test ($p<0.05\%$). Data analysis was performed using the SAS software (SAS, 2009).

Results and Discussion

Figure 1 shows the comparison of the substrates. For the three lots, higher values of germination and germination first count were obtained when using sand at temperatures of 25 and 30 °C, while the temperature of 20 °C and 20-30 °C generally resulted in higher germination on the paper roll.

For germination temperatures (Figure 1), for the three

lots which used the sand substrate, higher germination potential and first germination count occurred at 25 and 30 °C, while on paper roll, values obtained for these variables at 25, 30 and 20-30 °C did not differ and were higher than those observed at 20 °C. The lower values for both germination and first count were obtained at 20 °C, regardless of substrate. The lower germination observed in tests conducted at 20 °C can be attributed to the low metabolic activity of seeds at a lower temperature (Socolowski and Takaki, 2004;. Bewley et al, 2013), resulting in a lower percentage of normal seedlings at the end of the test.

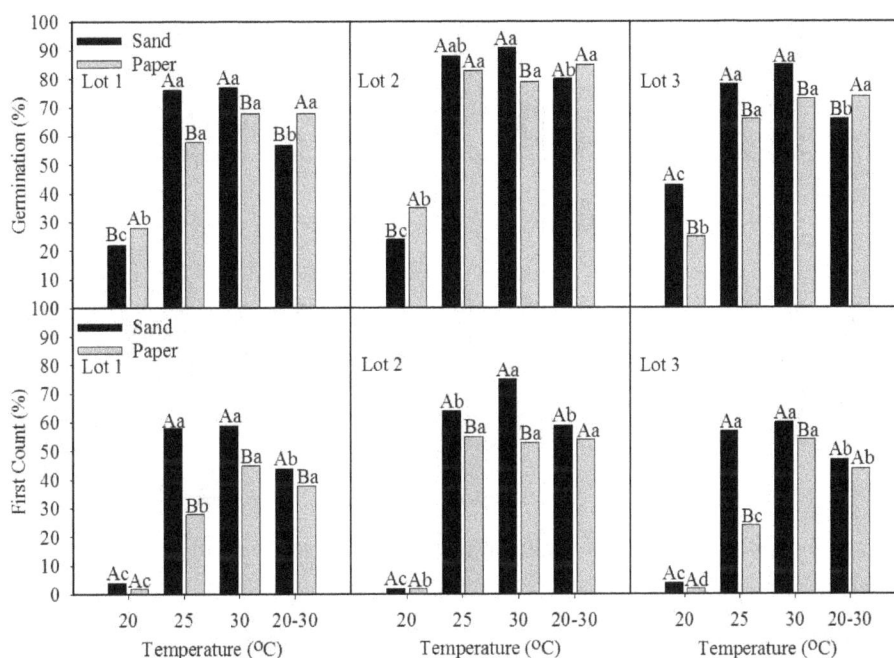

Figure 1.   Mean values of germination percentage and first germination count of three lots of *J. curcas* L. in sand and on paper roll under different temperatures.
*Capital letters compare substrates at each temperature (F, $p< 0.05$).
*Lowercase letters compare the temperatures in each substrate (Tukey, $p<0.05$).

The temperature of 25 °C was recommended for conducting the germination test of physic nut seeds in sand and on paper roll by Neves et al.(2009) and Vanzolini et al. (2010). However, Fogaça et al. (2007) observed that the maximum values of germination were not found at this temperature, which was also observed by Martins et al. (2008), and ranging the temperature between 20 and 30 °C best expressed seed germination, both in sand and on paper. In contrast, Dias et al. (2007), found higher values when the germination test was conducted at 30 and 35 °C. These authors also reported that physic nut seeds did not germinate at temperatures below 20 °C or above 45 °C. These results reinforce that, to date, there is no consensus as to the most suitable temperature for the germination of these seeds.

Figure 2 shows that a lower germination rate was obtained at 20 °C for all lots, a result which confirms the worse performance of the seeds at this temperature. According to Godoi and Takaki (2005), temperature affects water absorption rate, and according to Höes et al. (2004) and Bewley et al. (2013), it also affects the biochemical reactions that regulate the metabolism of the germination process. Therefore, it affects both the rate and uniformity of germination and total germination. According to Carvalho and Nakagawa (2012), germination occurs only within certain limits of temperature, and there is an optimum temperature at which the process occurs with maximum efficiency, yielding maximum germination in the shortest possible period.

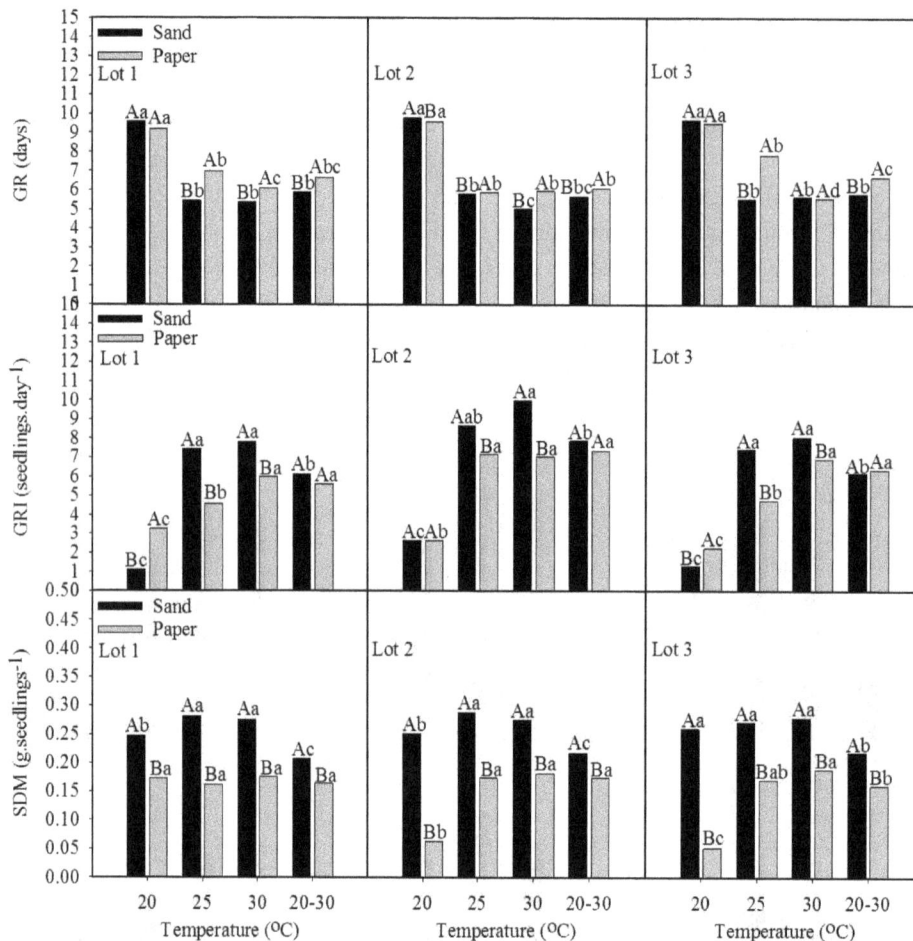

Figure 2. Mean values of germination rate (GR), germination rate index (GRI) and seedling dry matter (SDM) of three lots of
J. curcas L. in sand and on paper roll, under different temperatures.
*Capital letters compare substrates at each temperature (F, $p< 0.05$).
*Lowercase letters compare temperatures in each substrate (Tukey, $p<0.05$).

As observed in the germination test (Figure 1), a higher germination rate (Figure 2), both by GRI and GR, was obtained at 25 and 30 °C in the sand substrate. In general, a lower germination rate was observed at 20 °C, with no significant difference in GR between the substrates for lots 1 and 3. Neves et al. (2009) also observed a reduction in the germination rate index in physic nut seeds germinated at 20 °C.

In general, it is observed that the physic nut seeds from the three lots expressed their maximum germination potential at 25 and 30 °C in sand, where values were greater than those for paper roll (Figure 1). It was also observed that both GRI (Figure 2) and DM (Figure 3) were higher at 25 and 30 °C also in sand substrate. Similar results were found by Martins et al. (2008). Importantly, there was higher contamination by microorganisms on paper roll than in sand, which explains the higher values of germination in the latter substrate. There is no direct contact between substrate and seedling throughout the test in sand, as opposed to paper roll. Thus, seedlings

may be able to avoid microorganisms that can affect their development, resulting in higher germination.

However, the advantages of using the paper substrate over sand should be taken into account: for example, ease of test setup, less space used in the germinators, and no need for sterilization.

The substrate is a physical medium whose purpose is to maintain the right conditions for germination and seedling development. According to Martins et al. (2008), the substrate directly affects germination and, hence, germination rate, because of its structure, aeration, water holding capacity, degree of pathogen infestation, among others; thus, it may or may not increase the rate and percentage of seed germination.

The highest rates of seedling dry matter (Figure 2) were observed at 25 and 30 °C in the sand substrate, possibly because this substrate allows better seedling development after germination compared with the paper roll, which poses some physical restraint to seedling growth. At all temperatures, there was a higher development of seedlings in sand. Seedling growth influences the definition of times for performing the first and final count of the germination test, and also the interpretation of the test results, because seedlings should reach a minimum level of development so that their basic structures can be assessed and they can be classified as normal or abnormal.

Figure 3 shows the seedlings characterized as normal. These seedlings have a well-developed hypocotyl, and their root system comprises a well developed taproot and four basal roots in its periphery, or a short taproot with well-developed basal roots. Considering the criteria for abnormality of seedlings established by RST (Brasil, 2009), abnormal seedlings were considered as such when they showed atrophy of the entire root system, stunted and necrotic root, negative geotropism, twisting or folding of the hypocotyl, short and thick hypocotyl, damage at the point where the hypocotyl connects with the endosperm, and seeds in early germination without characterizing a seedling (Figure 4).

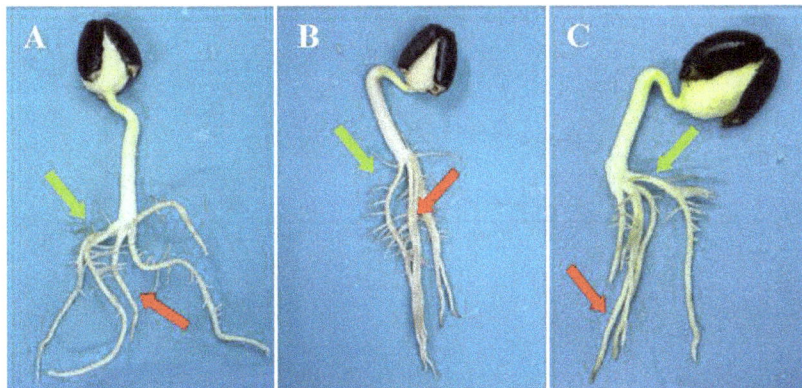

Figure 3. Structural characteristics of normal seedlings of *J. curcas* L.: A, B, C - well-developed hypocotyl and root system comprised of a well developed root (red arrow) and four secondary roots in its periphery (green arrow).

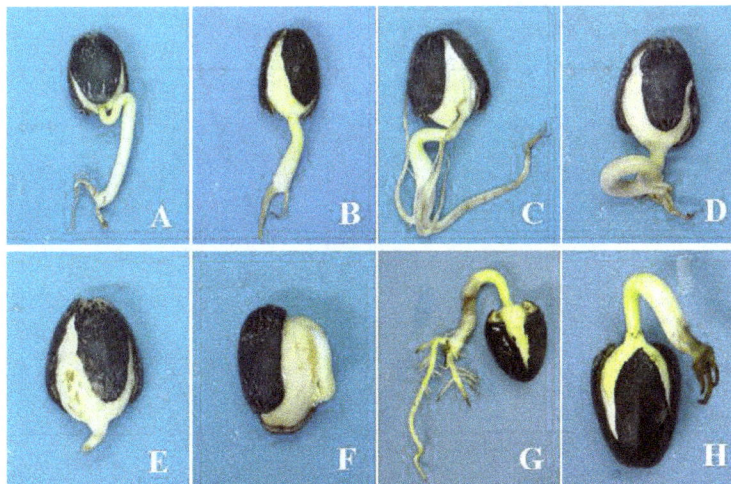

Figure 4. Structural characteristics of abnormal seedlings of *J. curcas* L.: A, B - stunted root; C- negative geotropism; D - twisted hypocotyl and stunted root system; E, F - seed in early germination without characterizing a seedling; G - infected seedling; H - stunted and necrotic root system.

The data collected from the daily assessment of germination enabled the design of cumulative germination curves for each treatment (Figure 5). From the third day of conducting the test it was observed the appearance of normal seedlings stabilizing between the tenth and twelfth day following the start of the test suggest most if not already normal seedlings, leaving only non-germinated seeds, which have been classified how hard (did not absorb water) or dead (loose seeds and attacked by microorganisms, according to Brasil (2009).

Five days after sowing (Figure 5), it is also observed, for most treatments, that the values obtained for normal seedlings were below 50%, and the seedlings were not very developed yet, while on the seventh day, in the germination tests in sand at 25, 30 and 20-30 °C and on paper at 30 and 20-30 °C, the percentage of normal seedlings was above 50%. A different behavior was observed at 20 °C, in both substrates. Some seedlings were still underdeveloped and it was difficult to classify them as normal at both 10 and 12 days after sowing.

These results in both Figures 5 and 6 reinforce the periods of 7 and 12 days as the most appropriate for performing the first and final germination count for physic nut seeds. Figure 6 shows

that the temperature of 20 °C resulted in the worst performances in both count periods. Higher germination percentage was obtained in tests conducted at temperatures of 25, 30 and 20-30 °C, with counts performed at 7 (first count) and 12 (final count) days after sowing. Comparing these results with those of germination (Figure 1), where the temperatures were compared within each substrate, higher values of germination at 25 and 30 °C were observed in sand for the three lots. On the paper roll, there was no significant difference between germination at temperatures of 25, 30 and 20-30 °C.

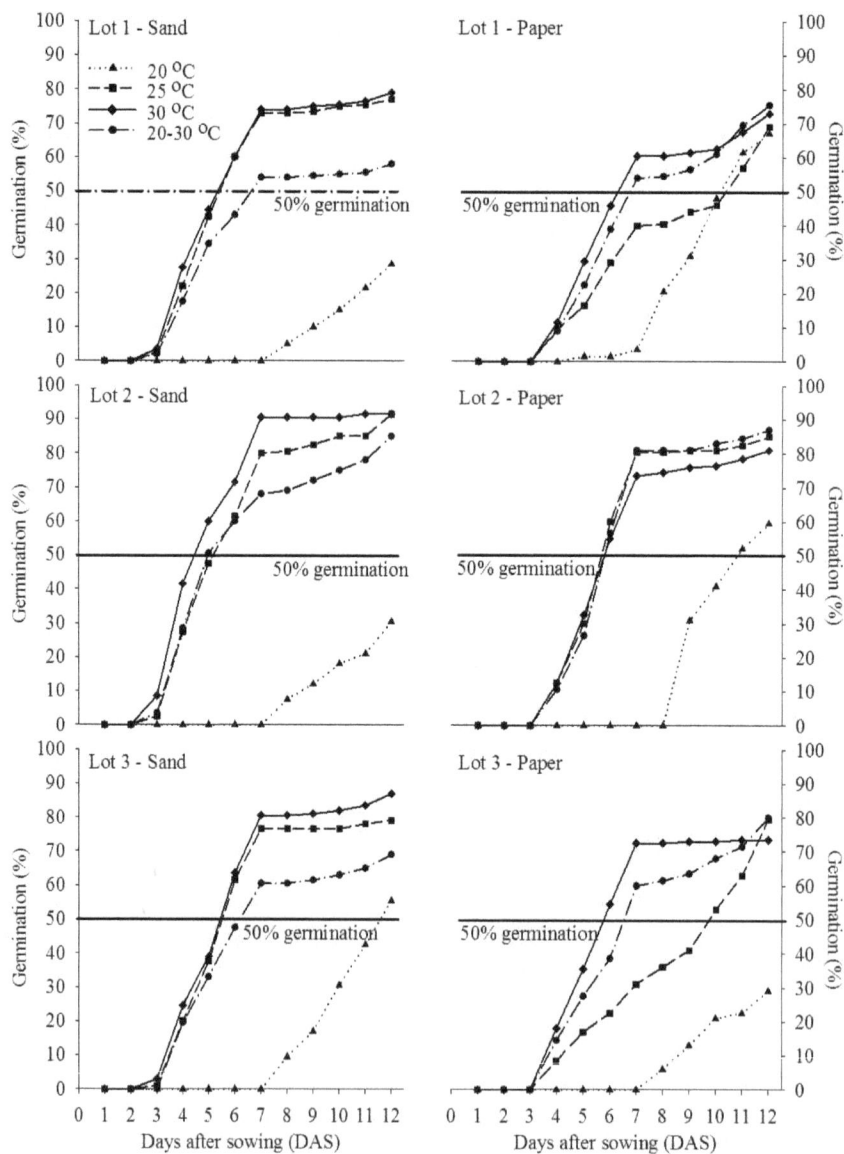

Figure 5. Cumulative germination of three lots of *J. curcas* L. in sand and on paper roll under different temperatures.

To date, there is no consensus in the literature regarding the most appropriate time for assessing the seed germination test with this crop. According to Martins et al. (2008), the

first count and the final count of germination in physic nut seeds should be performed at 5 and 10 days after sowing, respectively. Similar results were found by Silva et al.

(2008) who recommended using the temperature of 30 °C, on paper roll, with final count at 10 days after sowing, while Neves et al. (2009) indicated the 10th and 15th days are suitable for the first and final count of the germination test. The results obtained in this study showed that evaluations of the test must be made at 7 and 12 days after sowing, since there was stabilization of scores from the 12th day, and there was increased occurrence of microorganisms with the seedlings remaining on the paper roll for a few more days.

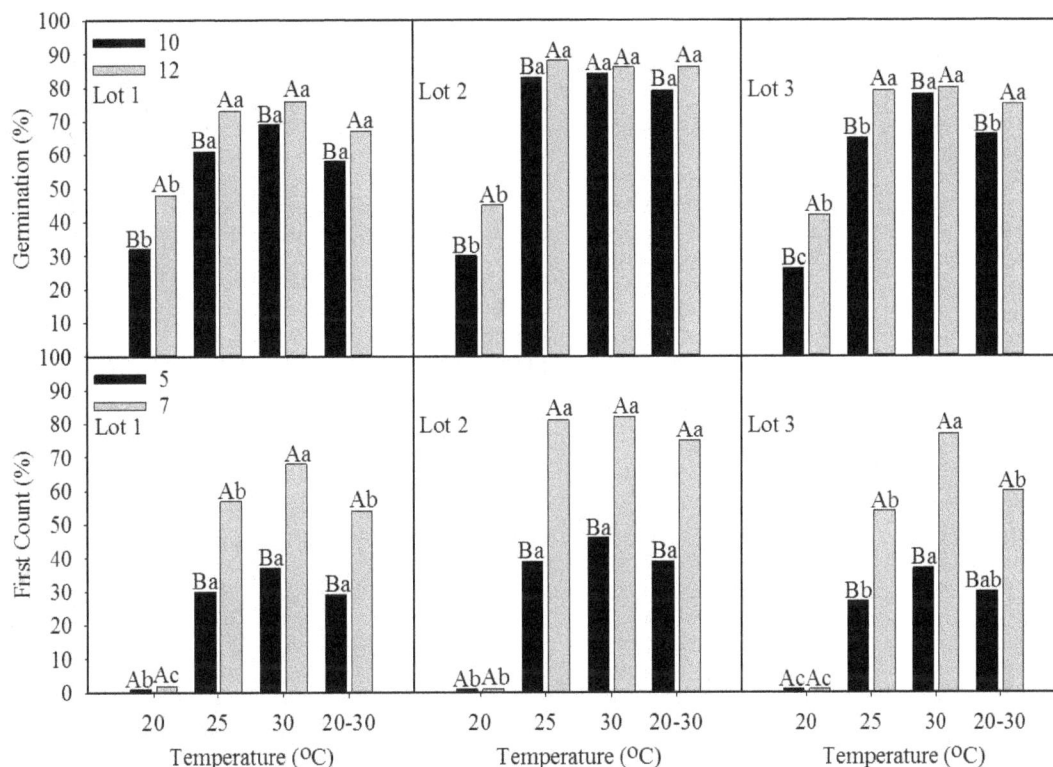

Figure 6. Mean values of germination at 10 and 12 days after sowing, and first germination count, at 5 and 7 days after sowing, obtained for the three seed lots of *J. curcas* L. at different temperatures.
*Capital letters compare assessment times at each temperature (F, $p < 0.05$).
*Lowercase letters compare temperatures at each assessment time (Tukey, $p < 0,05$).

Conclusions

For physic nut seeds, the germination test should be conducted at temperatures of 25 °C or 30 °C, with seeding in sand or paper roll.

The first and final count of the germination test should be performed at 7 and 12 days after sowing, respectively.

Acknowledgments

The authors are thankful to CNPq and FAPEMIG for their financial support and the graduate research grant given to the first author.

References

BEWLEY, J. D.; BRADFORD, K. J.; HILROST, H. W. M.; NONOGAKI, H. *Seeds:* physiology of development, germination and dormancy. 3 ed., New York: Springer, 381p. 2013.

BRASIL. Instrução normativa nº 4, de 14 de janeiro de 2008. Diário Oficial da União, Ministério da Agricultura, Pecuária e Abastecimento, Brasília, DF, 15 jan. 2008, Seção I, p.4.

BRASIL. Instrução normativa nº 45, de 17 de setembro de 2013. Diário Oficial da União, Ministério da Agricultura, Pecuária e Abastecimento, Brasília, DF, 18 set. 2013, Seção I, p.16.

BRASIL. Ministério da Agricultura, Pecuária e Abastecimento. *Regras para análise de sementes.* Ministério da Agricultura, Pecuária e Abastecimento. Secretaria de Defesa Agropecuária. Brasília: MAPA/ACS, 2009. 395p. http://www.agricultura.gov.br/arq_editor/file/2946_regras_analise__sementes.pdf

CARVALHO, N.M.; NAKAGAWA, J. *Sementes*: ciência, tecnologia e produção. 5 ed. Jaboticabal: FUNEP, 590 p. 2012.

DIAS, L.A.S. Biofuel plant species and the contribution of genetic improvement. *Crop Breeding and Applied Biotechnology*, v.1, p.16-26, 2011. http://dx.doi.org/10.1590/S1984-70332011000500004

DIAS, M.P.; DIAS, D.C.F.S.; DIAS, L.A.S. Germinação de sementes de pinhão manso (*Jatropha curcas* L.) em diferentes temperaturas e substratos. In: II Congresso da Rede Brasileira de Tecnologia de Biodiesel, 2007, Brasília. *Anais do II Congresso da Rede Brasileira de Tecnologia de Biodiesel*. Brasília: MCT/ABIPIT, p. 1-5, 2007.

FOGAÇA, C.A.; SILVA, L.L; POLIDORO, J.C.; BREIER, T.B.; LELES, P. S.S. Metodologia para a condução do teste de germinação em sementes de *Jatropha curcas* L. In: 4°Congresso Brasileiro de Plantas Oleaginosas, Óleos, Gorduras e Biodiesel. *Anais*. Varginha, p.1351-1357, 2007. http://oleo.ufla. br/anais_04/artigos/pdfs/a4206.pdf

GODOI, S.; TAKAKI, M. Efeito da temperatura e a participação do fitocromo no controle da germinação de sementes de embaúba. *Revista Brasileira de Sementes*, v.27, n.2, p.87-90, 2005. http://dx.doi.org/10.1590/S0101-31222005000200013

HÖES, A.; SCHULCH, L.O.B.; PESKE, S.T.; BARROS, A.C.S.A. Emergência e crescimento de plântulas de arroz em resposta à qualidade fisiológica de sementes. *Revista Brasileira de Sementes*, v.26, n.1, p.92-97, 2004. http://dx.doi.org/10.1590/S0101-31222004000100014

MARCOS- FILHO, J. *Fisiologia de sementes de plantas cultivadas*. FEALQ: Piracicaba, 2005. 495p.

MARTINS, C.C.; MACHADO, C.G.; CAVASINI, R. Temperatura e substrato para o teste de germinação de sementes de pinhão manso. *Revista Ciência e Agrotecnologia*, v.32, n.3, p.863-868, 2008. http://dx.doi.org/10.1590/S1413-70542008000300024

NAKAGAWA, J. Testes de vigor baseados no desempenho das plântulas. In: KRZYZANOWSKI, F.C.; VIEIRA, R.D.; FRANÇA-NETO, J.B. (Ed.). *Vigor de sementes:* conceitos e testes. Londrina: ABRATES, Cap.2, p.1-21, 1999.

NEVES, J.M.G.; SILVA, H.P.; BRANDÃO-JÚNIOR, D.S; MARTINS, E.R.; NUNES, U.R. Padronização do teste de germinação para sementes de pinhão-manso. *Revista Caatinga*, v.22, n.4, p.76-80, 2009. http://periodicos. ufersa.edu.br/revistas/index.php/sistema/article/view/1262

NOBRE, D.A.C; ANDRADE, J.A.; DAVID, A.M.S.; RESENDE, J.C.F.; FARIA, M.A.V.R.; DAVID, D.A. Germinação de sementes de pinhão-manso submetidas a diferentes condições de temperaturas. In: 4° Congresso Brasileiro de Plantas Oleaginosas, Óleos, Gorduras e Biodiesel. *Anais*. Varginha, p.1119- 1126, 2007.

PASCUALI, L.C.; SILVA, F.S.; PORTO, A.G.; SILVA FILHO, A.; MENEGHELLO, G.E. Germinação de sementes de pinhão manso em diferentes temperaturas, luz e substratos. *Semina: Ciências Agrárias*, v.33, n.4, p.1435-1440, 2012. http://dx.doi.org/10.5433/1679-0359.2012v33n4p1435

PINTO, T.L.F.; MARCOS FILHO, J.; FORTI, V.A.; CARVALHO, C.D.; GOMES JUNIOR, F.G. Avaliação da viabilidade de sementes de pinhão manso pelos testes de tetrazólio e de raios X. *Revista Brasileira de Sementes*, v.31, n.2, p.195-201, 2009. http://dx.doi.org/10.1590/S0101-31222009000200023

SAS- Program SAS - Getting started with the SAS Learning Edition. North Carolina: Cary SAS Publishing, 2009, 200p.

SILVA, H.P.; NEVES, J.M.G.; BRANDÃO JUNIOR, D.S.; COSTA, C.A. Quantidade de água do substrato na germinação e vigor de sementes de pinhão-manso. *Revista Caatinga*, v.21, n.5, p.178-184, 2008. http:// periodicos.ufersa.edu.br/revistas/index.php/sistema/article/view/666

SILVA, L.J.; DIAS, D.C.F.S.; MILAGRES, C.C.; DIAS, L.A.S. Relationship between fruit maturation stage and physiological quality of physic nut (*Jatropha curcas* L.) seeds. *Revista Ciência e Agrotecnologia*, v.36, n.1, p.39-44, 2012. http://dx.doi.org/10.1590/S1413-70542012000100005

SOCOLOWSKI, F.; TAKAKI, M. Germination of *Jacaranda mimosifolia* (D. Don-Bignoniaceae) seeds: effects of light, temperature and water stress. *Brazilian Archives of Biology and Technology*, v.47, p.785-792, 2004. http:// dx.doi.org/10.1590/S1516-89132004000500014

TIWARI, A.K.; KUMAR, A.; RAHEMAN, H. Biodiesel production from jatropha oil (*Jatropha curcas*) with high free fatty acids: An optimized process. *Biomass and Bioenergy*, v.31, n.8, p.569–575, 2007. http://dx.doi. org/10.1016/j.biombioe.2007.03.003

VANZOLINI, S.; MEORIN, E.B.K.; SILVA, R.A.; NAKAGAWA, J. Qualidade sanitária e germinação de sementes de pinhão-manso. *Revista Brasileira de Sementes*, v.32, n.4, p.9-14, 2010. http://dx.doi.org/10.1590/S0101-31222010000400001

Fruit processing and the physiological quality of *Euterpe edulis* Martius seeds

Patrícia Ribeiro Cursi[1]*, Silvio Moure Cicero[2]

ABSTRACT - Different pulping methods were analyzed for *Euterpe edulis*, with regard to its seed physiological quality, and the physiological performance of seeds submitted to continuous drying and monitored by radiographs was assessed. In order to do so, 2x2 factorial design treatments were carried out, using two different manners to store fruit prior to pulping, immersed in water and not. Also, two different methods for the pulping of fruit were used for each lot; namely, manual and mechanical. Seeds from fruit that had been immersed in water at 40 °C for 20 minutes showed greater physiological performance than those that had not been immersed in water previously. Immersion in water at 55 °C resulted in seed death. The pulp extraction method, whether manual or mechanical, did not affect seed physiological performance. Radiographs made it possible to observe that the volume occupied by embryos in the embryo cavity reduced with decreased water content. Mechanical pulping showed no reduction in percentage of seedling emergence for water content up to 33.3%. However, reduction in water content to 16.1% caused seed death. The drying of manually pulped seeds until reaching 39.0% water content did not adversely affect seedling emergence, whereas lower water contents than 25.6% caused seed death.

Index terms: juçara palm, tolerance to drying, pulped fruit, X - ray.

Métodos de despolpa e qualidade fisiológica de sementes de *Euterpe edulis* Martius

RESUMO - Foram estudados os efeitos de diferentes métodos de despolpa de frutos de *Euterpe edulis* sobre a qualidade fisiológica das sementes, e avaliado o desempenho fisiológico de sementes submetidas à secagem contínua assistida por tomada de imagens radiográficas. Para tanto, foram realizados tratamentos em arranjo fatorial 2x2, sendo duas formas de pré-condicionamento dos frutos, com e sem imersão em água, e dois métodos de despolpa, manual e mecânico, para cada lote. As sementes obtidas a partir de frutos imersos em água a 40 °C por 20 minutos apresentaram desempenho fisiológico superior. A temperatura da água a 55 °C para a imersão dos frutos foi prejudicial e ocasionou a morte das sementes. A forma de extração da polpa, não influenciou o desempenho fisiológico das sementes. As imagens de raios X possibilitaram observar a redução do volume ocupado na cavidade embrionária pelo embrião com a redução do teor de água. Para despolpa mecânica, não há redução na percentagem de emergência de plântulas para teor de água de até 33,3% e para 16,1% de água a desidratação é letal. A secagem das sementes, despolpadas manualmente, até 39,0% de água não prejudica a emergência de plântulas e a 25,6% é letal.

Termos para indexação: palmiteiro juçara, tolerância à secagem, despolpa, raios - X.

Introduction

Euterpe edulis Martius, popularly known as "juçara palm", is a native tree of the Atlantic Rainforest of Brazil that has great ecological and socioeconomic importance.

Palms of the genus *Euterpe* are the most commonly grown for the production of hearts of palm and the great market demand has resulted in its disorderly exploitation, adding it to the list of Brazilian endangered flora species.

The agribusiness of hearts of palm has significantly grown over the years and, according to the latest survey on the production of plant extraction and forestry, there has been a 13.1% increase in production, totaling 5563 tonnes of cultivated hearts of palm (IBGE, 2011).

The *Euterpe edulis* hearts of palm is one of the most exploited non-timber forest products of the Atlantic Forest (Fantini and Guries, 2007). Aiming at decreasing the predatory process, nonprofit organizations and government

[1]Coordenadoria de Assistência Técnica Integral - CATI, Caixa Postal 962, 13070-172 - Campinas, SP, Brasil.

[2]Departamento de Produção Vegetal, USP/ESALQ, Caixa Postal, 9, 13418-900 - Piracicaba, SP, Brasil.
*Corresponding author < patrícia.cursi@cati.sp.gov.br>

agencies have encouraged other options for the sustainable exploitation of this species, and the marketing of its fruit pulp to be used in human consumption has great potential for that (Cembraneli et al., 2009).

The production of native forest seed currently grows due to the increased demand for seedlings to be used in the recovery of degraded areas and the preservation of protected natural reserves (Smith et al., 2007). Especially in the state of São Paulo, a large number of companies from the agribusiness industry have undertaken reforestation projects in order to obtain environmental certification and access to financial subsidies (Brancalion et al., 2011). Thus, residues from the *Euterpe edulis* fruit agribusiness processing to extract pulp are "clean" seeds, with no epicarp nor mesocarp, which can be traded for the production of seedlings in nurseries for native forest species.

Considering that this species can only propagate through seed dispersal, the production of good quality seeds - taking into account their genetic, physical, physiological, and hygienic aspects - is critical to ensuring healthy and vigorous seedlings for the success of reforestation, enrichment of native woodlands and commercial crops.

E. edulis seeds are round and may range from 8 to 15 mm in diameter (Fleig and Rigo, 1998; Pizo et al., 2006; Martins et al., 2009b) and from 0.35 to 1.81 grams of dry matter (Brancalion et al., 2011). When freshly harvested, seeds have high water content (Brancalion et al., 2011; Roberto et al., 2011) and as shown by research studies on their storage conditions, are recalcitrant due to their sensitivity to dehydration and short life period when stored (Bovi and Cardoso, 1978; Queiroz and Cavalcante, 1986; Reis et al., 1999; Martins et al., 2000; Andrade, 2001). The very small embryo accounts for only 0.54% of the total seed mass and it has only a single prominent cotyledon, a short hypocotyl-radicle axis and epicotyl, storing protein and lipid substances as sources of reserve (Panza et al., 2004).

As with most palms, germination of *E. edulis* is unbalanced and slow (Bovi and Cardoso, 1976; Panza et al., 2004; Roberto and Habermann, 2010; Roberto et al., 2011), and may take from 60 to 90 days, not because the embryo is undeveloped, but because of a physical impediment that hinders the penetration of water (Bovi and Cardoso, 1976). Their thick mesocarp and hard endocarp could act as a barrier to seed imbibition, causing dormancy (Roberto et al., 2011).

The sowing of the whole fruit, if compared to that of shelled seed, presents lower germination values and increased mean time to germinate (Bovi and Cardoso, 1976). The pulping of fruits was favorable when considering germination speed and uniformity (Bovi et al., 1987). The pulping of fruit,

which consisted of soaking it in water for three to four days, caused a stimulating effect of the pre-soaking treatment on the final percentage and speed of germination (Bovi, 1990). Furthermore, Martins et al. (2004) found that storing the whole fruit was unfavorable to the physiological quality of seed compared with storing pulped seeds.

Thus, the potential of having seed production combined with the production of pulp becomes a great technical, economic, and environmental move.

Given the above, this study aimed to assess how different methods of pulping *Euterpe edulis* fruit may affect the physiological performance of seeds and its tolerance to drying seeds, accompanied by radiographic images.

Material and Methods

Ripe fruits of *Euterpe edulis* were collected manually at two locations and in two different moments, forming two lots. Lot 01 was collected on December 1, 2011 from mother plants located in Parque das Neblinas, in the city of Mogi das Cruzes (Lat 23° 44' 52" S, Log 46° 09' 44" W); and Lot 02 was collected on May 24th, 2012 from mother plants near by the Experimental Station of the Agronomic Institute of Campinas, in the city of Ubatuba (Lat 23° 24' 38" S, Log 45° 06' 59" W). Both lots were transported to Piracicaba, SP, and stored in a cold chamber at 10 °C until the start of treatment, for a period of 24 hours.

Initially, ripe fruit was selected, cleaned and washed in tap water. Then it was sanitized in a 200 ppm solution of crop sanitizer (Sumaveg®) for 15 minutes.

After that, treatments were performed as follows: Treatment 1 (T1) - fruits were immersed in water and mechanically pulped; Treatment 2 (T2) - fruits were not immersed in water and were mechanically pulped; Treatment 3 (T3) - fruits were immersed in water and manually pulped; Treatment 4 (T4) - fruits were not immersed in water and were manually pulped.

For lot 01, immersion in water at 40 ± 2 °C was conducted for 20 min, and this temperature was strictly kept under control and it was found to be the most appropriate in previous trials. For Lot 02, the immersion temperature used was 55 °C, for a period of 20 minutes, to simulate a less strict control with respect to temperature, considering the reality of the locations where fruit processing is made, such as traditional communities and rural populations. It was not possible to conduct any of these two treatments for neither lot due to the limited availability of seeds.

Pulp from processing was discarded and the seeds placed in plastic trays and kept in lab environment to remove excess water from the processing of fruit, for 24 hours.

Seed initial water content was determined in the oven at 105 ± 3 °C for 24 h (Brasil, 2009).

A seedling emergence test was conducted to assess its physiological performance. A sterilized vermiculite moistened with water content equivalent to 60% of its water holding capacity was used as substrate. The test was conducted in a location with no control of temperature and relative humidity. The substrate was moistened and normal seedlings were weekly counted up until 90 days, in accordance with the Rules for Seed Testing (Brasil, 2009); and the final count was made at 135 days, period for which seedling emergence should stabilize.

Tests for the first count of seedling emergence, speed of seedling emergence, shoot length and dry weight of seedling, were performed simultaneously to the emergency test.

Seedling emergence was first counted when at least one treatment showed 10% of normal seedlings emerged. Seedling emergence speed was calculated based on the count of normal seedlings, every seven days for each replication, by applying the equation described by Maguire (1962). At the end of the emergency test, normal seedlings were removed from the substrate and shoot length was measured and expressed in mm. Once that had been determined, normal seedlings, separated from the rest of the seed, were stored in paper bags and placed in an oven at 70 °C for 48 hours prior to weighing. The dry weight of seedlings was expressed in grams.

Aiming to assess tolerance to drying and the evolution of damage caused by it, seeds from treatment 1, considered to be the recommended treatment for obtaining pulp, as well as seeds from treatment 4, in this work assessed as control treatment, were subjected to continuous drying in an oven with air circulation at temperature of 30 ± 2 °C at intervals of approximately 3% from initial water content, until exhaustion of available seed. For each water content, determined through the oven method at 105 °C (Brasil, 2009), the seeds were radiographed by a MX -20 Faxitron X- ray digital equipment, coupled to a Core 2 Duo computer (3.16 GHz, 2 GB RAM and 160 GB Hard Disk) with a MultiSync LCD 1990SX 17-inch monitor. Seeds were distributed in plastic sheets and numbered according to their position, allowing for their identification to carry out subsequent measurements.

The experimental design was completely randomized with five replicates of 20 seeds. The analysis of variance for the pulping factors was conducted separately for each test, in a 2 x 2 factorial arrangement (manual and mechanical pulping, with and without prior immersion in water). Treatment mean comparison was made using Tukey's test at 5% probability, with the aid of a 9.1 SAS software. Seedling emergence first count data were transformed into Log (x+1), for not having

normal distribution.

For tolerance to drying, the experimental design used for all water contents obtained was the completely randomized type; and mean comparison for each water content was conducted using the Tukey test at 5% probability, with the aid of a 9.1 SAS statistical software.

Results and Discussion

Pulping Methods

Table 1 shows seed physiological performance for each pulping method. Seed initial water content ranged from 43.0% to 45.0% for Lot 01 and from 45.9% to 46.7% for Lot 02, between treatments.

Mechanical impact caused by pulp extractor did not cause damage to the embryo. No significant differences were found between the manual and mechanical pulping methods used for the two lots analyzed (Table 1). The process of mechanical pulping to remove parts of the fruit had already been recommended to other genera of family Arecaceae (Ferreira and Gentil, 2006; Costa and Marchi, 2008) with the intent to accelerate and standardize the germination process.

Regardless of the pulping method, whether manual or mechanical, seeds from treatments that included fruit immersed in water at 40 °C for 20 minutes showed greater physiological performance, according to results found for Lot 01. These treatments resulted in an increase of 19% and 14% in the percentage of seedling emergence, in each type of pulping, mechanical and manual, respectively, at 90 days after test application (Table 1). As for the final count, made at 135 days, an increase of 8% was noted for the mechanical pulping and 5% for manual pulping. Similarly, differences found in speed of seedling emergence at 90 days were greater than at 135 days. Thus, although there is an increase in the percentage of seedling emergence at 135 days, the 90-day assessment was enough to show differences in the physiological performance of treatments, confirming the assessment time period recommended by the Rules for Seed Testing (Brasil, 2009).

Moreover, for the first count of seedling emergence, at 39 days after the application of test, a difference of up to 16% was noted for treatments with and without prior seed immersion in Lot 01. Thus, differences in vigor for each treatment were more obvious at the beginning of tests, up to 90 days, and reduced over time.

The slow germination of the seeds of this palm is due to a physical impairment that hinders the penetration of water (Bovi and Cardoso, 1976; Roberto et al., 2011). When ripe, the fruit of the palm has a fully formed seed, ready to start the seed germination process. The embryo, although small, is fully developed. The only two impediments to germination are fruit pulp and the presence of a serous operculum that hinders the penetration of water in the seed (Bovi, 1990).

Table 1. Results obtained for water content (WC), seedling emergence at 90 days (E 1) and at 135 days (E 2), emergence speed rate at 90 days (ESR 1) and at 135 days (ESR 2), first count of seedling emergence (FC), shoot length of seedling (SL), and dry mass of seedling (DM) for pulping treatments applied to *Euterpe edulis* fruit.

Parameter	Pulping	Lot 01		CV (%)	Lot 02		CV (%)
		With Immersion (40 °C/20')	Without Immersion		With Immersion (55 °C/20')	Without Immersion	
WC (%)	Mechanical	44.6	45.0	-	46.2	46.7	-
	Manual	43.0	45.0	-	46.5	45.9	-
E 1 (%)	Mechanical	91 aA[1]	72 aB	8.53	0 aA[1]	72 aB	31.27
	Manual	86 aA	72 aB		0 aA	71 aB	
E 2 (%)	Mechanical	96 aA	88 aB	6.26	0 aA	82 aB	18.48
	Manual	96 aA	91 aB		0 aA	72 aB	
ESR 1	Mechanical	1.658 aA	1.100 aB	13.76	0 aA	1.04 aB	26.88
	Manual	1.476 aA	1.146 aB		0 aA	0.95 aB	
ESR 2	Mechanical	1.700 aA	1.214 aB	11.18	0 aA	1.125 aB	23.79
	Manual	1.560 aA	1.300 aB		0 aA	1.906 aB	
FC (%)	Mechanical	18 aA[2]	2 aB[2]	99.07	0 aA[2]	10 aB[2]	103.93
	Manual	8 aA[2]	1 aB[2]		0 aA[2]	5 aB[2]	
SL (mm)	Mechanical	20.06 aA	17.12 aB	7.34	0 aA	14.67 aB	9.36
	Manual	19.70 aA	17.27 aB		0 aA	14.75 aB	
DM (g)	Mechanical	3.21 aA	2.60 aB	10.16	0 aA	2.01 aB	21.07
	Manual	3.19 aA	2.60 aB		0 aA	1.77 aB	

[1]For each examined characteristic, means followed by the same lower case letter in the column and capital on the line do not differ statistically among themselves by Tukey test at 5% probability.
[2]The figures are the original data. For statistical analysis, the values were transformed into Log (x+1), for not having normal distribution.

Removing or simple perforating the operculum in palm seeds has proven to be efficient, as it increases the percentage of germinated seeds, according to Myint et al. (2010) and Goudel et al. (2013). Bovi and Cardoso (1976) have found for *E. edulis* that a slight scarification in the germ pore causes increased germination. In that same study, they observed that fruit immersed in water at room temperature, for a period of three days, and which had their pulp removed manually, had seeds with higher performance than those that were not pulped. Similar results were found by Bovi et al. (1987) for *E. oleracea* and *E. edulis*, where pulped fruit showed improvement in the speed and uniformity of germination. Increased germination speed rate for seeds of forest species represents an advantage, because it proves the important relationship of germination ability with the establishment of seedling in natural environment (Gomes et al., 2006; Brancalion and Marcos- Filho, 2008).

Treatments that included prior fruit immersion in water at 40 °C for 20 min, showed differences of 2.94 and 2.43 cm for shoot length, and 0.61 and 0.59 g for seedling dry matter, for each type of pulping, mechanical and manual, respectively,

demonstrating the beneficial effects that immersion has provided to seeds from Lot 01 (Table 1).

When assessing the consequences of pre-imbibition at room temperature for a period of three days, Bovi (1990) observed detrimental effects on fruit due to the presence of pathogenic microorganisms easily developed in high pulp humidity, affecting the seed. After fruit pre-imbibition, pulp was removed and "clean" seeds, exclusively, were sown with the intent to obtain better results. Similarly, two-day pre-imbibition of seeds clear of oleaginous pulp has positively affected the final germination percentage.

Thus, fruit immersion in water, in addition to improving both manual and mechanical extraction of pulp, it has probably eased the protrusion of plant bud. Immersion temperature of 40 °C may have decreased operculum mechanical strength, caused by changes in the physical properties of its serous layer.

As for seeds from fruit immersed in water at 55 °C for 20 minutes, immersion was not favorable to physiological performance, and this temperature has proven to be lethal for this species. Seeds from the same lot, which were not subjected to immersion, had greater performance than the immersed seeds, for all parameters studied.

Costa and Marchi (2008) have proposed that seed hydration at high temperatures can positively affected palm seed germination. The immersion of seeds from species of genera *Acrocomia* e *Astrocaryum* in a water bath at 65 and 70 °C, for 2-3 weeks, was favorable for germination.

Wolkers et al. (1998) have reported that protein denaturation temperature is directly related to tolerance to seed drying. Studies on *Arabdopsis thaliana*, showed that for wild species, tolerant to drying, proteins did not denature when subjected to 150 °C. As for mutants sensitive to water loss, there has been reduced temperature of protein denaturation for values from 68 to 87 °C.

Considering that *Euterpe edulis* seeds are highly sensitive to water loss, it is possible to assume that the temperature of 55 °C, used for the fruit, may have caused the denaturation of proteins that are essential for germination, which justifies the death of seeds in that treatment.

Tolerance to drying

The evolution of the drying process made it possible to monitor the reduction of the volume occupied by the embryo within the embryo cavity (Figure 1); thus, the embryo has been associated with the physiological performance of seeds (Tables 2, 3 and 4). When mapping images to test results, it was possible to note that treatments with the largest number of embryos occupying all or most of the embryo cavity have matched the ones that showed better physiological performance. Reduced water content resulted in decreased volume of embryo, appearance of void spaces, and increased percentage of dead seeds for treatments 01 and 04 in Lot 01, and for treatment 04 in Lot 02 (Figure 1).

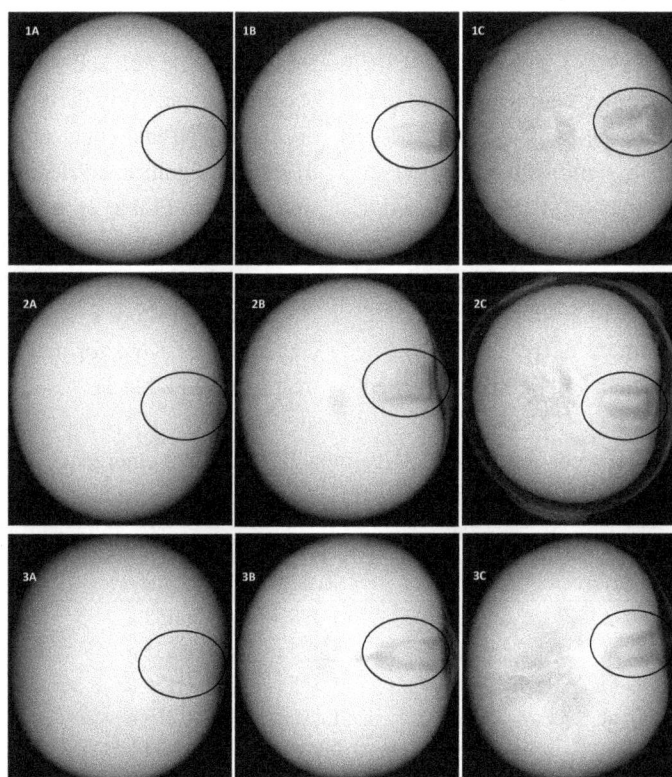

Figure 1. Representative image of the reduction of the volume occupied by the embryo within the embryo cavity throughout the drying process. 1 - Radiographic images that refer to treatment 01, Lot 01. 1A - initial water content of 44.9%; 1B - critical water content of 33.3%; 1C - lethal water content of 16.1%. 2 - Radiographic images that refer to treatment 04, Lot 01. 2A - initial water content of 44.9%; 2B - critical water content of 39.0%; 2C - lethal water content of 20.4%. 3 - Radiographic images that refer to treatment 04, Lot 02. 3A - initial water content of 45.9%; 3B - critical water content of 37.0%; 3C - lethal water content of 27.1%. Embryo is highlighted.

For treatment 01 of Lot 02, in which fruit were immersed in water at 55 °C for 20 minutes, it was observed that even for initial water contents, embryos did not occupy the whole embryo cavity, and presented reduced volume compared with other treatments, yielding dead seeds for all water contents assessed (Figure 2). Protein denaturation caused by high temperature may have caused the disorganization and breakdown of membranes, resulting in the death of the embryo, detected by tests that assessed the physiological performance of different methods of pulping (Table 1) and shown in radiographs (Figure 2).

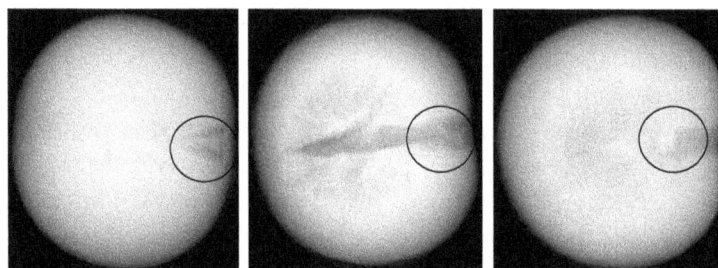

Figure 2.   Representative image of the reduction of the volume occupied by the embryo within the embryo cavity during the drying process for treatment 01 of Lot 02 (in water at 55 °C for 20 minutes), which yielded dead seeds. A - initial water content of 46.2%; B - intermediate water content of 27.8%; C - water content at the end of the drying process 13.6%. Embryo is highlighted.

The progressive dehydration of seeds intensified the process of deterioration, evidenced by the reduction of all parameters for all treatments (Tables 2, 3 and 4), except for treatment 01 of Lot 02, which performance was canceled due to the high immersion temperature used.

For Lot 01, there has been a significant reduction in the percentage of seedling emergence at 90 days from water contents 33.3% and 39.0%; and at 135 days for water contents 33.3% and 35.3% in treatments 01 e 04 (Tables 2 and 3), respectively. The emergence of seedlings was nulled for treatment 01, when water content decreased to 16.1% (Table 2); and for treatment 04, when water content reached 20.4% (Table 3).

Regarding the speed of seedling emergence, a significant reduction to 35.7% and 39.0% for treatments 01 and 04 has occurred, respectively (Tables 2 and 3), as well as for seedling dry mass. The first count of seedling emergence for treatment 01 showed significant difference of 42.7% compared with initial water content (Table 2), and this was the test that showed the greatest sensitiveness to water loss. No seedling emergence was noted on the first count at 39 days for treatment 04 (Table 3). A reduction in shoot length of seedlings was noted with water content from 18.6% for treatment 01 (Table 2) and 22.4% for treatment 04 (Table 3).

Table 2.   Results obtained for drying seeds from Lot 01, Treatment 01 - seedling emergence and emergence speed rate (ESR) at 90 and 135 days, first count of seedling emergence (FC), shoot length of seedling (SL) and dry mass of seedling (DM).

T 01	Water Content (%)	Emergence (%)		ESR		FC (%)	SL (cm)	DM (g)
		90 days	135 days	90 days	135 days			
U0	44.9	80 ab[1]	87 ab	1.388 a	1.452 a	7 a	18.70 a	2.67 ab
U1	42.7	88 a	95 a	1.495 a	1.555 a	2 b	20.08 a	3.07 a
U2	39.6	87 a	94 a	1.385 a	1.456 a	2 b	20.47 a	2.68 ab
U3	35.7	76 ab	83 ab	1.110 b	1.216 b	0 b	18.56 a	2.39 bc
U4	33.3	71 b	75 bc	0.908 b	1.034 bc	0 b	18.30 a	1.77 cd
U5	29.8	55 c	67 c	0.639 c	0.858 c	0 b	18.25 a	1.87 cd
U6	27.0	45 c	46 d	0.549 c	0.631 d	0 b	19.14 a	1.29 d
U7	23.2	16 d	16 e	0.209 d	0.224 e	0 b	17.78 ab	0.48 e
U8	18.6	4 de	5 ef	0.044 d	0.047 ef	0 b	9.5 b	0.13 e
U9	16.1	0 e	0 f	0.000 d	0.000 f	0 b	0.00 c	0.00 e
U10	15.2	0 e	0 f	0.000 d	0.000 f	0 b	0.00 c	0.00 e
U11	11.5	0 e	0 f	0.000 d	0.000 f	0 b	0.00 c	0.00 e
U12	8.8	0 e	0 f	0.000 d	0.000 f	0 b	0.00 c	0.00 e
CV (%)		15.21	15.77	17.32	14.57	148.59	31.78	24.44

[1]Means followed by the same letters, within each column, do not differ by Tukey test at 5% probability.

Table 3. Results obtained for drying seeds from Lot 01, Treatment 04 - seedling emergence and emergence speed rate (ESR) at 90 and 135 days, first count of seedling emergence (FC), shoot length of seedling (SL) and dry mass of seedling (DM).

T 04	Water Content (%)	Emergence (%)		ESR		FC (%)	SL (cm)	DM (g)
		90 days	135 days	90 days	135 days			
U0	44.9	73 ab[1]	87 a	1.175 a	1.294 a	0 a	17.84 a	2.39 ab
U1	42.0	76 a	86 a	1.206 a	1.286 a	0 a	19.24 a	2.73 a
U2	39,0	50 cd	77 ab	0.718 b	0.946 b	0 a	13.90 a	2.01 bc
U3	35.3	53 c	67 b	0.742 b	0.886 b	0 a	16.66 a	1.79 c
U4	33.0	56 bc	64 b	0.697 b	0.806 b	0 a	16.61 a	1.55 c
U5	29.7	33 de	40 c	0.362 c	0.491 c	0 a	16.06 a	0.93 d
U6	25.8	16 ef	17 d	0.204 cd	0.225 d	0 a	16.27 a	0.39 de
U7	22.4	5 f	5 de	0.049 d	0.063 d	0 a	13.21 a	0.12 e
U8	20.4	0 f	0 e	0.000 d	0.000 d	0 a	0.00 b	0.00 e
U9	17.7	0 f	0 e	0.000 d	0.000 d	0 a	0.00 b	0.00 e
U10	14.7	0 f	0 e	0.000 d	0.000 d	0 a	0.00 b	0.00 e
U11	10.8	0 f	0 e	0.000 d	0.000 d	0 a	0.00 b	0.00 e
U12	8.1	0 f	0 e	0.000 d	0.000 d	0 a	0.00 b	0.00 e
CV (%)		28.82	20.99	32.40	23.93	-	27.82	27.34

[1]Means followed by the same letters, within each column, do not differ by Tukey test at 5% probability.

For Lot 02 seeds, there has been a significant reduction in percentage of seedling emergence at 90 and 135 days with water content 37.0% for treatment 04 (Table 4). For that same water content, a reduction in seedling dry mass was also found in relation to initial water content (Table 4). Seedling emergence was nulled from water content of 24.1% (Table 4). The speed rate of seedling emergence, as well as its first count, showed significant difference of 38.9% compared with initial water content. Seedling shoot length only showed the effects of drying for water content of 25.6%. For treatment 01, no seedling emergence was noted for any of the water contents analyzed, since prior immersion of fruit in water at 55 °C for 20 minutes has caused the death of embryos.

Table 4. Results obtained for drying seeds from Lot 02, Treatment 04 - seedling emergence and emergence speed rate (ESR) at 90 and 135 days, first count of seedling emergence (FC), shoot length of seedling (SL) and dry mass of seedling (DM).

T 04	Water Content (%)	Emergence (%)		ESR		FC (%)	SL (cm)	DM (g)
		90 days	135 days	90 days	135 days			
U0	45.9	88 a[1]	89 a	1.290 a	1.305 a	60 a	14.99 a	2.26 a
U1	38.9	70 ab	78 ab	0.923 b	0.999 b	0 b	14.60 a	1.97 a
U2	37.0	52 b	63 bc	0.651 c	0.759 bc	0 b	14.03 ab	1.47 b
U3	33.0	29 c	45 cd	0.352 d	0.512 cd	0 b	12.18 ab	0.93 c
U4	31.6	26 c	33 de	0.310 de	0.381 de	0 b	11.06 ab	0.62 c
U5	27.1	0 d	13 ef	0.000 f	0.132 ef	0 b	8.24 abc	0.19 d
U6	25.6	7 d	9 f	0.103 ef	0.103 f	0 b	7.25 bc	0.17 d
U7	24.1	0 d	1 f	0.000 f	0.010 f	0 b	2.30 cd	0.00 d
U8	19.8	0 d	0 f	0.000 f	0.000 f	0 b	0 d	0.00 d
U9	14.6	0 d	0 f	0.000 f	0.000 f	0 b	0 d	0.00 d
U10	10.1	0 d	0 f	0.000 f	0.000 f	0 b	0 d	0.00 d
CV (%)		34.38	32.97	32.49	30.43	24.35	41.79	28.79

[1]Means followed by the same letters, within each column, do not differ by Tukey test at 5% probability.

The effects of dehydration on the physiological performance of *Euterpe edulis* seeds were also examined by Panza et al. (2007), Martins et al. (2009a, 2009b) and Roberto et al. (2011), indicating that the reduction in water content to values between 35-30% has hindered seed physiological performance with initial water content ranging from 50-45%. These same authors have pointed out that water content from 18 to 29% is lethal for seeds of *Euterpe edulis*.

Water loss in recalcitrant seeds may trigger several deteriorative processes, such as protein denaturation, changes in enzyme activity, and damage to the system of membranes, resulting in complete loss of seed feasibility (Nautiyal and Purohit, 1985; Wang et al., 2012). Furthermore, it may lead to a reduction in the speed of the metabolic processes, slowing the development of embryo during the pre-germination stage.

The physiological performance of seeds from fruit immersed in water at 40 °C for 20 minutes during the drying process was greater than that of seeds obtained from fruit that were not previously immersed in water. Besides the fact that the temperature of immersion has favored the protrusion of plant buds, as discussed above, the impact of the propellers in mechanical pulping has detached the fibrous portion of the mesocarp; while through manual pulping only the epicarp and the oleaginous portion of the mesocarp were removed, leaving the fibrous layer of the mesocarp adhered to the seed, making germination hard to occur and increasing the capacity of water retention when compared to the "clean" seeds. Thus, the mechanically pulped seeds have no physical barrier that hinders the germination process, leading to a faster germination, as evidenced by the first count of seedling emergence and speed of seedling emergence. Therefore, the mechanical pulping method may be considered advantageous for seedling emergence and for the subsequent seedling establishment in the soil.

Conclusions

Pulping methods where fruit are immersed in water at 40 °C for 20 minutes result in greater seed physiological performance when compared with methods that do not include prior fruit immersion in water.

Water temperature at 55 °C for fruit immersion for 20 minutes is harmful and causes seed death.

The method of pulp extraction, whether mechanical or manual, does not affect the physiological performance of seeds.

In mechanical pulping, no percentage reduction in seedling emergence is noted for water content of up to 33.3%. Reduced water content to 16.1% causes seed death.

The dehydration of manually pulped seeds, of up to 39.0% water, does not affect seedling emergence. Water content lower than 25.6% causes seed death.

References

ANDRADE, A.C.S. The effect of moisture content and temperature on the longevity of heart of palm seeds (*Euterpe edulis*). *Seed Science and Technology*, v.29, p.171-182, 2001. http://www.academia.edu/881988/The_ effect_of_moisture_content_and_temperature_on_the_longevity_of_heart_ of_palm_Euterpe_edulis_seeds

BOVI, M.L.A.; CARDOSO, M. Seed germination of *Euterpe edulis* Mart. *Bragantia*, v.35, n.1, p.23-29, 1976. http://www.scielo.br/pdf/brag/v35n1/26.pdf

BOVI, M.L.A.; CARDOSO, M. Conservação de sementes de palmiteiro (*Euterpe edulis* Mart.). *Bragantia*, v.37, n.1, p.65-71, 1978.

BOVI, M.L.A.; GODOY-JUNIOR, G.; SAES, L.A. Pesquisas com os gêneros *Euterpe* e *Bactris* no Instituto Agronômico de Campinas. *Agronômico*, v.39, n.2, p.129-174, 1987.

BOVI, M.L.A. Pré-embebição em água e porcentagem e velocidade de emergência de sementes de palmiteiro. *Bragantia*, v.49, n.1, p.11-22, 1990. http://www.scielo.br/pdf/brag/v49n1/02.pdf

BRANCALION, P.H.S.; MARCOS-FILHO, J. Distribuição da germinação no tempo: causas e importância para a sobrevivência das plantas em ambientes naturais. *Informativo ABRATES*, v.18, p.11-17, 2008.

BRANCALION, P.H.S.; NOVEMBRE, A.D.L.C.; RODRIGUES, R.R. Seed development, yield and quality of two palm species growing in different tropical forest types in SE Brazil: implications for ecological restoration. *Seed Science and Technology*, v.39, n.2, p.412-424, 2011. http://www. ingentaconnect.com/content/ista/sst/2011/00000039/00000002/art00013

BRASIL. Ministério da Agricultura, Pecuária e Abastecimento. *Regras para análise de sementes*. Ministério da Agricultura, Pecuária e Abastecimento. Secretaria de Defesa Agropecuária. Brasília: MAPA/ACS, 2009. 395p. http://www.agricultura.gov.br/arq_editor/file/laborat%c3%b3rio /sementes/ regras%20para%20analise%20de%20sementes.pdf

CEMBRANELI, F.; FISH, T.V.; CARVALHO, C.P. Exploração sustentável da palmeira *Euterpe edulis* Mart. no Bioma Mata Atlântica, Vale do Paraíba-SP. *Revista Ceres*, v.56, n.3, p.233-240, 2009. http://cncflora.jbrj.gov.br/ plataforma2/arquivos/biblio/4fc3c155aef18_V56N003P51809.pdf

COSTA, C.J.; MARCHI, E.C.S. *Germinação de sementes de palmeiras com potencial para a produção de agroenergia*. Planaltina-DF: Embrapa Cerrados. 2008. 35p.

FANTINI, A.C.; GURIES, R.P. Forest structure and productivity of palmiteiro (*Euterpe edulis* Martius) in the Brazilian Mata Atlântica. *Forest Ecology and Management*, v.242, p.185-194, 2007. http://cmq.esalq.usp.br/wiki/lib/exe/ fetch.php?media=biometria:palmito:fantini-guries-2007.pdf

FERREIRA, S.A.N.; GENTIL, D.F.O. Extração, embebição e germinação de sementes de tucumã (*Astrocaryum aculeatum*). *Acta Amazonica*, v.36, n.2, p.141-146. 2006. http://www.scielo.br/pdf/%0D/aa/v36n2/v36n2a02.pdf

FLEIG, F.D.; RIGO, S.M. Influência do tamanho dos frutos do palmiteiro *Euterpe edulis* Mart. na germinação das sementes e crescimento das mudas. *Ciência Florestal*, v.8, n.1, p.35-41, 1998.

GOMES, P.B.; VÁLIO, I.F.M.; MARTINS, F.R. Germination of *Geonoma brevispatha* (Arecaceae) in laboratory and its relation to the palm spatial distribution in a swamp. *Aquatic Botany*, v.85, p.16-20, 2006. http://www. sciencedirect.com/science/article/pii/S0304377006000155

GOUDEL, F.; SHIBATA, M.; COELHO, C.M.M.; MILLER, P.R.M. Fruit biometry and seed germination of *Syagrus romanzoffiana* (Cham.) Glassm. *Acta Botanica Brasilica*, v.27, n.1, p.147-154, 2013. http://www.scielo.br/ pdf/abb/v27n1/v27n1a15.pdf

IBGE. Instituto Brasileiro de Geografia e Estatística. *Produção da Extração Vegetal e da Silvicultura*, v.26, 2011. Accessed on Apr. 15th. 2013. ftp://ftp. ibge.gov.br/Producao_Agricola/Producao _da_Extracao_Vegetal_e_da_ Silvicultura_%5Banual%5D/2011/pevs2011.pdf.

MAGUIRE, J.D. Speed of germination: aid in selection and evaluation for seedling emergence and vigor. *Crop Science*, v.2, n.2, p.176–177, 1962.

MARTINS, C.C.; BOVI, M.L.A.; NAKAGAWA, J.; GODOY JÚNIOR, G. Temporary storage of jussara palm seeds: effects of time, temperature and pulp on germination and vigor. *Horticultura Brasileira*, v.22, n.2, p.271-276, 2004. http://www.scielo.br/pdf/hb/v22n2/21029.pdf

MARTINS, C.C.; BOVI, M.L.A.; NAKAGAWA, J.; MACHADO, C.G. Secagem e armazenamento de sementes de juçara. *Revista Árvore*, v.33, p.635-642, 2009a. http://www.scielo.br/pdf/rarv/v33n4/v33n4a06.pdf

MARTINS, C.C.; MACHADO, C.G.; NAKAGAWA, J.; OLIVEIRA, S.S.C. Tamanho e secagem de sementes de palmeira Jussara sobre a germinação e o vigor. *Revista Caatinga*, v.22, p.117-120, 2009b. http://periodicos.ufersa.edu.br/revistas/index.php/sistema/article/view/1132/584

MARTINS, C.C.; NAKAGAWA, J.; BOVI, M.L.A. Desiccation tolerance of four seed lots from *Euterpe edulis* Mart. *Seed Science and Technology*, v.28, p.101-113, 2000.

MYINT, T.; CHANPRASERT, W.; SRIKUL, S. Germination of seed of oil palm (*Elaeis guianeensis* Jacq.) as affected by different mechanical scarification methods. *Seed Science and Technology*, v.38, n.3, p.635-645, 2010. http://www.ingentaconnect.com/content/ista/sst/2010/00000038/00000003/art00011

NAUTIYAL, A.R.; PUROHIT, A.N. Seed viability in sal. II. Physiological and biochemical aspects of ageing in seeds of *Shorea robusta*. *Seed Science and Technology*, v.13, n.1, p.69-76, 1985.

PANZA, V.; LÁINEZ, V.; MALDONADO, S. Seed structure and histochemistry in the palm *Euterpe edulis*. *Botanical Journal of the Linnean Society*, v.145, n.4, p.445-453, 2004. http://onlinelibrary.wiley.com/doi/10.1111/j.1095-8339.2004.00293.x/pdf

PANZA, V.; LÁINEZ, V.; MALDONADO, S.; MARODER, H.L. Effects of desiccation on *Euterpe edulis* Martius seeds. *Biocell*, v.31, n.3, p.383-390, 2007. http://www.scielo.org.ar/pdf/biocell/v31n3/v31n3a04.pdf

PIZO, M.A.; ALLMEN VON, C.; MORELLATO, L.P.C. Seed size variation in the palm *Euterpe edulis* and the effects of seed predators on germination and seedling survival. *Acta Oecologica*, v.29, p.311-315, 2006. http://www.sciencedirect.com/science/article/pii/S1146609X05001414

QUEIROZ, M.H.; CAVALCANTE, M.D.T.H. Efeito do dessecamento das sementes de palmiteiro na germinação e no armazenamento. *Revista Brasileira de Sementes*, v.8, n.3, p.121-125. 1986.

REIS, A.; SILVEIRA PAULILO, M.T.; NAKAZONO, E.M.; VENTURI, S. Effect of different level of desiccation in the seed germination of *Euterpe edulis* Martius – Arecaceae. *Insula*, v.28, p.31-42, 1999.

ROBERTO, G.G.; HABERMANN, G. Morphological and physiological responses of recalcitrant *Euterpe edulis* seeds to light, temperature and gibberellins. *Seed Science and Technology*, v.38, p.367-378, 2010. http://www.ingentaconnect.com/content/ista/sst/2010/00000038/00000002/art00010

ROBERTO, G.G.; COAN, A.I.; HABERMANN, G. Water content and GA$_3$-induced embryonic cell expansion explain *Euterpe edulis* seed germination, rather than seed reserve mobilization. *Seed Science and Technology*, v.39, n.3, p.559-571, 2011. http://www.ingentaconnect.com/content/ista/sst/2011/00000039/00000003/art00003

SMITH, S.L.; SHER, A.A.; GRANT III, T. Genetic diversity in restoration materials and the impacts of seed collection in Colorado's restoration plant production industry. *Restoration Ecology*, v.15, n.3, p.369-374, 2007. http://onlinelibrary.wiley.com/doi/10.1111/j.1526100X.2007.00231.x/pdf

WANG, W.Q.; CHENG, H.Y.; SONG, S.Q. The role of recovery of mitochondrial structure and function in desiccation tolerance of pea seeds. *Physiologia Plantarum*, v.144, p.20-34, 2012. http://sourcedb.ib.cas.cn/cn/ibthesis/201202/P020120216528870574542.pdf

WOLKERS, W.F.; ALBERDA, M.; KOORNNEEF, M.; LÉON-KLOOSTERZIEL, K.M.; HOEKSTRA, F.A. Properties of proteins and the glassy matrix in maturation defective mutant seeds of *Arabdopsis thaliana*. *The Plant Journal*, v.16, n.2, p.133-143. 1998.

ERRATUM

The Figure 1 of the article "**Fruit processing and the physiological quality of *Euterpe edulis* Martius seeds**", published on Journal of Seed Science, v. 36, n. 2, p. 134 - 142, 2014, is not correct. The corrected Figure can be find below. The authors regret the error.

http://dx.doi.org/10.1590/2317-1545v32n2847

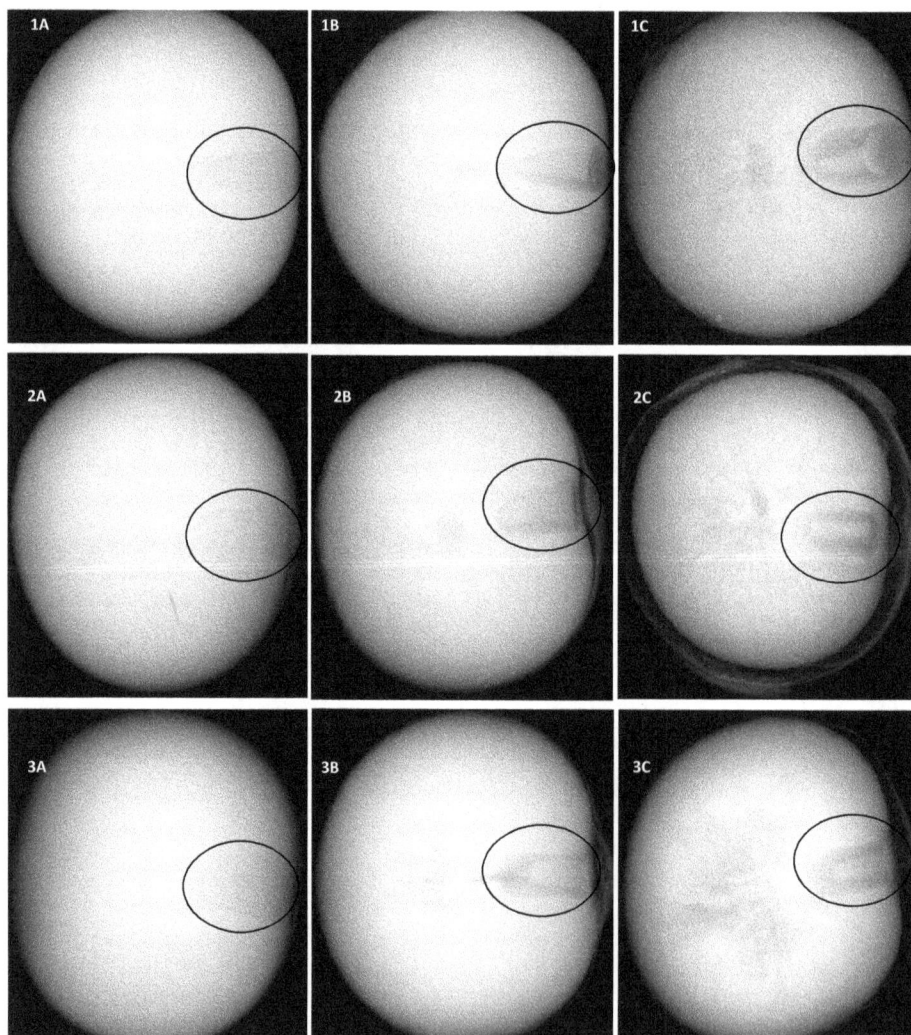

Figure 1. Representative image of the reduction of the volume occupied by the embryo within the embryo cavity throughout the drying process. 1 - Radiographic images that refer to treatment 01, Lot 01. 1A - initial water content of 44.9%; 1B - critical water content of 33.3%; 1C - lethal water content of 16.1%. 2 - Radiographic images that refer to treatment 04, Lot 01. 2A - initial water content of 44.9%; 2B - critical water content of 39.0%; 2C - lethal water content of 20.4%. 3 - Radiographic images that refer to treatment 04, Lot 02. 3A - initial water content of 45.9%; 3B - critical water content of 37.0%; 3C - lethal water content of 27.1%. Embryo is highlighted.

ERRATA

A Figura 1 do artigo "**Fruit processing and the physiological quality of *Euterpe edulis* Martius seeds**", publicado no Journal of Seed Science, v. 36, n. 2, p. 134 - 142, 2014, continha erros na formatação que haviam passado despercebidos pelos autores. A figura com a formatação correta segue abaixo.

http://dx.doi.org/10.1590/2317-1545v32n2847

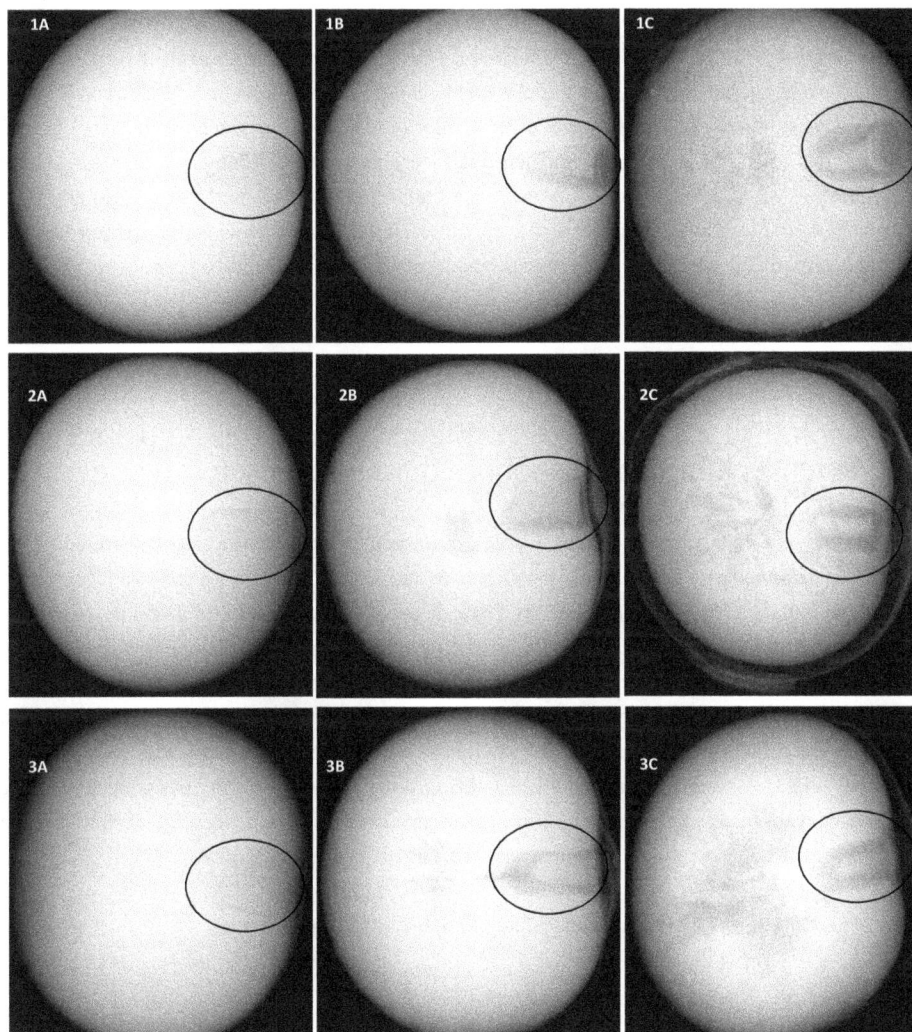

Figura 1. Imagem representativa da redução do volume ocupado pelo embrião dentro da cavidade embrionário durante a evolução do processo de secagem. 1- Imagens radiográficas referentes ao tratamento 01, lote 01. 1A – teor de água inicial de 44,9%; 1B – teor de água crítico de 33,3%; 1C – teor de água letal de 16,1%. 2- Imagens radiográficas referentes ao tratamento 04, lote 01. 2A – teor de água inicial de 44,9%; 2B – teor de água crítico de 39,0%; 2C – teor de água letal de 20,4%. 3- Imagens radiográficas referentes ao tratamento 04, lote 02. 3A – teor de água inicial de 45,9%; 3B – teor de água crítico de 37,0%; 3C – teor de água letal de 27,1%. Em destaque o embrião.

Physiological analysis and heat-resistant protein (LEA) activity in squash hybrid seeds during development

Patricia Pereira da Silva[1], Antônio Carlos Souza Albuquerque Barros[2],
Edila Vilela de Resende Von Pinho[3], Warley Marcos Nascimento[4*]

ABSTRACT – This study has aimed to evaluate the best time to harvest squash seeds and verify the effect of fruit storage on protein activity and in the physiological quality of 'Jabras' squash hybrid seeds. The fruits were harvested at 15, 30, 45, 60 and 75 days after anthesis (DAA). In every period, thirty fruits were harvested and fifteen had their fruit extracted immediately after harvesting, and the other fifteen fruits were stored for twenty days in plastic boxes under shade conditions, and only after this period the seeds were extracted. Seed quality was evaluated for the following parameters: moisture content, germination, first count, germination rate, seedling emergence, emergence rate, seed mass and seedling dry matter. Also, the electrophoretic profile analysis of heat-resistant proteins (Late Embryogenesis Abundant – LEA) was performed. Seed physiological maturity occurred in fruits harvested at 60 days after anthesis and stored for 20 days. During this period, seeds reached the maximum dry matter, maximum germination and vigor, and a high concentration of LEA proteins.

Index terms: *Cucurbita maxima*, *Cucurbita moschata*, seed quality.

Análise fisiológica e atividade de proteínas resistentes ao calor (LEA) em sementes híbridas de abóbora durante o desenvolvimento

RESUMO – O objetivo nesse trabalho foi avaliar a melhor época para a realização da colheita e verificar o efeito do armazenamento dos frutos na atividade de proteínas e na qualidade fisiológica de sementes do híbrido de abóbora 'Jabras'. Os frutos foram colhidos aos 15, 30, 45, 60 e 75 dias após a antese (DAA). Em cada época, foram colhidos trinta frutos, sendo que quinze frutos tiveram suas sementes extraídas imediatamente após a colheita e os outros quinze frutos foram armazenados por vinte dias em caixas plásticas em condições de galpão, e somente após este período, tiveram suas sementes extraídas. Avaliou-se a qualidade das sementes por meio do grau de umidade, germinação, primeira contagem, índice de velocidade de geminação, emergência, índice de velocidade de emergência, massa de 1000 sementes e massa seca de plântulas, além da análise do perfil eletroforético de proteínas resistentes ao calor (*Late Embryogenesis Abundant* – LEA). A maturidade fisiológica das sementes ocorreu em sementes oriundas de frutos colhidos aos 60 dias após a antese e armazenados por 20 dias. Nesse período, as sementes encontravam-se com o máximo de matéria seca, máxima germinação e vigor e alta concentração de proteínas LEA.

Termos para indexação: *Cucurbita maxima*, *Cucurbita moschata*, qualidade das sementes.

Introduction

Embrapa Hortaliças [Embrapa Vegetables (Brazilian Corporation for Agricultural Research (EMBRAPA – Empresa Brasileira de Pesquisa Agropecuária), a state-owned research corporation affiliated with the Brazilian Ministry of Agriculture, Livestock and Food Supply)] released in 1992 a squash interspecific F1 hybrid ('Jabras') with national technology, with the aim of providing to the market a similar product to the imported hybrid. It is a hybrid resulting from crossing between *Cucurbita maxima* Duch (female parent) and *Cucurbita moschata* Duch (male parent), with characteristics such as higher precocity and more uniform fruits when compared to open pollination regional cultivars (Nascimento et al., 2011).

A major factor in this hybrid seed production process is to

[1]Universidade Federal de Pelotas, 70351970 – Pelotas, RS, Brasil.
[2]Universidade Federal de Pelotas, Caixa Postal 354, 96010-900 – Pelotas, RS, Brasil.
[3]Departamento de Agricultura, UFLA, Caixa Postal 3037, 37200-000 – Lavras, MG, Brasil.
[4]Embrapa Hortaliças, Caixa Postal 218, 70359-970 – Brasília, DF, Brasil.
*Corresponding author <warley.nascimento@embrapa.br>

determine the period of seed physiological maturity to avoid incorrect harvest that could impair their quality, since there is not always the need for full fruit maturity in the field. After harvesting the fruits and subsequent storage period, the unripe seeds (or not fully ripe) can complete their development within the fruit, reaching maximum rates of germination and vigor.

In the early stages of development, the seeds are not able to tolerate desiccation. Tolerance acquisition is related to two protecting mechanisms which may be installed before or during the seed dehydration stage: the soluble sugars synthesis and the LEA (*Late Embryogenesis Abundant*) protein synthesis (Hoekstra et al., 2001). The LEA proteins are hydrophilic, stable, do not denature at high temperatures, are synthesized in seeds during development, when the abscisic acid (ABA) content is still high, and during the rapid dehydration stage, which occurs late in maturation process (Han et al.,1997). During seed maturity, the acquisition of desiccation tolerance may coincide with physiological maturity. The ability to germinate after the harvesting precedes the ability development to germinate after harvest and rapid artificial drying. Some studies report that the LEA proteins may also be accumulated in response to water stress, low temperature or salinity (Hong-Bo et al., 2005).

Physiological seed maturity is closely related to the ideal time to harvest, promoting the preservation of its physiological quality after harvesting. Thus, obtaining seeds with high physiological quality depends on the precise identification of the ideal time of harvesting, which often corresponds to the time when physiological maturity is reached, also coinciding with the moment of maximum dry matter accumulation, high vigor and high germination potential (Carvalho and Nakagawa, 2000).

Identifying the physiological maturity point to determine the correct time for fruit harvesting and extracting the seeds is of utmost importance in establishing efficient systems for commercial production. The unnecessary stay of fruits in field conditions, for example, results in progressive seed deterioration influenced by environmental conditions. In the maturity of fleshy fruit seeds, including cucurbitaceae, an outstanding feature is the continuation of the seed maturation even after the fruit harvesting (Vidigal et al., 2006). Unripe seeds from fleshy unripe fruits may have physiological quality comparable to ripe fruit seeds, provided that they are properly stored (Carvalho and Nakagawa, 2000). The possibility of holding early fruits, with subsequent storage, may be an interesting alternative for the grower of cucurbitaceae seeds (Barbedo et al., 1994). This issue has been extensively studied in seeds from fleshy fruits, as is the case of squash (Araújo et al., 1982), watermelon (Alvarenga et al., 1984) and cucumber (Barbedo et al., 1999), among others. In the specific case of hybrid squash, there are reports

indicating that fruit storage is essential to ensure the seed physiological quality (Costa et al., 2006).

Thus, the aim of this study was to evaluate the best time for harvest and check the fruit storage effect on physiological quality of 'Jabras' squash hybrid seeds by means of physiological studies and heat-resistant proteins (LEA) activity during seed development and maturation.

Material and Methods

Hybrid squash seeds derived from cross between *Cucurbita maxima* (female parent) and *Cucurbita moschata* (male strain) were produced in greenhouse in an experimental field at Embrapa Hortaliças, in Brasília, DF, from May to October 2012 period. Manual pollination was used, held in the early hours of the day, followed by labeling of pollinated flowers, with an average ratio of one male flower to three female flowers.

Fruit were harvested at 15, 30, 45, 60 and 75 days after anthesis (DAA). In every period, thirty fruits were harvested, with fifteen fruits being stored for twenty days in plastic boxes of a shaded and airy conditions. The remaining fruit had their seeds extracted immediately after harvesting, using hydrated lime for mucilage removal. Then the seeds were washed in tap water and dried at room temperature for 24 hours. In the next step, seeds were transferred to a chamber with ventilation and temperature of 32 °C for 48 hours. The dried seeds were then blowed in a pneumatic machine. The same procedure was performed with the seeds from the stored fruits. Seeds were subjected to the following analyses:

Determination of moisture content: the oven method at 105 °C for 24 hours was used, according to the Regras para Análise de Sementes (RAS; Rules for Seed Testing) (Brasil, 2009). About 2 g of seeds were used in each container, with two replicates for each treatment.

Germination: four replications of 50 seeds of each treatment were placed on a germitest-type paper roll moistened with distilled water at a ratio of 2.0 times the dry paper mass and incubated in a germination chamber with alternating temperatures of 20 °C (16 h, dark) and 30 °C (8 h, light). A count at eight days after the test establishment was carried out and evaluations were done according to criteria established by the Regras para Análise de Sementes (RAS; Rules for Seed Testing) (Brasil, 2009).

First count: it was held together with the germination test, counting the number of normal seedlings present on the fourth day after beginning the test. The results were expressed as percentages (Brasil, 2009).

Germination rate index: obtained by means of daily

counts of germinated seeds until the eighth day after sowing and calculated by the formula proposed by Labouriau, (1983), with the results being expressed in days.

Seedling emergence in a greenhouse: four replicates of 100 seeds of each treatment were used, which were sown in polystyrene multicellular trays expanded with 200 cells containing a commercial substrate. Irrigation was performed daily. The evaluation was performed 15 days after sowing and the results were expressed as a percentage.

Emergence rate index: obtained by daily counts of germinated seeds until the emergence stabilization, i.e.,15 days after sowing and calculated according to the germination rate described above.

Seed mass: eight subsamples of 100 dry seeds from each treatment were weighed and the average of the results was expressed in grams (g) (Brasil, 2009).

Seedling dry matter: held on 50 seedlings (shoot) obtained at 20 days after emergence, in the previously mentioned test. The samples were placed in brown paper envelopes which were weighed, identified and oven dried (65 °C) for a period of 48 hours. After this period, seedlings were kept in a desiccator for 15 minutes, determining the dry matter on a precision balance in mg.seedling^{-1}.

Heat-resistant proteins analysis – LEA (Late Embryogenesis Abundant): performed at Laboratório Central de Sementes (Central Laboratory of Seeds), Lavras Federal University, from February to May 2013. 100 mg of embryonic axes were macerated in crucibles in the presence of liquid nitrogen and then, extraction buffer was added at a ratio of 10 parts of buffer to 1 part of the sample. The samples were centrifuged at 14,000 rpm for 30 minutes at 4 °C. The supernatant was removed and incubated in a water bath at 85 °C for 15 minutes. Then, centrifugation was repeated as mentioned above. Subsequently, 70 μL of the extract were collected, put into microtubes and 40 μL of sample buffer were added (Tris HCl 50 mM pH 7.5). Subsequently, a hole was made in the lid of each microtube and they were placed in their own support, which was placed in boiling water for 5 minutes. At the end of this period, electrophoretic run described by Alfenas (2006) was carried out. The gels staining was performed with the use of 0.05% *Coomassie Blue* solution for 12 hours, and discoloration, until visualization of the bands, with 10% acetic acid solution.

The tests were carried out in a completely randomized design with four replications in a 5 x 2 (five harvest seasons and two fruits storage periods) factorial arrangement. Data were subjected to analysis of variance, and the means comparison was done by Tukey's test (p ≤ 0.05). After analysis of variance, regression analysis was performed by the PROC

REG procedure of the computer software SAS (SAS 9.1.3, 2000-2004). Analysis interpretation of the LEA proteins was based on visual analysis of gel electrophoresis, taking into account the presence/absence and the intensity of each of the electrophoretic bands.

Results and Discussion

The seed moisture content decreased continuously as the fruit age increased. For non-stored fruits, moisture content was 84% in seeds extracted from fruits at 15 DAA, and 45% in those extracted from fruits at 75 DAA. As for the fruits stored for 20 days, the moisture content was 76% in seeds extracted from fruits at 15 DAA, and 40% in those extracted from fruits at 75 DAA (Figure 1). This moisture is considered high and there is the need for quick drying to prevent possible fermentation and formation of products which result in immediate damage to the seed quality (Marcos-Filho, 2005). After drying, seeds presented moisture between 6% and 9%. It is recommended moisture for squash seed storage in waterproof packaging is 6% (Nascimento et al., 2008).

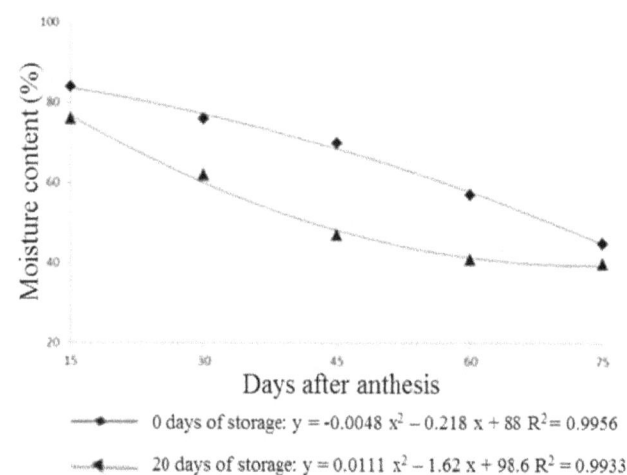

Figure 1. Moisture content of squash hybrid seeds before drying from fruits harvested at 15, 30, 45, 60 and 75 days after anthesis (DAA) and stored for 0 and 20 days.

Similar results were observed by other authors in fleshy fruits, such as by Alvarenga et al. (1984), with Italian squash, whose seed moisture content decreased from 89% to 42% throughout the harvesting period. As in pepper seeds, Vidigal et al. (2008) have found that seed moisture content ranges from 96% at 20 DAA to 46% at 70 DAA, at which time the seed has reached physiological maturity. In fleshy fruit species, osmotic equilibrium occurs between rich solute pericarp and seeds, resulting in stabilization of their moisture content at the end of maturation. Although it is used in determining physiological maturity, moisture content from fleshy fruits is not a suitable indicator because it is affected by environmental and genetic influences (Welbaum and Bradford, 1988).

The high initial moisture content found in the seeds of the first harvesting and its subsequent decrease are related to the importance of water in the seeds during and maturation filling and maturation processes. For the photosynthates products in the leaves to be deposited on the seed formation as a construction material and subsequently reserve, it is necessary that this moisture remain high, which occurs until the maximum accumulation of dry matter, when rapid dehydration begins (Carvalho and Nakagawa, 2000).

Significant differences were also found among harvesting periods regarding seed germination in fruits stored for 20 days. Seed germination increased, and relative stability was observed between 60 and 75 DAA, this period being considered as a possible indication of seed physiological maturation, which have attained maximum germination (97%) at 75 DAA (Figure 2). At this point, seeds produced were able to meet the commercial standards for squash seeds (germination = 80%), established by Ministry Administrative Regulation no. 457, of 12/18/1986 (Brasil, 1986). It was also found that the seeds from fruits stored for 20 days numerically showed higher germination compared with those seeds extracted immediately after fruit harvesting (Figure 2). Importantly, fruit storage for 20 days contributed to seed quality improvement, as may be observed in fruits harvested at 75 DAA, being 84% for non-stored and 97% for those stored. Similar results were obtained by Vidigal et al. (2009b) in pepper seeds, where higher physiological quality was observed in seeds from fruits harvested at 75 DAA and stored for seven days.

Generally, the harvesting period depends on the seed

physiological maturity, which, in most cases, coincides with the maximum accumulation of dry matter, and when the seeds reach this stage, usually their potential for germination and vigor increases (Duarte and Carneiro, 2009). The seed vigor modifications occur in parallel with the transfer progress of plant dry matter for the seeds, i.e., the proportion of vigorous seeds increases during maturation, reaching a maximum at a time that is very close to or coincident with the maximum accumulation of reserves (Marcos-Filho, 2005).

Seed vigor, determined by the first count and germination rate index tests, increased with age and fruit storage (Figure 3). In these two tests, increase in the old vigor obtained from fruits harvested at 60 DAA and stored for 20 days was observed. Similar results were observed by Barbedo et al. (1997) in cucumber seeds. In Italian squash seeds, storing for nine days fruits harvested at 55, 65 and 75 DAA was also beneficial to the vigor, measured by the germination first count (Alvarenga et al., 1984).

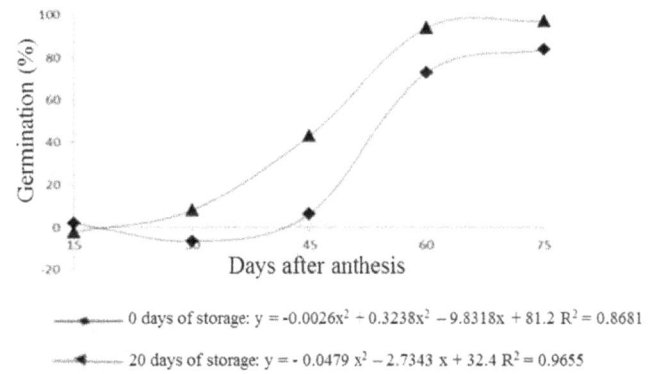

0 days of storage: y = -0.0026x² + 0.3238x² − 9.8318x + 81.2 R² = 0.8681

20 days of storage: y = - 0.0479 x² − 2.7343 x + 32.4 R² = 0.9655

Figure 2. Germination of squash hybrid seeds from fruits harvested at 15, 30, 45, 60 and 75 days after anthesis (DAA) and stored for 0 and 20 days.

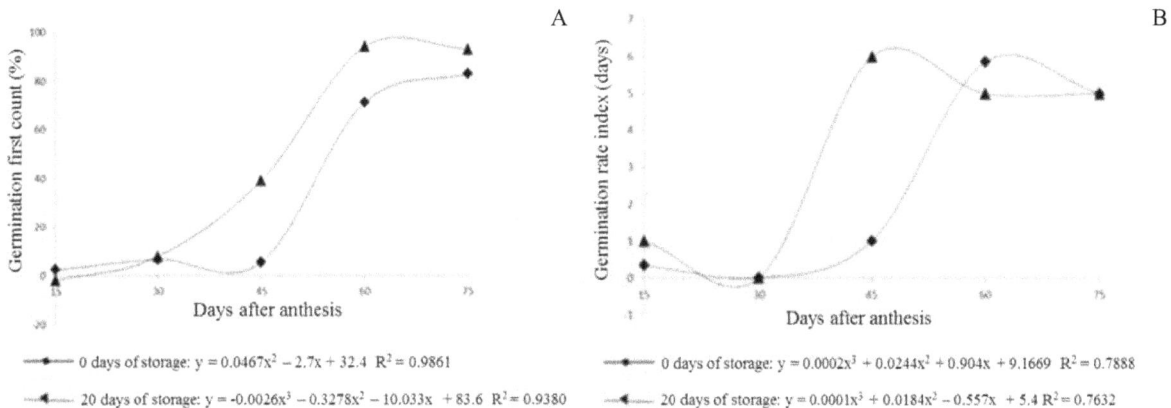

A

0 days of storage: y = 0.0467x² − 2.7x + 32.4 R² = 0.9861

20 days of storage: y = -0.0026x³ − 0.3278x² − 10.033x + 83.6 R² = 0.9380

B

0 days of storage: y = 0.0002x³ + 0.0244x² + 0.904x + 9.1669 R² = 0.7888

20 days of storage: y = 0.0001x³ + 0.0184x² − 0.557x + 5.4 R² = 0.7632

Figure 3. Germination first count (A) and germination rate index (B) of squash hybrid seeds from fruits harvested at 15, 30, 45, 60 and 75 days after anthesis (DAA) and stored for 0 and 20 days.

Seedling emergence in greenhouse and emergence rate index increments were also observed with advancing age and fruit storage (Figure 4). Seedling emergence was 94% in seeds from fruits harvested at 45 DAA and stored for 20 days and 97% in seeds from fruits harvested at 60 DAA and stored for 20 days. Higher emergence rate was obtained in seeds from fruits harvested at 45 days and stored for 20 days (Figure 4).

In the Figure 5 are the results for the seed mass (M1000), which were maximum for those from fruits harvested at 75 DAA and stored for 20 days, averaging 21.53 g. It was observed

that the seeds from stored fruits showed higher mass compared to seeds from non-stored fruits throughout the harvesting period. Seed mass is one factor that may influence germination and seedling vigor. Higher weight (or size) seeds usually have well-formed embryos and larger amounts of reserves, being potentially the most vigorous and having higher potential for survival (Carvalho and Nakagawa, 2000). For seedling shoots, growth was observed according to fruit age, and seeds from fruits harvested between 60 and 75 DAA, stored or not, showed similar results (Figure 6).

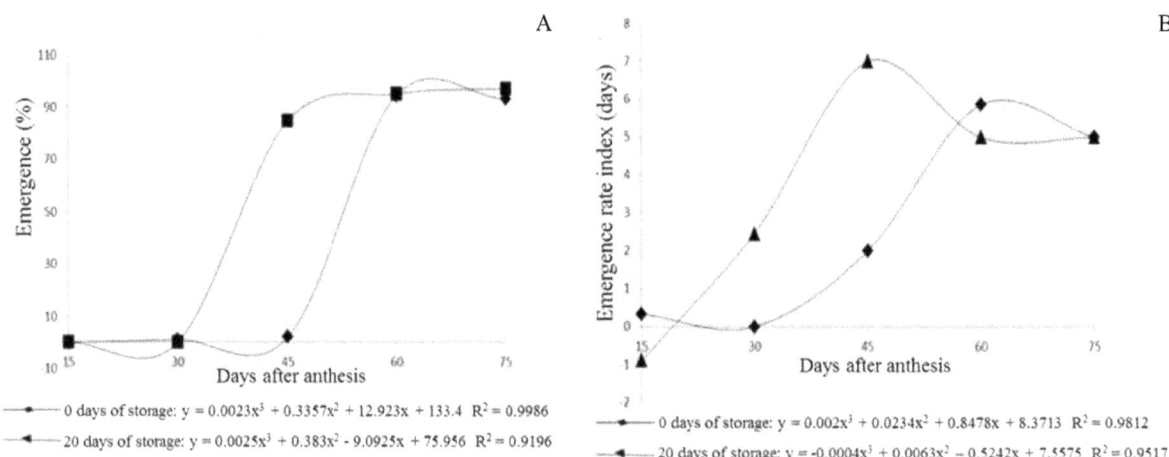

Figure 4. Seedling emergence (A) and seedling emergence rate in a greenhouse (B) of squash hybrid seeds from fruits harvested at 15, 30, 45, 60 and 75 days after anthesis (DAA) and stored for 0 and 20 days.

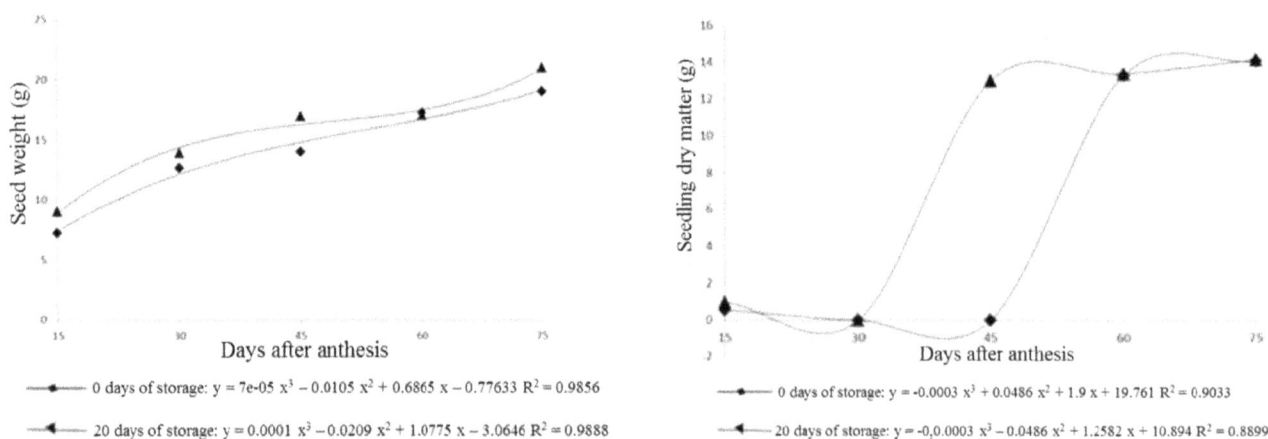

Figure 5. Squash hybrid seed mass from fruits harvested at 15, 30, 45, 60 and 75 days after anthesis (DAA) and stored for 0 and 20 days.

Figure 6. Squash hybrid seedling dry matter from fruits harvested at 15, 30, 45, 60 and 75 days after anthesis (DAA) and stored for 0 and 20 days.

Seeds from fruits from the 60 DAA are well formed and with high physiological quality, as seen in the results of germination, vigor and seed mass (Figures 2, 3 and 5).

Seedlings from high physiological potential seeds have higher efficiency in the production of dry biomass, and reduced differences in plant development (Mondo et al., 2012). These results corroborate those found in this study.

The electrophoretic profile of "*Late Embryogenesis Abundant*" (LEA) proteins demonstrates the absence of bands in the early stages of seed maturation, i.e., until the 30 DAA without fruit storage (Figure 7). Protein expression was initiated in the seeds from fruits harvested at 30 DAA and stored for 20 days; its higher thickness being visible when the maturation stages became more advanced, showing higher activity from 60 DAA, with 20 days of fruit storage. This indicates that from this time seeds probably acquire tolerance to drying, a characteristic that is associated with the time when physiological maturity is reached, which generally coincides with the seeds highest quality (Bewley and Black, 1994). The LEA proteins low activity during the early stages of seed maturation show that, probably, during this phase the seeds were unripe and thus still showed no efficiency in their free radical removal mechanism, promoting their low physiological quality. This result coincides with those obtained in the germination and first count tests (Figures 5 and 6) when it was possible to observe better seed physiological quality with the advance of the maturation process. Similar results were observed in cucumber by Nakada et al. (2011), who found a higher expression of LEA proteins in seeds that had higher physiological maturity. The band electrophoretic pattern analysis of these proteins may facilitate the timely detection of physiological maturation. In another study, the absence of LEA proteins in the early stages of maize seed development was also observed (Faria et al., 2004). During the seed drying process, changes may occur in the relative composition of membrane phospholipids (Dussert et al., 2006), heat-resistant protein synthesis, and the seeds ability to prevent, tolerate or repair damage caused by free radicals. In this process, heat-resistant proteins such as LEA proteins have an important role in preventing the damage caused by drying (Vidigal et al., 2009a), accumulating in response to the abscisic acid (ABA) in the late stages of seed maturation. Thus, seed development process determines the most appropriate procedures for obtaining high quality seeds. It is undeniable that the seed maximum physiological potential is achieved almost simultaneously to maturation (Carvalho and Nakagawa, 2000).

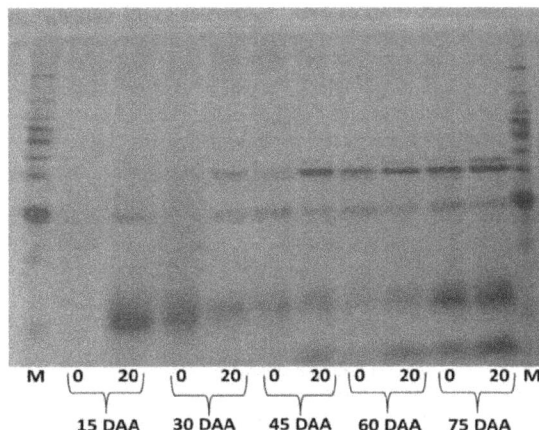

Figure 7. LEA proteins electrophoresis profile in squash hybrid seeds obtained from fruits harvested at 15, 30, 45, 60 and 75 DAA and stored for 0 and 20 days.

Conclusions

Physiological maturity of 'Jabras' squash seeds began after 60 DAA, and the seeds from fruits stored for 20 days showed higher physiological quality than those from non-stored fruit.

References

ALFENAS, A.C. *Eletroforese e marcadores bioquímicos em plantas e microrganismos*. Viçosa: UFV, 2006. 627p.

ALVARENGA, E.M.; SILVA, R.F.; ARAÚJO, E.F.; CARDOSO, A.A. Influência da idade e armazenamento pós-colheita dos frutos na qualidade de sementes de melancia. *Horticultura Brasileira*, v.2, n.2, p.5-8. 1984. http://www.scielo.br/scielo.php?script=sci_nlinks&ref=000092&pid=S0101-31222006000100018000001&lng=em

ARAÚJO, E.F.; MANTOVANI, E.C.; SILVA, R.F Influência da idade e armazenamento dos frutos na qualidade de sementes de abóbora. *Revista Brasileira de Sementes*, v.4, n.1, p.77-87, 1982. http://www.scielo.br/scielo.php?script=sci_nlinks&ref=000094&pid=S0101-31222006000100018000003&lng=en

BARBEDO, C.J.; BARBEDO, A.S.C.; NAKAGAWA, J.; SATO, O. Efeito da idade e do repouso pós-colheita de frutos de pepino na semente armazenada. *Pesquisa Agropecuária Brasileira*, v.35, p.839-847, 1999. http://dx.doi.org/10.1590/S0100-204X1999000500015

BARBEDO, C.J.; NAKAGAWA, J.; BARBEDO, A.S.C.; ZANIN, A.C.W. Influência da idade e do período de repouso pós-colheita de frutos de pepino cv. Rubi na qualidade fisiológica de sementes. *Horticultura Brasileira*, v.12, p.14-18, 1994. http://www.scielo.br/scielo.php?script=sci_nlinks&ref=000066&pid=S0100-204X199900050001500005&lng=en

BARBEDO, C.J.; NAKAGAWA, J.; BARBEDO, A.S.C.; ZANIN, A.C.W. Qualidade fisiológica de sementes de pepino cv. Pérola, em função da idade e do tempo de repouso pós-colheita dos frutos. *Pesquisa Agropecuária Brasileira*, v.32, p.905-913, 1997. http://www.scielo.br/scielo.php?script=sci_nlinks&ref=000071&pid=S141370542008000500020000004&lng=en

BEWLEY, J.D.; BLACK, M. *Seeds:* physiology of development and germination. 2.ed, New York: Plenum Press, 1994. 445 p.

BRASIL. Ministério da Agricultura, Pecuária e Abastecimento. *Regras para análise de sementes.* Ministério da Agricultura, Pecuária e Abastecimento. Secretaria de Defesa Agropecuária. Brasília: MAPA/ACS, 2009. 395p. http://www.agricultura.gov.br/arq_editor/file/2946_regras_analise__sementes.pdf

BRASIL. Ministério da Agricultura, Pecuária e Abastecimento. Portaria no. 457, de 18 de dezembro de 1986. Estabelece os padrões de sementes olerícolas para distribuição. *Diário Oficial da União*, Poder Executivo, Brasília, DF, p. 19653, 23 dez. 1986. 395 p.

CARVALHO, N.M.; NAKAGAWA, J. *Sementes:* ciência, tecnologia e produção. 4.ed. Jaboticabal: FUNEP, 2000. 588 p.

COSTA, C.J.; CARMONA, R.; NASCIMENTO, W.M. Idade e tempo de armazenamento de frutos e qualidade fisiológica de sementes de abóbora híbrida. *Revista Brasileira de Sementes*, v.28, p.127-132, 2006. http://www.scielo.br/scielo.php?script=sci_nlinks&ref=000068&pid=S0101-312220100003000200005&lng=en

DUARTE, E.F.; CARNEIRO, I.F. Qualidade fisiológica de sementes de *Dyckia goehringii* Gross & Rauh (Bromeliaceae) em função do estádio de maturação dos frutos. *Bioscience Journal*, v.25, p.161-171, 2009. http://dx.doi.org/10.1590/S0101-31222012000300011

DUSSERT, S.; DAVEY, M.W.; LAFFARGUE, A.; DOULBEAU, S.; SWENNEN, R.; ETIENNE, H. Oxidative stress, phospholipids loss and lipid hydrolysis during drying and storage of intermediate seeds. *Physiologia Plantarum*, v.127, p.192-204, 2006. http://onlinelibrary.wiley.com/doi/10.1111/j.1399-3054.2006.00666.x/full

FARIA, M.A.V.R., VON PINHO, R.G., VON PINHO, E.V.R., GUIMARÃES, R.M.; FREITAS, F.E.O. Germinabilidade e tolerância à dessecação em sementes de milho colhidas em diferentes estádios de maturação. *Revista Brasileira de Milho e Sorgo*, v.3, 276-289, 2004. http://www.scielo.br/scielo.php?script=sci_nlinks&ref=000073&pid=S0101312220090004000120 0012&lng=en

HAN, B.; HUGHES, W.; GALAU, G.A.; BEWLEY, J.D.; KERMODE, A.R. Changes in late-embryogenesis – abundant (LEA) messenger RNAs and dehydrins during maturation and premature drying of *Ricinus communis* L. seeds. *Planta*, v.201, p.27-35, 1997. link.springer.com/article/10.1007%2FBF01258677

HOEKSTRA, F.A.; GOLOVINA, E.A.; BUITINK, J. Mechanisms of plant desiccation tolerance. *Trends in Plant Science*, v.6. p.431–438, 2001. http://www.scielo.br/scielo.php?script=sci_nlinks&ref=000139&pid=S010390162008000600012 00019&lng=pt

HONG-BO, S.; ZONG-SUO, L; MING-AN, S. LEA proteins in higher plants: Structure, function, gene expression and regulation. Colloids Surf. B. *Biointerfaces*, v.42, p.107-113, 2005. www.ncbi.nlm.nih.gov/pubmed/16199145

LABOURIAU, L.G. *A germinação das sementes*. Washington: Secretaria Geral da Organização dos Estados Americanos, 1983. 174 p.

MARCOS-FILHO, J. *Fisiologia de sementes de plantas cultivadas.* Piracicaba: FEALQ, 2005. 495 p.

MONDO, V.H.V.; CICERO, S.M.; DOURADO-NETO, D.; PUPIM, T.L.; DIAS, M.A.N. Vigor de sementes e desempenho de plantas de milho. *Revista Brasileira de Sementes*, v.34, n.1, p. 143 – 155, 2012. http://dx.doi.org/10.1590/S0101-31222012000100018

NAKADA, P.G.; OLIVEIRA, J.A.; MELO, L.C.; GOMES, L.A.A.; VON PINHO, E.V.R. Desempenho fisiológico e bioquímico de sementes de pepino nos diferentes estádios de maturação. *Revista Brasileira de Sementes*, v.33, p.113-122, 2011. http://dx.doi.org/10.1590/S0101-31222011000100013

NASCIMENTO, W.M.; PESSOA, H.B.S.V.; SILVA, P.P. Produção de sementes híbridas de abóbora do tipo tetsukabuto. *Palestra XI Curso Sobre Tecnologia de Produção de Sementes de Hortaliças.* 8. ed. Porto Alegre: 2011. CD-ROM.

NASCIMENTO, W.M.; FREITAS, R.A.; CRODA, M.D. *Conservação de sementes de hortaliças na agricultura familiar.* Brasília: Comunicado Técnico 54, 2008. p.1-5.

VIDIGAL, D.S.; DIAS, D.C.F.S.; NAVEIRA, D.S.P.; ROCHA, F.B.; BHERING, M.C. Qualidade fisiológica de sementes de tomate em função da idade e do armazenamento pós-colheita dos frutos. *Revista Brasileira de Sementes*, v.28, p.87-93, 2006. http://www.scielo.br/scielo.php?script=sci_nlinks&ref=000119&pid=S0101-312220090002000150 0027&lng=pt

VIDIGAL, D.S.; DIAS, D.C.F.S.; VON PINHO, E.V.R.; DIAS, L.A.S. Sweet pepper seed quality and Lea protein activity in relation to fruit and post-harvest storage. *Revista Brasileira de Sementes*, v.31, p.192-201, 2009a. http://www.scielo.br/scielo.php?script=sci_nlinks&ref=000150&pid=S0100-204X201300120000300030&lng=en

VIDIGAL, D.S.; DIAS, D.C.F.S.; VON PINHO, E.V.R.; DIAS, L.A.S. Alterações fisiológicas e enzimáticas durante a maturação de sementes de pimenta (*Capsicum annuum* L.). *Revista Brasileira de Sementes*, v.31, p.129-136, 2009b. http://dx.doi.org/10.1590/S0101-31222009000200015

VIDIGAL, D.S.; LIMA, J.S.; BHERING, M.C; DIAS, D.C.F.S. Teste de condutividade elétrica para sementes de pimenta. *Revista Brasileira de Sementes*, v.30, p.168-174, 2008. http://www.scielo.br/scielo.php?script=sci_nlinks&ref=000108&pid=S0101-312220090003000080 0017&lng=en

WELBAUM, G.E.; BRADFORD, K.J. Water relations of seed development and germination in muskmelon (*Cucumis melo* L.). In. Water relations of seeds and fruit development. *Plant Physiology*, v.86, p.406-411, 1988. http://www.scielo.br/scielo.php?script=sci_nlinks&ref=000104&pid=S0101-312220020001000220 0021&lng=en

Treating sunflower seeds subjected to ozonization

Vitor Oliveira Rodrigues[1], Fabiano Ramos Costa[1], Marcela Carlota Nery[1], Sara Michelly Cruz[2*], Soryana Gonçalves Ferreira de Melo[1], Maria Laene Moreira de Carvalho[2]

ABSTRACT – Sunflower crops (*Helianthus annuus* L.) are a source of pathogens that can be transmitted by means of the seeds. An alternative for the treatment of seeds can be the application of an ozone compound, which has been used in various segments to eliminate microorganisms. Three lots of the Embrapa 122 variety have been used to assess the efficacy of ozone to control pathogens in sunflowers seeds. To typify the profile of the lots we have determined the moisture content, the first germination counting, the germination, the germination speed index, emergence, initial stand, emergence speed index and health. The lots have been stored in an ozone-rich environment for periods of 20, 60 and 120 minutes and without ozone (control). After the treatments, the seeds were tested to assess their health and physiological quality through the said tests as well as a test for accelerating aging, tetrazolium and incidence and severity of fungi. It has been concluded that the treatment for *H. annuus* seeds, Embrapa 122 variety, with ozone concentration of 1741 ppmv (0.24 g/h), for 60 minutes, reduces the fungal presence of *Alternaria* sp., *Fusarium* sp., *Aspergillus* sp. and *Penicillium* sp. without affecting their physiological potential.

Index terms: *Helianthus annuus*, vigor, viability, health, ozone.

Tratamento de sementes de girassol submetidas à ozonização

RESUMO – A cultura do girassol (*Helianthus annuus* L.) é hospedeira de inúmeros patógenos que podem ser transmitidos via sementes. Uma alternativa para o tratamento das sementes pode ser a aplicação de ozônio composto que tem sido usado em diversos segmentos na eliminação de microrganismos. Para avaliar a eficiência do ozônio no controle de patógenos em sementes de girassol foram utilizados três lotes da cultivar Embrapa 122. Para caracterização do perfil dos lotes determinou-se o grau de umidade, a primeira contagem de germinação, germinação, índice de velocidade de germinação, emergência, estande inicial, índice de velocidade de emergência e sanidade. Os lotes foram mantidos em ambiente rico em ozônio pelos períodos de 20, 60 e 120 minutos e sem ozônio (controle). Após os tratamentos as sementes foram submetidas à avaliação da qualidade fisiológica e sanitária, por meio dos testes já mencionados, além dos testes de envelhecimento acelerado, tetrazólio e incidência e severidade de fungos. Concluiu-se que o tratamento de sementes de *H. annuus*, cultivar Embrapa 122, com ozônio na concentração de 1741 ppmv (0.24 g/h), por 60 minutos, reduz a população fúngica de *Alternaria* sp., *Fusarium* sp., *Aspergillus* sp. e *Penicillium* sp. sem afetar o seu potencial fisiológico.

Termos para indexação: *Helianthus annuus*, vigor, viabilidade, sanidade, ozônio.

Introduction

Helianthus annuus L. belongs to the *Asteraceae* family. It is an annual dicotyledonous and it is used as animal feed, production of oil for human consumption and biodiesel production (Porto et al., 2007).

The spread of *H. annuus* is essentially seminiferous and it is important to use high quality seeds to obtain high productivity. According to Gomes et al. (2006), the spread of *H. annuus* culture can be hampered by the seeds low physiological and health quality. Most of the etiologic disease agents is transmitted by seeds, especially those caused by fungi that reduce germination power and can be disseminated by establishing primary infestation focuses on growing areas (Machado, 1994).

The control of diseases and pests in agriculture, including the control of pathogens in seeds, is accomplished primarily by means of synthetic products that generate high costs and environmental and toxicological risks (Campanhola and Bettiol, 2003).

[1]Departamento de Agronomia, Universidade Federal dos Vales do Jequitinhonha e Mucuri, 39100-000 – Diamantina, MG, Brasil.

[2]Departamento de Agricultura, Universidade Federal de Lavras, Caixa Postal 3037, 37200-000 – Lavras, MG, Brasil.
*Corresponding author <saramichellycruz@gmail.com>

As an alternative to chemicals, ozone has been used to control insect pests and fungi in stored grains (Mendez et al., 2003; Pereira et al., 2008), in postharvest fruits (Simão and Rodríguez, 2008), in minimally processed fruits (Ponce et al., 2010), in food ripening rooms (Pinto et al., 2007; Serra et al., 2003), and in the degradation of mycotoxins (Akbas e Ozdemir, 2006). An advantage is the possibility of being produced at the site of use and not requiring special packaging and transport of goods for the treatment (Mendez et al., 2003).

The effects of seeds exposure to ozone have been studied and a method that benefits the seed germination has been patented by Yvin and Coste (1995), and the another one by Klaptchuk (2004) to sterilize the seeds and degrade residual herbicides with use of ozone. It has been found that exposure to ozone favors the acceleration of tomato seed germination by breaking dormancy (Sudhakar et al., 2011); however, according to Wu et al. (2006) the prolonged exposure to gas can reduce the germination of wheat seeds. Studies by Violleau et al. (2008) have suggested that long periods of ozonization may be harmful for the quality of the seeds; however, which would a long period be has not been specified.

Given the potential of ozone use in the inactivation of fungi and the little knowledge about its effects on germination, the aim of this study was to verify the effectiveness of ozone use in the control of plant phytopathogens and its possible effects on seed physiological quality of *H. annuus*.

Material and Methods

Three lots of *Helianthus annuus* L. seeds from cultivar Embrapa 122 were used, acquired from the seed bank at "Embrapa Produtos e Mercado", harvested in the Brazilian city of Dourados, MS, 2013 harvest. These lots were ranked by sieve screen, with lot 1 of 3.5 mm small sieve, lot 2 of 4.0 mm average sieve and lot 3 of 5.5 mm large sieve, so as to ensure greater uniformity of the gas passage by the seeds.

The seeds were kept in a rich atmosphere in ozone by periods of 20, 60 and 120 minutes and without ozonization (control). Soon after, determinations and physiological and health tests were carried out.

Equipment to apply the ozone treatment: ozone was obtained by means of the Electrochemical Ozone Generator (Gerador de Ozônio Eletroquímico) which is under patent Leonardo and Jardim, 2011 F. "Gerador de Ozônio Eletroquímico – INPI – PI1101892-5.2011, Brasil." The seeds were placed in a continuous flow chamber made with a PVC (polyvinyl chloride) pipe, which has a valve at each end; inside the tube the seeds were placed on individual cage rings in a single layer to avoid overlapping seeds,

ensuring the contact of ozone with all of them (Figure 1). The ozone concentrations were measured directly by means of a spectrophotometer attached to the apparatus throughout the ozonization period, thus evaluating the average ozone concentration in each tested period. The average production of ozone in the ozonization time interval for the three periods (20 minutes, 60 minutes and 120 minutes) was 1741 ppmv (parts per million by volume) and 0.24 g/h.

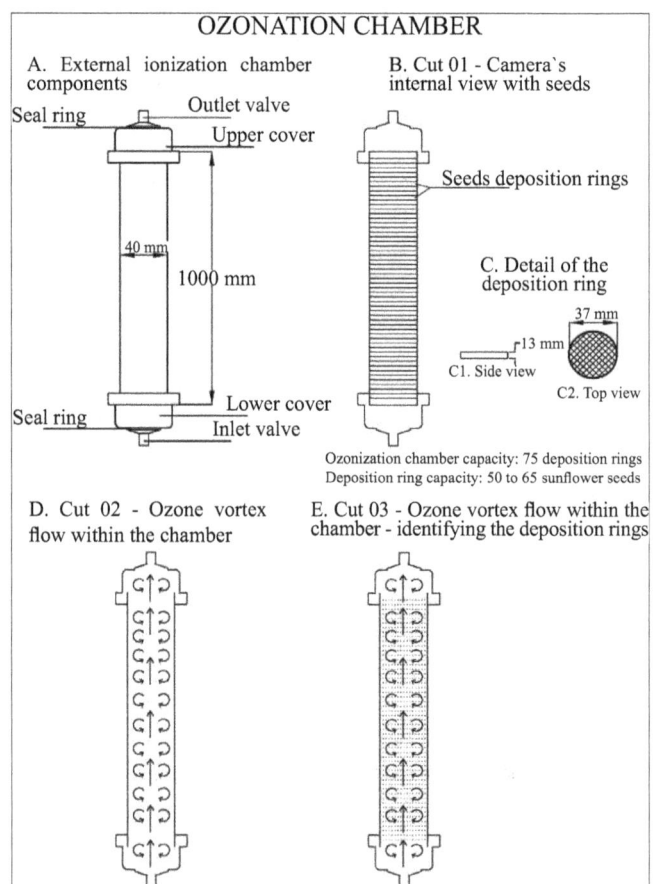

Figure 1. Outline of the ozonization chamber.

The characterization of the lots was carried out by determining the moisture content (M), first germination counting (FGC), germination (G), germination speed index (GSI), emergence (E), initial stand (IS) emergence speed index (ESI) and health. The moisture content was determined by the oven method at 105 °C for 24 hours (Brasil, 2009b).

The following tests were carried out with four repetitions of 50 seeds per lot.

Germination test: roll paper substrate was used at a constant temperature of 25 °C and a 12-hour photoperiod. Evaluations were performed on the fourth day, first germination counting, and on the tenth day, final counting (Brasil, 2009b). The germination speed index (GSI) was calculated using the formula proposed by Maguire (1962).

Seedling emergence test: they were seeded in plastic boxes containing earth and sand in the ratio 2:1, with the substrate sterilized and moistened with distilled water, and the boxes kept in a growth room at 25 °C with constant photoperiod. Evaluations of the number of seedlings emerged from the initial stand were held at four days and of the final stand at ten days after sowing. The results were expressed in percentage. For the emergence speed index (ESI), the number of emerged seedlings was daily computed from the beginning of the emergence and the calculation was performed according to Maguire (1962).

Health test: the lots were divided into ten replicates of 20 seeds. The method of the blotter paper in gerbox box was used, with a thin layer of agar-agar at 1.0%. The blotter papers were moistened with distilled water and 2,4-dichlorophenoxyacetic acid. The test was associated with the freezing method (Goulart, 2009). The gerboxes with the seeds were incubated under white fluorescent light bulbs and a 12-hour photoperiod for seven days at a temperature of 20 °C (Brasil, 2009a). After the incubation period, the seeds were individually analyzed in a stereoscopic magnifier and optical microscope for the identification and quantification of fungi. The results were expressed as percentage of infected seeds, by computing the incidence of the fungus.

After the treatments with ozone, the lots were submitted to the germination tests, first germination counting, GSI, emergence, ESI and health, as described above, and by the accelerated aging and tetrazolium tests.

Accelerated aging test: 50 seeds from each lot and each treatment were placed on an aluminum screen inside a gerbox plastic box with 40 mL of distilled water at the bottom, at 42 °C for 72 hours (Aguiar et al., 2001). After this period, the germination test was performed as previously described.

Tetrazolium test: 50 seeds from each lot and each treatment were placed to soak on paper for 16 hours in a B.O.D. (Biochemical Oxygen Demand)-type germination chamber at 25 °C. Subsequently to the removal of the pericarp and the integument, the seeds were then placed to soak in tetrazolium solution at the concentration of 0.5% and packed in B.O.D. at 30 °C for one hour (Brasil, 2009b). After color development, the percentage of viable, unviable and vigorous seeds were evaluated.

To assess the ozone potential as a method of disinfection of *H. annuus* seeds, the tests described below were performed.

Infestation severity index (ISI): a grading scale based on the percentage of each seed covered by typical surface structures of fungi was defined, attributing grades (0) for the absence of fungal structures, (1) for up to 25% of the seed covered by fungal structures, (2) for up to 50% of the seed

covered by fungal structures, (3) for up to 75% of the seed covered by fungal structures, and (4) for more than 75% of the seed covered by fungal structures. This analysis was performed along the health check.

To determine the seed infestation severity index, an adaptation of a methodology proposed by McKinney (1923) was used.

$$ISI = ISI = \frac{\Sigma (f \times n) \times 100}{F \times N}$$

Where:

(ISI): infestation severity index by seed evaluated; (f): a grade of scale attributed to the evaluated seed; (n): number of seeds which have received the grade; (F): maximum scale grade, and (N): total number of seeds evaluated by replicate.

The experiment was conducted in a completely randomized design with four replications, and the analyses of variance were performed using the statistical software SISVAR (Ferreira, 2011). The data were transformed into arcsin to approach the normal curve and subjected to analysis of variance and the averages compared by Scott-Knott test at 5% probability.

Results and Discussion

The lots showed good physiological quality before being ozonized (Table 1). The germination of all the *H. annuus* seeds lots was in accordance with the marketing standard in Brazil for S1 and S2 seeds, which is 70% (Brasil, 2013), and the water content was within the range for storage, transport and marketing, ranging from 5% to 10%, according to Leite et al. (2005) and showed no statistical differences.

For the germination test, first germination counting, GSI, emergence and ESI (Table 1) it was observed that there were no significant differences among the lots. In the case of the initial stand, the greatest vigor was observed for lot 2 in relation to the other lots.

A higher percentage of emergence compared to the germination test was observed (Table 1). This fact can be explained by the health quality of seeds and substrate difference among tests. There was a high incidence of fungus *Alternaria* sp. (Table 2) in the lots, which, according to Salustiano et al. (2005), causes loss of viability in the initial stand of the development. In the germination test in paper roll, the integument remains associated to the cotyledons, and the pathogens associated with it cause the seed deterioration, since, upon the emergence in soil and sand mixture of the emergence test, the seedling releases the infected integument in the soil (Juliatti et al., 2011).

Table 1. Results of the characterization of lots of *Helianthus annuus* L. seeds, of cultivar Embrapa 122, according to moisture content (M), germination (G), first germination counting (FGC), germination speed index (GSI), emergence (E), initial stand (IS) and emergence speed index (ESI).

Lots	M (%)	G (%)	FGC (%)	GSI	E (%)	IS (%)	ESI
L1	6.45 A	72 A	45 A	37.29 A	83 A	75 B	14.75 A
L2	7.27 A	80 A	49 A	37.80 A	92 A	85 A	16.04 A
L3	7.17 A	75 A	47 A	37.88 A	87 A	79 B	16.12 A
CV (%)	4.20	8.23	10.28	7.15	5.17	5.33	8.87

Means followed by the same letter in the column do not differ by the Scott Knott at 5% probability.

In the health check (Table 2) were identified, besides *Alternaria* sp., the fungi *Fusarium* sp., *Aspergillus* sp. and *Penicillium* sp., without any statistical difference among the lots. Gomes et al. (2006) have found the same pathogens for cultivar Embrapa 122 and these are harmful fungi for the quality of sunflower seeds (Grisi et al., 2009).

After 120 minutes of ozonization of the seeds, inferior results were observed in the germination tests and first counting for all lots (Table 3). This reduction can be attributed to phytotoxicity, because this treatment resulted in abnormalities in the root system, affecting the growth of sunflower seedlings.

There was a reduction of moisture content only in the period of 20 minutes of ozonization and only for lot 2, statistically differing from the other lots. For the other ozonization times, there was no difference among the moisture values of seeds of each lot (Table 3).

By the first germination counting data (Table 3) it was observed that the smallest percentages of germination occurred in the control seeds without ozonization, and in the period of 120 minutes. For lot 1, the 60-minute ozonization period showed a percentage of first germination counting as higher than in the other periods of ozonization; for lot 2 the percentage was higher in the 60-minute period, followed by the 20-minute period. As for lot 3, the 20-minute and 60-minute ozonization periods were higher, not differing from each other.

In the periods of 20 minutes of exposure of sunflower seeds to ozone, higher percentage of the first counting for lots 2 and 3 was observed relative to lot 1, and in the 60-minute ozonization period, superiority of lot 2 was observed in relation to the others. In the control and after 120 minutes of exposure there was no statistical difference among the lots (Table 3).

As for the germination, it was observed that there was no effect of different periods of ozone exposure for lot 1. As for lot 2, the germination percentage was higher after 60 minutes of ozonization and, for lot 3, there was higher germination after 20 minutes and 60 minutes of ozonization compared to the other treatments. Similar results were observed by Violleau et al. (2008) with maize seeds, where there was an increase of germination after 20 minutes of exposure to ozone.

Table 2. Results of the incidence of *Fusarium* sp. (FUS), *Penicillium* sp.(PEN), *Aspergillus* sp.(ASP) and *Alternaria* sp.(ALT) in the seeds of *Helianthus annuus L,* cultivar Embrapa 122, in the characterization of the lots.

Lots	FUS (%)	PEN (%)	ASP (%)	ALT (%)
L1	76 A	53 A	67 A	71 A
L2	74 A	35 A	56 A	74 A
L3	83 A	50 A	60 A	73 A
CV (%)	22.15	42.22	30.44	20.98

Means followed by the same letter in the column do not differ by the Scott Knott at 5% probability.

Table 3. Results of moisture content (M), first germination counting (FGC), germination (G) and germination speed index (GSI) of the seed lots of *Helianthus annuus L.,* cultivar Embrapa 122, after different times (minutes) of ozonization.

Ozonization time	M (%)			FGC (%)			G (%)			GSI		
	L1	L2	L3	L1	L2	L3	L1	L2	L3	L1	L2	L3
0	6.54 Aa	7.53 Aa	6.69 Aa	47 Ab	53 Ac	52 Ab	70 Aa	77 Ab	71 Ab	40.46 Aa	44.61 Aa	42.49 Aa
20	7.52 Aa	6.17 Ba	7.72 Aa	53 Bb	63 Ab	60 Aa	74 Aa	80 Ab	75 Aa	40.68 Aa	42.55 Aa	39.17 Ba
60	7.12 Aa	6.80 Aa	7.80 Aa	61 Ba	75 Aa	64 Ba	77 Ba	88 Aa	78 Ba	39.35 Aa	43.91 Aa	42.00 Aa
120	7.34 Aa	6.32 Aa	7.20 Aa	51 Ab	51 Ac	50 Ab	69 Aa	73 Ab	67 Ab	37.84 Aa	40.48 Aa	38.38 Ba
CV (%)	8.43			8.77			7.66			5.54		

Means followed by the same lowercase letter in the column and uppercase letter in the row do not differ by the Scott Knott test at 5% probability.

In the control periods, 20 minutes and 120 minutes, there was no difference among the percentages of seed germination. After 60 minutes of ozonization it was possible to classify the lots into different groups, as was observed for the data of the first germination counting (Table 3).

Regarding the germination speed index (GSI) (Table 3), the control treatment and the other treatments belong to the same group for all three lots. In studies using chemical treatment for sunflower seeds sterilization, reduction in the germination speed index is observed in the control treatment seeds (Medeiros et al., 2013; Grisi et al., 2009; Gomes et al., 2006). It is observed that in the ozonization times of 20 and 120 minutes, the germination speed index was higher for lots 1 and 2 in relation to lot 3.

By the results of the vigor tests (Table 4) there was no effect of ozonization time on seedling emergence. In the initial stand test, 120-minute period, lot 2 had higher percentage of normal seedlings compared to the other lots. However, for the same lot, statistical differences after ozonization were not observed. As for the emergence speed index, there were no differences among the lots; however, there was a significant interaction among the ozonization times within lots, with the highest rates for lot 2 occurring in seeds without ozonization, and after 120 minutes of treatment, as well as for untreated lot 3 and after 60 minutes of ozonization. By means of the accelerated aging test, marked reduction in vigor of all lots of sunflower seeds was observed after treatment with ozone. However, only in the seeds without ozonization (control) differences were detected vigor among lots, it is possible to separate the lots 1 and 2 in a group with higher quality than lot 3.

Table 4. Emergence results (E), initial stand (IS), emergence speed index (ESI), accelerated aging (AA), of the lots of the seeds of *Helianthus annuus L.,* cultivar Embrapa 122, after different times (minutes) of ozonization.

Ozonization time	IS (%)			E (%)			ESI			AA (%)		
	L1	L2	L3	L1	L2	L3	L1	L2	L3	L1	L2	L3
0	77 Aa	85 Aa	78 Aa	80 Aa	89 Aa	86 Aa	21.52 Aa	22.86 Aa	23.28 Aa	27 Aa	34 Aa	20 Ba
20	77 Aa	85 Aa	79 Aa	81 Aa	86 Aa	82 Aa	15.19 Aa	17.43 Ab	15.45 Ab	27 Aa	17 Ab	21 Aa
60	78 Aa	84 Aa	83 Aa	81 Aa	82 Aa	88 Aa	17.73 Aa	16.59 Ab	20.25 Aa	13 Ab	27 Aa	8 Ab
120	73 Ba	80 Aa	75 Ba	81 Aa	91 Aa	84 Aa	19.00 Aa	20.67 Aa	17.83 Ab	17 Ab	17 Ab	10 Ab
CV (%)	9.16			8.18			17.77			33.47		

Means followed by the same lowercase letter in the column and uppercase letter in the row do not differ by the Scott Knott test at 5% probability.

By the results of viability of sunflower seeds obtained in the tetrazolium test (Table 5) there was no statistical difference among lots after the different ozonization periods tested. This fact is positive since one of the desirable traits in seed treatment is that the product used do not affect viability (Machado, 2000). It was observed that for the 60-minute ozonization time there was a differentiation among the lots, lot 2 being higher than lot 1 and lot 3, and these results are similar to those observed in the germination test (Table 3).

As for the results of the vigorous seeds percentage (Table 5), obtained in the tetrazolium test, there was no statistical difference among the different ozonization periods. Similar results were observed in the data of GSI, emergence, initial stand (Table 4) and viability tetrazolium (Table 5). There was a distinction of lot 2 as the one of greater vigor in the seeds without ozonization (control) and in treatments of 20 minutes and 120 minutes of ozonization. In the treatment of 60 minutes of ozonization there was no statistical difference among lots.

As occurs after the use of chemical fungicides (Grisi et al., 2009), ozone did not reduce the quality of sunflower seeds.

Table 5. Viability and vigor results by the tetrazolium test, in lots of sunflower seeds, cultivar Embrapa 122, in the different times (minutes) of ozonization.

Ozonization time	Viability (%)			Vigor (%)		
	L1	L2	L3	L1	L2	L3
0	80 Aa	86 Aa	80 Aa	58 Ba	70 Aa	59 Ba
20	80 Aa	87 Aa	82 Aa	63 Ba	72 Aa	62 Ba
60	84 Ba	89 Aa	83 Ba	68 Aa	75 Aa	66 Aa
120	80 Aa	85 Aa	78 Aa	64 Ba	70 Aa	57 Ba
CV (%)	5.08			9.56		

Means followed by the same lowercase letter in the column and uppercase letter in the row do not differ by the Scott Knott test at 5% probability.

The incidence of *Alternaria* sp. (Table 6, Figure 2) in all lots was highest in seeds without the treatment and after 20 minutes of ozonization. After 60 minutes and 120 minutes, there was a reduction in the incidence of this fungus, and therefore there was a tendency of reducing the incidence of *Alternaria* sp. with the increased time of exposure of the seeds to ozone (Figure 2). Only for the time of 60 minutes there was a difference among lots separating them into two groups, and in lots 1 and 3 the reduction of the incidence was greater than for lot 2.

In the assessment of *Fusarium* sp. (Table 6, Figure 2) there was a higher incidence in the seeds without ozonization and after 20 minutes. The lowest incidence was found after 60 and 120 minutes of ozonization for lot 1 and lot 2 and for lot 3 the lowest incidence was observed only after 120 minutes

of ozonization. There was no statistical difference among lots without ozonization and in the 20-minute period of ozonization for incidence of this fungus. After the 60-minute period, the seeds in lots 1 and 2 had a lower incidence of *Fusarium* sp. and for the 120-minute period, the lowest incidence was observed in lot 1.

Table 6.   Incidence Index of *Alternaria* sp., *Fusarium* sp., *Aspergillus* sp and *Penicillium* sp., in the seed lots of *Helianthus annuus L.*, cultivar Embrapa 122, due to the times (minutes) of ozonization.

Ozonization time	*Alternaria* sp.			*Fusarium* sp.			*Aspergillus* sp			*Penicillium* sp.		
	L1	L2	L3	L1	L2	L3	L1	L2	L3	L1	L2	L3
0	75 Aa	78 Aa	80 Aa	96 Aa	92 Aa	98 Aa	73 Aa	59 Aa	68 Aa	64 Aa	47 Ba	61 Aa
20	71 Aa	78 Aa	68 Aa	94 Aa	92 Aa	94 Aa	65 Aa	63 Aa	56 Aa	53 Ab	44 Aa	49 Aa
60	40 Bb	60 Ab	38 Bb	74 Bb	77 Bb	92 Aa	34 Ab	29 Ab	38 Ab	25 Ac	29 Ab	32 Ac
120	42 Ab	43 Ab	50 Ab	49 Bc	74 Ab	76 Ab	45 Ab	24 Ab	34 Ab	23 Ac	21 Ab	15 Ad
CV (%)	27.13			16.65			28.26			26.57		

Means followed by the same lowercase letter in the column and uppercase letter in the row do not differ by Scott Knott at 5% probability.

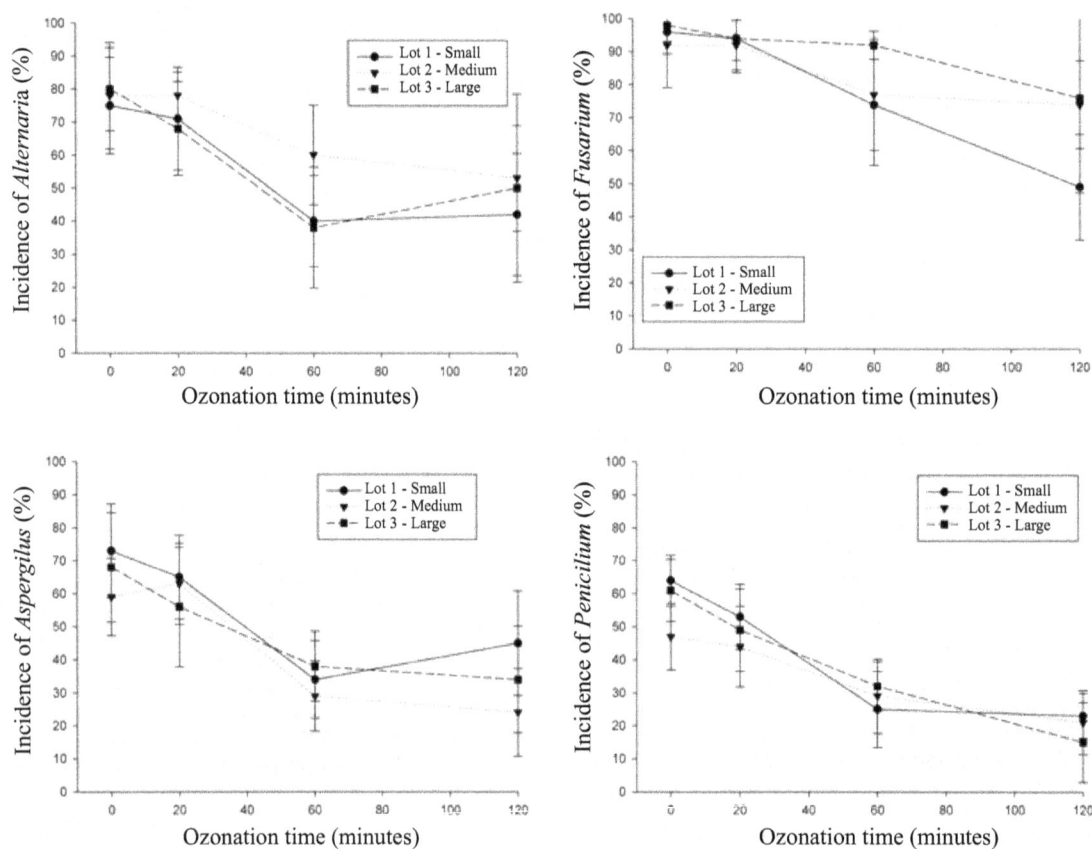

Figure 2.   Incidence of the fungi *Alternaria* sp., *Fusarium* sp., *Aspergillus* sp *and Penicillium* sp. (mean ± standard deviation) in three seed lots of *Helianthus annuus L.,* cultivar Embrapa 122, due to the times of ozonization.

The incidence of *Aspergillus* sp. (Table 6, Figure 2) was reduced from 60 minutes of ozonization, as observed for *Alternaria* sp. and *Fusarium* sp. According to Freitas-Silva

and Venâncio (2008), depending on the ozone concentration injected into cultivated colonies, the inactivation of *Aspergillus* sp. can reach 80% to 100%. In all ozonization times the lots were similarly ranked.

There was a reduction in the incidence of *Penicillium* sp. (Table 6, Figure 2) from 20 minutes of ozonization for lots 1 and from 60 minutes for 2 lot. Only without ozonization lot 2 statistically differentiated from the other lots with lower incidence of the fungus. Palou et al. (2002) have obtained similar results reducing the incidence of *Penicillium* sp. using ozone to sanitize pear fruits after harvest.

In the severity assessments (Table 7, Figure 3) in sunflower seeds that were not treated with ozone, there was a prevalence of the fungus *Fusarium* sp., followed by *Alternaria* sp., *Aspergillus* sp. and *Penicillium* sp. When subjected to treatment with ozone, reduction of the infestation severity index for all fungi was found.

Table 7. Severity Index of *Alternaria* sp., *Fusarium* sp., *Aspergillus* sp and *Penicillium* sp., in the lot of seeds of *Helianthus annuus L.*, cultivar Embrapa 122, due to the times (minutes) of ozonization.

Ozonization time	*Alternaria* sp.			*Fusarium* sp.			*Aspergillus* sp.			*Penicillium* sp.		
	L1	L2	L3	L1	L2	L3	L1	L2	L3	L1	L2	L3
0	60 Aa	67 Aa	65 Aa	70 Aa	72 Aa	79 Aa	34 Aa	29 Aa	33 Aa	26 Aa	19 Ba	26 Aa
20	55 Aa	65 Aa	55 Aa	71 Aa	71 Aa	78 Aa	31 Aa	29 Aa	25 Ab	21 Ab	18 Aa	20 Ab
60	31 Ab	41 Ab	28 Ab	49 Bb	51 Bb	71 Aa	11 Ab	11 Ab	13 Ac	8 Ac	10 Ab	11 Ac
120	32 Ab	40 Ab	40 Ab	32 Bc	54 Ab	56 Ab	17 Ab	9 Ab	12 Ac	7 Ac	8 Ab	5 Ad
CV (%)	39.98			23.96			33.18			30.63		

Means followed by the same lowercase letter in the column and uppercase letter in the row do not differ by Scott Knott test at 5% probability.

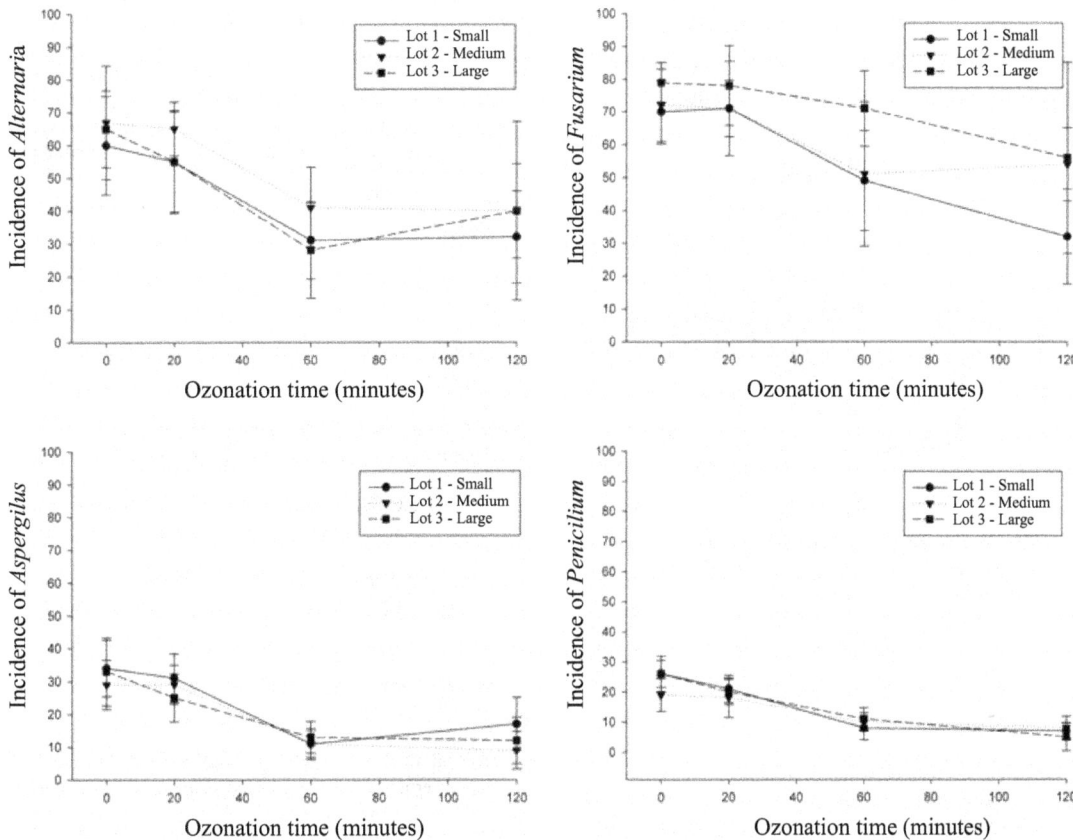

Figure 3. Severity Index of *Alternaria* sp., *Fusarium* sp., *Aspergillus* sp and *Penicillium* sp. (mean ± standard deviation) of three seed lots of *Helianthus annuus L.*, cultivar Embrapa 122, due to the times (minutes) of ozonization.

There was a reduction of the infestation severity index (ISI) of *Alternaria* sp. from 60 minutes of ozonization for all lots of sunflower seeds. In all ozonization times the lots belonged to the same data group. For *Fusarium* sp., from 60 minutes of ozonization there was a reduction of the infestation on seeds of lots 1 and 2 and after 120 minutes for lot 3. There were differences among lots from 60 minutes, when lower infestation severity index was observed for lots 1 and 2, and in the 120-minute period when lot 1 obtained lower ISI compared to the other lots.

For *Aspergillus* sp. the same tendency of reduction occurred in the evaluation of the incidence of fungus (Table 6, Figure 2) for lot 1 and lot 2, wherein the reduction was observed from 60 minutes of ozonization. As for lot 3, the ISI reduction of *Aspergillus* sp. occurred from 20 minutes of ozonization. All lots corresponding to the control and the tested ozonization treatments were ranked in the same group.

The results of the *Penicillium* sp severity index observed had the same tendency of incidence results (Table 6, Figure 2) for the same fungus, where ISI was reduced from 20 minutes for lots 1 and 3, and for lot 2 the reduction was observed from 60 minutes of ozonization.

It can be seen that the ozonization for 60 minutes has caused no negative effect on the viability and vigor of sunflower seeds of cultivar Embrapa 122 and has promoted reduction of incidence and severity of all the pathogens found, indicating ozone potential as a sunflower seed sanitizer.

Conclusions

The treatment of sunflower seeds, cultivar Embrapa 122 with ozone in the concentration of 1741 ppmv (0.24 g/h), for 60 minutes, reduces the fungi population of *Alternaria* sp., *Fusarium* sp., *Aspergillus* sp. and *Penicillium* sp., without affecting their physiological potential.

References

AGUIAR, R.H.; FANTINATTI, J.B.; GROTH, D.; USBERTI, R. Qualidade física, fisiológica e sanitária de sementes de girassol de diferentes tamanhos. *Revista Brasileira de Sementes*, v.23, n.1, p.134-139, 2001. http://www.abrates.org.br/revista/artigos/2001/v23n1/artigo19.pdf

AKBAS, M.Y.; OZDEMIR, M. Effect of different ozone treatments on aflatoxin degradation and physico-chemical properties of pistachios. *Journal of the Science of Food and Agriculture*, v.86, n.13, p.2099–2104, 2006. 10.1002/jsfa.2579

BRASIL. Ministério da Agricultura Pecuária e Abastecimento. *Manual de Análise Sanitária de Sementes (Handbook on Seed Health Testing)*. Ministério da Agricultura e Abastecimento. Brasília: MAPA-ACS, 2009a. 200p. http://www.agricultura.gov.br/arq_editor/file/12261_sementes_-web.pdf

BRASIL. Ministério da Agricultura, Pecuária e Abastecimento. *Regras para análise de sementes*. Ministério da Agricultura e Abastecimento. Secretaria de Defesa Agropecuária. Brasília: MAPA-ACS, 2009b. 395p. http://www.agricultura.gov.br/arq_editor/file/2946_regras_analise__sementes.pdf

BRASIL. Ministério da Agricultura, Pecuária e Abastecimento. Padrões para produção e comercialização de sementes de girassol. *Instrução Normativa* nº 45, de 17 de setembro de 2013. Brasília, DF: MAPA/SCS.

CAMPANHOLA, C.; BETTIOL, W. *Métodos alternativos de controle fitossanitário*. Jaguariúna: Embrapa Meio Ambiente, 2003. 279 p.

FERREIRA, D.F. Sisvar: a computer stastistical analysis system. *Ciência e Agrotecnologia*, v.35, n.6, p.1039-1042, 2011. http://dx.doi.org/10.1590/S1413-70542011000600001

FREITAS-SILVA, O.; VENÂNCIO, A. Supressão de *Aspergillus* produtores de aflatoxina e ácido ciclopiazonico por ozônio. *Revista Ciência e Vida*, v.28, p.128-200, 2008. http://ainfo.cnptia.embrapa.br/digital/bitstream/item/65812/1/2008-078.pdf

GOMES, D.P.; BRINGEL, J.M.M.; MORAES, M.F.H.; GOMES, J.J.A.; LEITE, R.M.V.B.C. Qualidade fisiológica e sanitária de sementes de girassol produzidas na região de Timon, Maranhão. *Summa Phytopathologica*, v.32, n.3, p.291-292, 2006. http://dx.doi.org/10.1590/S0100-54052006000300016

GOULART, A.C.P. *Detecção e controle químico de Colletotrichum em sementes de soja e algodão*. Dourados: Embrapa Agropecuária, 2009. 33p.

GRISI, P.U.; SANTOS, C.M.; FERNANDES, J.J.; SÁ JÚNIOR, A. Qualidade das sementes de girassol tratadas com inseticidas e fungicidas. *Bioscience Journal*, v.25, n.4, p.28-36, 2009. http://www.seer.ufu.br/index.php/biosciencejournal/article/viewArticle/6948

JULIATTI, F.C.; DEL BIANCO JUNIOR, R.; MARTINS, J.A.S. Qualidade fisiológica e sanitária de sementes de algodoeiro produzidas nas regiões do Triângulo Mineiro e sul de Goiás. *Bioscience Journal*, v.27, n.1, p.24-31, 2011. http://www.seer.ufu.br/index.php/biosciencejournal/article/view/7370

KLAPTCHUK, P. Method of destroying seed. World Patent n. WO089078, 7 abr. 2004, 21 out. 2004.

LEITE, R.M.V.B.; BRINGHENTI, A.M.; CASTRO, C. (eds.). *Girassol no Brasil*. Londrina: Embrapa Soja, 2005. 641p.

LEONARDO, M.S; JARDIM, W.F. Gerador de Ozônio Eletroquímico. Brasil INPI n. PI1101892-5, 2011.

MACHADO, J.C. Padrões de tolerância de patógenos associados às sementes. *Revisão Anual de Patologia de Plantas*, v.2, p. 229-263, 1994.

MACHADO, J.C. *Tratamento de sementes no controle de doenças*. Lavras: Editora UFLA, 2000. 138p.

MAGUIRE, J.D. Speeds of germination-aid selection and evaluation for seedling emergence and vigor. *Crop Science*, v.2, p.176-177, 1962.

MCKINNEY, H.H. Influence of soil, temperature and moisture on infection of wheat seedlings by *Helminthosporium sativum*. *Journal of Agricultural Research*, v.26, p.195-217, 1923.

MEDEIROS, J.G.F.; NETO, A.C.A; MENEZES, N.P.C.; NASCIMENTO, L.C. Sanidade e germinação de sementes de *Clitoria fairchildiana* Howard tratadas com extratos de plantas. *Pesquisa Florestal Brasileira*, v.33, n.76, p.403-408, 2013. http://pfb.cnpf.embrapa.br/pfb/index.php/pfb/article/viewFile/541/339

MENDEZ, F.; MAIER, D.E.; MASON, L.J.; WOLOSHUK, C.P. Penetration of ozone into columns of stored grains and effects on chemical composition and performance. *Journal of Stored Products Research*, v.39, p.33-44, 2003. http://www.sciencedirect.com/science/article/pii/S0022474X02000152

PALOU, L.; CRISOTO, C.H.; SMILANICK, J.L.; ADASKAVEG, J.E.; ZOFFOLI, J.P. Effects of continuous 0.3 ppm ozone exposure on decay development and physical responses of peaches and table grapes in cold storage. *Postharvest Biology and Technology*, v.24, p.39–48, 2002. http://handle.nal.usda.gov/10113/19498

PEREIRA, A.M.; FARONI, L.R.D.; SILVA JUNIOR, A.G.; SOUZA, A.H.; PAES, J.L. Viabilidade econômica do gás ozônio como fumigante em grãos de milho armazenados. *Engenharia na Agricultura*, v.16, n.2, p.144-154, 2008. http://www.seer.ufv.br/seer/index.php/reveng/article/viewFile/12/2

PINTO, A.T.; SCHMIDT, V.; RAIMUNDO, S.A.; RAIHMER, F. Uso de ozônio no controle de fungos em sala de maturação de queijos. *Acta Scientiae Veterinariae*, v.35, n.3, p.333-337, 2007. http://hdl.handle.net/10183/20606

PONCE, A.R.; BASTIANI, M.I.D.; MINIM, V.P.; VANETTI, M.C.D. Características físico-químicas e microbiológicas de morango minimamente processado. *Ciência Tecnologia Alimentos*, v.30, n.1, p.113-118, 2010. http://www.scielo.br/pdf/cta/v30n1/aop_3254.pdf

PORTO, W.S.; CARVALHO, C.G.P.; PINTO, R.J.B. Adaptabilidade e estabilidade como critérios para seleção de genótipos de girassol. *Pesquisa Agropecuária Brasileira*, v.42, n.4, p.491-499, 2007. http://www.scielo.br/pdf/pab/v42n4/06.pdf

SALUSTIANO, M.E.; MACHADO, J.C.I.; PITTIS, J.E. Patogenicidade de *Alternaria helianthi* (Hansf.) e *Alternaria zinniae* (Pape) ao girassol a partir de sementes. *Revista Brasileira de Sementes*, v.27, n.1, p.138-143, 2005. http://dx.doi.org/10.1590/S0101-31222005000100017

SERRA, R.; ABRUNHOSA, L.; KOZAKIEWICZ, Z.; VENANCIO, A.; LIMA, N. Use of ozone to reduce molds in a cheese ripening room. *Journal of Food Protection,* v.66, n.12, p.2355 – 2358, 2003. http://hdl.handle.net/1822/1654

SIMÃO, R.; RODRÍGUEZ, T.D.M. Utilização do ozônio no tratamento pós-colheita do tomate (*Lycopersicon esculentum* Mill.). *Revista de Estudos Sociais*, v.2. n.22, p.115-124, 2008. http://periodicoscientificos.ufmt.br/index.php/res/article/view/246/235

SUDHAKAR, N.; NAGENDRA-PRASAD, D.; MOHAN, N.; HILL, B.; GUNASEKARAN, M.; MURUGESAN, K. Assessing influence of ozone in tomato seed dormancy alleviation. *American Journal of Plant Sciences,* v.2 n.3, p.443-448, 2011. http://www.scirp.org/journal/PaperInformation.aspx?PaperID=7608

VIOLLEAU, F.; HADJEBA, K.; ALBET, J.; CAZALIS, R.; SUREL, O. Increase of corn seeds germination by oxygen and ozone treatment. *Ozone Science & Engineering Journal*, v.30, n.6, p.418 – 422, 2008. 10.1080/01919510802474631

WU, J.; DOAN, H.; CUENCA, M.A. Investigation of gaseous ozone as an anti-fungal fumigant for stored wheat. *Journal of Chemical Technology and Biotechnology*, v.81, n.7, p.1288–1293, 2006. http://onlinelibrary.wiley.com/doi/10.1002/jctb.1550/abstract

YVIN, J.C.; COSTE, C. Method and system for the treatment of seeds and bulbs with ozone. World Patent n. WO1995009523 A1, 4 out. 1994, 13 abr. 1995.

Seed germination of Brazilian *Aldama* species (Asteraceae)

Aline Bertolosi Bombo[1], Tuane Santos de Oliveira[1], Beatriz Appezzato-da-Glória[1*], Ana Dionísia da Luz Coelho Novembre[2]

ABSTRACT - Brazilian samples from the *Aldama* genus (Asteraceae) could not propagate vegetatively despite their thickened underground system; thus, this study on sexual propagation is critical given the lack of data on reproductive strategies for such species. The aim for this research was to assess the optimal temperature for *Aldama arenaria*, *A. filifolia*, *A. linearifolia*, *A. robusta* and *A. trichophylla* seed germination. Seed germination was evaluated at the constant temperatures 20, 25 and 30 °C and the alternating temperatures 15-35, 20-30 and 20-35 °C with an 8-h daily photoperiod, using fluorescent-lamp. The ungerminated seeds were evaluated for embryo viability. The *A. filifolia* seed health was also evaluated. The optimal temperatures for germination are 20 and 25 °C for *Aldama arenaria*, *A. filifolia*, *A. robusta* and *A. trichophylla* seeds and 20 °C for *A. linearifolia* seeds. The alternating temperature 15-35 °C is not recommended for germinating seeds from these species. The six fungi taxa studied herein did not affect *A. filifolia* seed germination.

Index terms: Compositae, sexual propagation, *Viguiera*, cerrado.

Germinação de sementes de espécies brasileiras de *Aldama* (Asteraceae)

RESUMO - Amostras de espécies brasileiras do gênero *Aldama* (Asteraceae) não têm capacidade de se propagar vegetativamente, apesar do seu sistema subterrâneo espessado; assim, esse estudo da propagação sexuada é importante, devido à falta de informações sobre as estratégias de reprodução dessas espécies. O objetivo da pesquisa foi determinar a faixa ideal de temperatura para a germinação das sementes de *Aldama arenaria*, *A. filifolia*, *A. linearifolia*, *Aldama robusta* e *A. trichophylla*. A germinação das sementes foi avaliada em temperaturas constantes de 20, 25 e 30 °C e alternadas de 15-35, 20-30 e 20-35 °C sob fotoperíodo diário de 8 h, sob lâmpada fluorescente. As sementes não germinadas foram avaliadas quanto à viabilidade do embrião. A sanidade das sementes de *A. filifolia* também foi avaliada. As temperaturas ideais para a germinação das sementes de *Aldama arenaria*, *A. filifolia*, *A. robusta* e *A. trichophylla* são 20 e 25 °C e, para as de *A. linearifolia* é 20 °C. A temperatura alternada de 15-35 °C não é recomendada para a germinação das sementes dessas espécies. Os seis táxons de fungos que foram determinados não influenciam na germinação das sementes de *A. filifolia*.

Termos para indexação: Compositae, propagação sexuada, *Viguiera*, cerrado.

Introduction

The genus *Aldama* La Llave belongs to the Asteraceae family and includes South American species previously classified as *Viguiera sensu lato* (Schilling and Panero, 2011). Brazilian species from this genus occur mainly in the Cerrado regions in climates that have dry winters (Magenta et al., 2010), and these species are most often observed under such conditions that may be related to the type of thickened underground system, which supports the plants' water needs during dormancy (Zaidan and Carreira, 2008; Magenta et al., 2010) and is essential for regenerating the vegetative aerial portions (Oliveira et al., 2013;

Bombo et al., 2014). The thickened underground systems, which can bear buds, were observed in other Asteraceae species, and these systems may be morphologically diverse, such as xylopodia, rhizophores, root buds and diffuse underground systems (Vilhalva and Appezzato-da-Glória, 2006; Hayashi and Appezzato-da-Glória, 2007; Appezzato-da-Glória et al., 2008; Appezzato-da-Glória and Cury, 2011). However, it has been observed that some species cannot propagate vegetatively even bearing a thickened underground system, and the sexual reproduction ensures maintenance of the species (Cury et al., 2010). The underground system for Brazilian species from the *Aldama* genus is characterized by a

[1]Departamento de Ciências Biológicas, USP/ESALQ, Caixa Postal 9, 13418-900 – Piracicaba, SP, Brasil.

[2]Departamento de Produção Vegetal, USP/ESALQ, Caixa Postal 9, 13418-900 – Piracicaba, SP, Brasil..
*Corresponding author:< bagloria@usp.br>

xylopodium emitting tuberous roots (Oliveira et al., 2013; Bombo et al., 2014); therefore, these species do not use vegetative propagation as a form of reproduction.

Few studies have investigated seed germination for Brazilian Asteraceae species (Ferreira et al., 2001; Gomes and Fernandez, 2002; Velten and Garcia, 2005; Garcia et al., 2006; Cury et al., 2010), and about the propagation strategies of Brazilian *Aldama* species. Germination behavior has only been reported for *Aldama robusta* (Gardner) E.E.Schill. & Panero (= *Viguiera robusta*), and for this species the germination was characterized as slow process (Ruggiero and Zaidan, 1997).

Research on the phytochemistry of several species of *Aldama* genus have highlighted the pharmacological importance of this genus (Costa et al., 1996; Marquina et al., 2001; Spring et al., 2003; Ambrosio et al., 2004); therefore, reproductive studies on plant species are important for proposing strategies to ensure a sustainable development (Gomes and Fernandez, 2002) and conservation of such species, since *Aldama filifolia* and *A. linearifolia* are on the red list as species threatened with extinction (Brasil, 2014). Thereby, the aim of this research was to study germination and determine the optimal temperature for five species from the *Aldama* genus.

Material and Methods

Cypsela collection and storage

According to Marzinek et al. (2008), Asteraceae plants have cypsela-type fruits. Seeds could not be removed from the fruits in that species; thus, the propagating unit that comprised the fruit and seed (referred as the seed in the text) was evaluated in the experiments.

Cypselas used for germination tests were collected from plants of natural populations of *Aldama arenaria* (Baker) E.E.Schill. & Panero (= *Viguiera arenaria*), *A. filifolia* (Sch.Bip. ex Baker) E.E.Schill. & Panero (= *V. filifolia*), *A. linearifolia* (Chodat) E.E.Schill. & Panero (= *V. linearifolia*), *Aldama robusta* (Gardner) E.E.Schill. & Panero (= *V. robusta*) and *A. trichophylla* (Dusén) Magenta (= *V. trichophylla*) in Cerrado in Itirapina - Estação Ecológica de Itirapina, SP, Brazil, Alto Paraíso de Goiás - Goiás (GO), Ponta Porã - Mato Grosso do Sul (MS), Altinópolis - São Paulo (SP) and Palmeira – PR, respectively, between April and November 2010, along the fruiting period for each species. The plants were identified by an expert on this genus and included in the ESA herbarium, 'Luiz de Queiroz' College of Agriculture (Escola Superior de Agricultura 'Luiz de Queiroz') under the

numbers 111847, 111848, 113164, 114255 and 111850.

The entire capitula were collected, spread on paper, maintained at 22 °C until dried and stored for three months in paper bags at the same temperature.

Experiment set-up and seed germination monitoring

Seeds were manually removed from the capitula, homogenized in a tray and visually selected in order to eliminate the wilted or predated seeds, which were discarded, and just the apparently undamaged and healthy ones were manually selected for germination test. However, it was impossible to assess whether they contained an embryo based on their external physical appearance or by pressure.

Five replicates with at least 16 seeds each were used for each temperature evaluated (See Table 1). The amount of seeds in each repetition varied among the species because it was determined according to the availability of seeds at collection, since these seeds were obtained from natural populations. However, the amount was always the same among the replicates within a species. The seeds were randomly distributed on two sheets of blotting paper, hydrated with 13 mL of distilled water and packed in previously sterilized transparent plastic boxes (13 x 13 x 3.5 cm). The boxes were distributed into six germination chambers (Marconi brand, MA 402 model) at the constant temperatures of 20, 25 and 30 °C with a 8-h daily photoperiod and alternating temperatures 15-35, 20-30 and 20-35 °C, with a 8-h daily photoperiod in the top temperature, using fluorescent-lamp lighting with 40.32 mmol. $m^{-2}s^{-1}$ irradiation.

The test was performed until no germination was observed anymore, and this occurred after 32 days. When it was necessary, the paper was moistened to ensure the water supply to the seeds during the test. The seeds were considered germinated when they originated normal seedlings with two cotyledons, hypocotyl and primary root and the non-germinated seeds were also counted (Brasil, 2009).

With the obtained results, the germination speed index (GSI, using the formula proposed by Maguire (1962), mean germination time (Laboriau, 1983) and germination percentage were calculated for each species. The values generated were assessed through statistical analysis for the normal distribution using the Kolmogorov-Smirnov test. The data were transformed into an $arcsen \sqrt{x/100}$ and into the $\sqrt{x + 0.5}$ when the parameter values were equal to zero. Next, an analysis of variance was used for the results, and the means were compared through the Tukey test ($p < 0.05$) using the Statistical Analysis System (SAS) software (Sas Institute, 2003).

Table 1.   Germination (G, %), germination speed index (GSI) and mean germination time (MT, days) for *Aldama arenaria*, *A.*
filifolia, *A. linearifolia*, *A. robusta* and *A. trichophylla* seeds evaluated at the constant temperatures 20, 25 and 30 °C
with a 8-h daily photoperiod and the alternating temperatures 15-35, 20-30 and 20-35 °C with a 8-h daily photoperiod
in the upper temperature. Values with the same letter in the same column do not differ statistically, within each species.

Species	Temperature (°C)	G (%)	GSI	MT (days)
Aldama arenaria (n=600)	20	92 a	2.70 bc	7.24 bc
	25	88 ab	3.81 a	5.01 a
	30	74 ab	2.69 bc	6.29 abc
	15-35	73 b	1.52 d	10.86 d
	20-30	82 ab	2.93 b	6.07 ab
	20-35	71 b	1.95 cd	8.02 c
Aldama filifolia (n = 750)	20	50 a	1.14 ab	11.81 a
	25	48 a	1.29 a	10.05 a
	30	33 bc	0.76 c	12.71 ab
	15-35	8 d	0.10 d	23.73 c
	20-30	41 ab	0.89 bc	12.63 ab
	20-35	22 c	0.34 d	17.40 b
Aldama linearifolia (n = 705)	20	29 a	0.49 a	14.73 a
	25	22 a	0.43 ab	12.79 a
	30	13 a	0.23 ab	12.55 a
	15-35	17 a	0.21 b	21.75 b
	20-30	28 a	0.49 a	14.54 a
	20-35	20 a	0.34 ab	15.09 a
Aldama robusta (n=480)	20	97 a	2.01 a	7.99 ab
	25	95 ab	2.21 a	7.55 a
	30	67 c	1.14 c	10.56 ab
	15-35	17 d	0.12 d	24.71 d
	20-30	97 a	1.68 b	10.96 b
	20-35	85 bc	0.89 c	17.72 c
Aldama trichophylla (n = 750)	20	47 a	1.68 ab	7.30 a
	25	54 a	2.54 a	5.85 a
	30	34 ab	1.49 b	6.98 a
	15-35	50 a	1.26 b	10.48 b
	20-30	42 ab	1.60 b	7.22 a
	20-35	22 b	0.77 b	7.60 a

Evaluation of ungerminated seeds

After the final evaluation, the seeds were kept under the same conditions for an additional 10-day period to ensure stabilization of the germination process. Thereafter, the non-germinated seeds were analyzed to verify the presence or absence of the embryo and its viability. For such analysis, a cross-sectional cut was done along the fruit just below the pappus insertion, and if the seed contained an embryo, it was removed and evaluated for viability through the tetrazolium test using the method reported by Cury et al. (2010) and the results expressed as the percentage of viable embryos.

Seed health test

During the experiment, the incidence of fungi was recorded just for seeds from *Aldama filifolia* species, and they were evaluated by the health test. Thus, 100 seeds were collected as described before, distributed on three sheets of filter paper soaked in water, placed in Petri dishes (25 seeds per plate) and maintained at 20 °C with a 12-h daily photoperiod for seven days. Thereafter, the fungi were identified from the colonies and the results were expressed as the percentage of fungi.

Results and Discussion

Seed germination

The results of the germination tests for the five studied species (Table 1) indicated that all of them presented some germination percentage in all the tested temperatures. Among them, *A. robusta*, followed by *A. arenaria*, were the species with higher percentage of germination while *A. linearifolia* had the lowest performance at all the temperatures. The test also indicated that the constants temperatures 20, 25 °C and the alternating temperature 20-30 °C were the best treatments with the significantly higher percentage of germination in *A. filifolia* and *A. robusta* and 20, 25, 30 °C and 20-30 °C in *A. arenaria*, while for *A. trichophylla* all the treatments responded at the same way, except the alternating temperature 20-35 °C, which presented significantly lower values of percentage of germination. A statistically significant difference was not found among the temperatures evaluated for *A. linearifolia* species.

The 15-35 °C and 20-35 °C alternating temperature produced significantly lower results of germination, GSI and mean time of germination for seeds of *Aldama arenaria* and *A. filifolia*; for *A. robusta* 15-35 °C was not a good treatment for these parameters; the results for GSI and mean time of germination in *A. linearifolia* seeds were lower despite the statistically similar results for germination rate. For *A. trichophylla* seeds, the 20-35 °C alternating temperature significantly reduced the percentage of germination while 15-35 °C reduced the speed of that process.

Aldama seeds germinated in a range of constant and alternating temperatures. This ability may have useful consequences, because at least some seeds will germinate, whatever the environmental conditions they are submitted (Silva et al., 2002). Moreover, seed germination studies on Asteraceae species (Ferreira et al., 2001; Gomes and Fernandez, 2002; Velten and Garcia, 2005; Garcia et al., 2006; Cury et al., 2010) indicate that different species germinate differently and respond to distincts alternating temperatures with variations in germination rates. The constant or alternating temperatures between 20 and 30 °C were most efficient despite the variation in optimal temperature for seed germination for the species studied. Albuquerque et al. (2003) reported that the seeds from certain medicinal species in the Cerrado typically germinate at the constant temperatures 25 or 30 °C, and Ruggiero and Zaidan (1997) had already indicated that 20-30 °C is the most suitable temperature for *Aldama robusta* (*Vigueira robusta*) seeds germination.

The study herein demonstrated that the lowest germination percentages were observed for temperatures which seed germination was slowest, except for *A. trichophylla* seeds. In the temperatures where the germination percentages were higher, the GSI values were also higher, and consequently, the mean germination times were lower. High GSI values and low mean germination times indicate fast germination,

which is a typical feature of species that can be established in the environment as quickly as possible, under favorable conditions (Velten and Garcia, 2005).

Among the five species investigated here, *Aldama arenaria*, followed by *A. trichophylla* were the species whose seeds germinated faster, while *A. linearifolia* presented the slowest germination process. The mean time or speed germination are good indicators of the speed for occupying a territory by a species at a given environment, i.e., its ability to propagate in its habitat (Ferreira et al., 2001). Gomes and Fernandez (2002) also highlighted the importance of rapid germination for a species to be able to recover degraded areas or occupy an environment. As proposed by Ferreira et al. (2001), species may be classified according to their seed germination speed: at temperatures with the highest germination rates, *Aldama arenaria*, *A. trichophylla* and *A. robusta* seeds were classified as intermediate germination, while *A. filifolia* and *A. linearifolia* seeds have slow germination. Such observations may explain the presence of the reduced populations of *Aldama* species studied here, which were observed specially on the sides of highways whose areas were often invaded by exotic grass that hamper seed germination and seedling establishment.

Evaluation of ungerminated seeds

The evaluation herein showed that the percentages for *Aldama arenaria*, *A. robusta*, *A. filifolia* and *A. trichophylla* seeds without embryos were 7, 8, 23 and 21%, respectively. The highest percentage of seeds without embryos (67%) was observed for *Aldama linearifolia* species. Dead seeds accounted for 6% of the seeds in *Aldama linearifolia*, 7% in *A. arenaria* and *A. robusta*, 32% in *A. filifolia* and 27% in *A. trichophylla* (Table 2).

The abnormal seedling morphology, which accounted for 5% seedlings in *A. arenaria*, 9% in *A. filifolia,* 5% in *A. linearifolia*, 3% in *A. robusta* and 10% in *A. trichophylla* (Table 2), was another factor that affected the germination results, in addition to the absent of embryos and dead seeds. The highest number of abnormal seedlings was observed under the constant temperature 30 °C and the alternating temperature 15-35 and 20-35 °C in *A. arenaria*, *A. filifolia* and *A. trichophylla*. In *A. robusta* the alternating temperature 15-35 °C and in *A. linearifolia* the constant temperature 20 °C produced lower values of abnormal seedlings.

The highest percentages of viable seeds that do not germinated was observed under the alternating temperature 15-35 °C, which was the less efficient condition for seed germination in *Aldama arenaria* along with 20-35 °C, *A. filifolia*, *A. linearifolia* and *A. robusta* species, while, for *A. trichophylla*, 20-35 °C was the least efficient temperature.

Although some species require daily temperature fluctuations to germinate (Silva et al., 2002), the observations here indicated that these two alternating temperatures are not

recommended for germination in *Aldama* species, and it may even delay the process, which would explain the viable embryos in the ungerminated seeds.

Table 2.   Analysis for ungerminated seeds or germinated seeds that produced abnormal seedlings for *Aldama arenaria, A. filifolia, A. linearifolia, A. robusta* and *A. trichophylla*. Percentage of abnormal seedlings, viable seeds, dead seeds and seeds without embryos. Values with the same letter in the same column do not differ statistically, within each species.

Species	Temperature (°C)	Abnormal seedlings (%)	Ungerminated seeds (%)		
			Viable	Dead	Seed without embryos
Aldama arenaria	20	0 c	0 b	3 a	5 a
	25	1 bc	0 b	6 a	5 a
	30	14 a	0 b	9 a	3 a
	15-35	4 abc	6 a	9 a	8 a
	20-30	1 bc	0 b	8 a	9 a
	20-35	9 ab	0 b	7 a	13 a
Aldama filifolia	20	0 c	0 b	24 a	26 ab
	25	4 bc	0 b	33 a	15 b
	30	11 ab	1 b	33 a	22 ab
	15-35	26 a	10 a	42 a	14 b
	20-30	5 bc	1 b	27 a	26 ab
	20-35	12 ab	2 b	28 a	36 a
Aldama linearifolia	20	1 b	0 b	8 a	62 a
	25	3 ab	0 b	3 a	72 a
	30	5 ab	0 b	11 a	71 a
	15-35	10 a	3 a	7 a	63 a
	20-30	3 ab	0 b	4 a	65 a
	20-35	8 ab	0 b	5 a	67 a
Aldama robusta	20	0 b	0 b	0 b	3 b
	25	0 b	0 b	0 b	5 ab
	30	0 b	3 b	19 a	11 ab
	15-35	18 a	23 a	25 a	17 a
	20-30	0 b	0 b	0 b	3 b
	20-35	3 b	1 b	0 b	11 ab
Aldama trichophylla	20	2 c	0 a	30 a	21 ab
	25	2 c	0 a	21 a	23 ab
	30	17 ab	0 a	37 a	12 b
	15-35	9 abc	2 a	26 a	13 b
	20-30	6 bc	0 a	21 a	31 a
	20-35	22 a	0 a	30 a	26 ab

Several factors may affect the seed germination rate, including the decrease in seed viability during storage (Sassaki et al., 1999; Gomes and Fernandez, 2002), the number of seeds without embryos (Sassaki et al., 1999; Velten and Garcia, 2005; Cury et al., 2010) and the collection of immature seeds (Albuquerque et al., 2003). The number of seeds without embryo in the species studied herein, excluding *Aldama linearifolia* seeds, was relatively low when compared with other Asteraceae species (Sassaki et al., 1999; Velten and Garcia, 2005; Cury et al., 2010). For example,

the higher percentage of seeds without embryos in *Chresta sphaerocephala* DC. (85%) compared with *Lessingianthus bardanoides* (Less.) H. Rob. (54%) is related to the ability of *C. sphaerocephala* to propagate vegetatively through root buds, while *L. bardanoides* representatives bear a xylopodium and the species propagates only by seeds (Cury et al., 2010). For *Aldama* species studied herein, which have xylopodium as underground organ (Oliveira et al., 2013; Bombo et al., 2014), the vegetative propagation is not common, and this could be related to the low percentage of seeds without embryos in such species. In turn, the high number of empty seed in *A. linearifolia* is related to the maturation degree of the seeds, since this species has the same sort of underground organ.

Seed health test

The results from the seed health evaluation for *Aldama filifolia* allowed identification of six fungi taxa associated with its seeds. *Alternaria alternata* were found in 98% of seeds, *Cladosporium* sp. in 90%, *Fusarium* sp. in 9%, *Alternaria* sp. in 1%, *Phoma* sp. in 1% and *Nigrospora* sp. in 1%. Fungi have been reported in seeds for several species, including Asteraceae family representatives (Lima et al., 2003; Reis et al., 2006). Fungal development during germination may increase levels of infected or dead seeds in addition to diseased seedlings; thus, seed contamination may affect their physiological quality and, in certain cases, inhibit the germination process (Castellani et al., 1996; Botelho et al., 2008). Although fungal infestation was observed for all treatments analyzed, the increase in abnormal seedlings coincided with the temperature which provided the lowest germination rate. Therefore, such infestation seems not to have affected seed germination process of the studied species, as already observed by Oliveira et al. (2003) for *Peltophorum dubium* (Sprengel) Taubert (Fabaceae) seeds.

Conclusions

Considering the levels of germinated seeds, germination speed index and mean germination time simultaneously, the best temperatures for seed germination are the constant temperatures 20 °C and 25 °C for *Aldama filifolia*, *A. robusta* and *A. trichophylla*, 20 °C for *A. areanaria* and for *A. linearifolia* just the alternating temperature 15-35 °C is not recommended. However, the five species responded well to almost all treatments used and other temperature ranges may also be used. The alternating temperature 15-35 °C is not recommended for such species, except to *A. trichophylla*, for which 20-35 °C is the temperature not recommend.

Acknowledgments

We thank The National Council for Scientific and Technological Development (CNPq) for grants (302776/2010-9) and the São Paulo Council for Research (FAPESP) (Thematic Project Proc. n° 2010/51454-3, Proc. no. 2010/01931-0) for providing financial support and grants to the first author. We are also grateful to Professor Mara Angelina Galvão Magenta for plant identification, the coordinator of the Parque Nacional da Chapada dos Veadeiros, GO, Brazil and the coordinator of the Estação Ecológica de Itirapina, SP, Brazil for granting permission and facilities to collect plant material for this study. Finally, we thank the Seed Pathology Laboratory, Department of Plant Pathology and Nematology, 'Luiz de Queiroz' College of Agriculture/University of São Paulo, in which the seed health test was performed.

References

ALBUQUERQUE, M.C.F.; COELHO, M.F.B.; ALBRECTH, J.M.F. Germinação de sementes de espécies medicinais do Cerrado. In: COELHO, M.F.B.; COSTA JÚNIOR, P.; DOMBROSKI, J.L.D. (Ed.). *Diversos olhares em etnobiologia, etnoecologia e plantas medicinais*. Cuiabá UNICEN, 2003, p. 157-181.

AMBROSIO, S.R.; SCHORR, K.; COSTA, F.B. Terpenoids of *Viguiera arenaria* (Asteraceae). *Biochemical Systematics and Ecology*, v.32, p.221-224, 2004. http://www.sciencedirect.com/science/article/pii/S030519780300139X

APPEZZATO-DA-GLÓRIA, B.; CURY, G. Morpho-anatomical features of underground systems in six Asteraceae species from the Brazilian Cerrado. *Anais da Academia Brasileira de Ciências*, v.83, n.3, p.981-992, 2011. http://www.scielo.br/scielo.php?pid=S0001-37652011000300017&script=sci_arttext

APPEZZATO-DA-GLÓRIA, B.; HAYASHI, A.H.; CURY, G.; SOARES, M.K.M.; ROCHA, R. Ocurrence of secretory structures in underground systems of seven Asteraceae species. *Botanical Journal of the Linnean Society*, v.157, n.4, p.789-796, 2008. http://onlinelibrary.wiley.com/doi/10.1111/j.1095-8339.2008.00823.x/abstract

BOMBO, A.B.; OLIVEIRA, T.S.; OLIVEIRA, A.S.S.; REHDER, V.L.G.; APPEZZATO-DA-GLÓRIA, B. Anatomy and essential oil composition of the underground systems of three species of *Aldama* La Llave (Asteraceae). *The Journal of the Torrey Botanical Society*, v.141, n.2, p.115-125, 2014. http://www.bioone.org/doi/abs/10.3159/TORREY-D-12-00053.1

BOTELHO, L.D.S.; MORAES, M.H.D.; MENTEN, J.O.M. Fungos associados às sementes de ipê-amarelo (*Tabebuia serratifolia*) e ipê-roxo (*Tabebuia impetiginosa*): incidência, efeito na germinação e transmissão para as plântulas. *Summa Phytopathologica*, v.34, n.4, p.343-348, 2008. http://www.scielo.br/scielo.php?script=sci_arttext&pid=S0100-54052008000400008

BRASIL. Ministério da Agricultura, Pecuária e Abastecimento. *Regras para análise de sementes*. Ministério da Agricultura, Pecuária e Abastecimento. Secretaria de Defesa Agropecuária. Brasília: MAPA/ACS, 2009. 395p. http://www.agricultura.gov.br/arq_editor/file/2946_regras_analise__sementes.pdf

BRASIL. Ministério do Meio Ambiente. 17 de dezembro de 2014. *Diário oficial da união*, Brasília, DF, 18 dez. 2014, p. 110. Portaria n° 443 Seção I.

CASTELLANI, E.D.; SILVA. A.; BARRETO, M.; AGUIAR, I.B. Influência do tratamento químico na população de fungos e na germinação de sementes de *Bauhinia variegata* L. var. *Variegata*. *Revista Brasileira de Sementes*, v.18, n.1, p.41-44, 1996. http://www.abrates.org.br/revista/artigos/1996/v18n1/artigo07.pdf

COSTA, F.B.; VICHNEWSKI, W.; HERZ, W. Constituents of *Viguiera aspillioides* and *V. robusta*. *Biochemical Systematics and Ecology*, v.24, n.6, p.585-587, 1996. http://www.sciencedirect.com/science/article/pii/0305197896000579

CURY, G.; NOVEMBRE, A.D.L.C.; APPEZZATO-DA-GLÓRIA, B. Seed germination of *Chresta sphaerocephala* DC. and *Lessingianthus bardanoides* (Less.) H. Rob. (Asteraceae) from Cerrado. *Brazilian Archives of Biology and Technology*, v.53, n.6, p.1299-1308, 2010. http://www.scielo.br/scielo. php?pid=S1516-89132010000600006&script=sci_arttext

FERREIRA, A.G.; CASSOL, B.; ROSA, S.G.T.; SILVEIRA, T.S.; STIVAL, A.L.; SILVA, A.A. Germinação de sementes de Asteraceae nativas no Rio Grande do Sul, Brasil. *Acta Botanica Brasilica*, v.15, n.2, p.231-242, 2001. http://www.sciencedirect.com/science/article/pii/0305197896000579

GARCIA, L.C.; BARROS, F.V.; LEMOS FILHO, J.P. Comportamento germinativo de duas espécies de canga ferrífera: *Baccharis retusa* DC. (Asteraceae) e *Tibouchina multiflora* Cogn. (Melastomataceae). *Acta Botanica Brasilica*, v.20, n.2, p.443-448, 2006. http://www.scielo.br/scielo. php?pid=S0102-33062006000200019&script=sci_abstract&tlng=pt

GOMES, V.; FERNANDEZ, G.W. Germinação de aquênios de *Baccharis dracunculifolia* D.C. (Asteraceae). *Acta Botanica Brasilica*, v.16, n.4, p.421-427, 2002. http://www.scielo.br/scielo.php?script=sci_ arttext&pid=S0102-33062002000400005

HAYASHI, A.H.; APPEZZATO-DA-GLÓRIA, B. Anatomy of the Underground System in *Vernonia grandiflora* Less. and *V. brevifolia* Less. (Asteraceae). *Brazilian Archives of Biology and Technology*, v.50, n.6, p.979-988, 2007. http://www.scielo.br/scielo.php?script=sci_ arttext&pid=S1516-89132007000700009

LABORIAU, L.G. *A germinação de sementes*. Washington: Secretaria Geral da OEA, 1983, 171p.

LIMA, M.L.P.; REIS, A.; LOPES, C.A. Patogenicidade de *Alternaria cichorii* sobre espécies da família Asteraceae no Brasil. *Fitopatologia Brasileira*, v.28, n.6, p.682-685, 2003. http://www.scielo.br/pdf/fb/v28n6/a16v28n6.pdf

MAGENTA, M.A.G.; NUNES, A.D.; MENDONÇA, C.B.F.; GONÇALVES-ESTEVES, V. Palynotaxonomy of Brazilian *Viguiera* (Asteraceae) species. *Boletin de la Sociedad Argentina de Botanica*, v.45, n.3-4, p.285-299, 2010. http://www.scielo.org.ar/scielo.php?script=sci_ arttext&pid=S1851-23722010000200008

MAGUIRE, J.D. Speed of germination-aid in selection and evaluation for seedling emergence and vigor. *Crop Science*, v.2, n.2, p.176-177, 1962. https://www.crops.org/publications/cs/pdfs/2/2/CS0020020176

MARQUINA, S.; MALDONADO, N.; GARDUÑO-RAMÍREZ, M.L.; ARANDA, E. VILLARREAL, M.L.; NAVARRO, V.; BYE, R.; DELGADO, G.; ALVAREZ, L. Bioactive oleanolic acid saponins and other constituents from the roots of *Viguiera decurrens*. *Phytochemistry*, v.56, p.93-97, 2001. http://www.sciencedirect.com/science/article/pii/S0031942200002831

MARZINEK, J.; DE-PAULA, O.C.; OLIVEIRA, D.M.T. Cypsela or achene? Refining terminology by considering anatomical and historical factors. *Revista Brasileira de Botânica*, v.31, n.3, p.549-553, 2008. http://www. scielo.br/scielo.php?pid=S0100-84042008000300018&script=sci_arttext

OLIVEIRA, T.; BOMBO, A.B.; APPEZZATO-DA-GLORIA, B. Anatomy of vegetative organs with an emphasis on the secretory structures of two species of *Aldama* (Asteraceae – Heliantheae). *Botany*, v.91, n.6, p.335-342, 2013. http:// www.nrcresearchpress.com/doi/abs/10.1139/cjb-2012-0271#.Ul7YR_msiSo

OLIVEIRA, L.M.; DAVIDE; A.C.; CARVALHO, M.L.O. Avaliação de métodos para quebra da dormência e para a desinfestação de sementes de canafístula (*Peltophorum dubium*) (Sprengel) Taubert. *Revista Árvore*, v.27, n.5, p.597-603, 2003. http://www.redalyc.org/articulo.oa?id=48827501

REIS, A.; SATELIS, J.F.; PEREIRA, R.S.; NASCIMENTO, W.M. Associação de *Alternaria dauci* e *A. alternata* com sementes de coentro e eficiência do tratamento químico. *Horticultura Brasileira*, v.24, n.1, p.107-111, 2006. http:// www.scielo.br/scielo.php?pid=S0102-05362006000100022&script=sci_ abstract&tlng=pt

RUGGIERO, P.G.C.; ZAIDAN, L.B.P. Estudos de desenvolvimento de *Viguiera robusta* Gardn., uma Asteraceae do Cerrado. *Revista Brasileira Botânica*, v.20, n.1, p.1-9, 1997. http://www.scielo.br/scielo.php?pid=S0100-84041997000100001&script=sci_arttext

SAS INSTITUTE. *Sas System: SAS/STAT version 9.1* (software). Cary, 2003.

SASSAKI, R.M.; RONDON, J.N.; ZAIDAN, L.B.P.; FELIPPE, G.M. Germination of seeds from herbaceous plants artificially stored in Cerrado soil. *Revista Brasileira de Biologia*, v.59, n.2, p.271-279, 1999. http://www. scielo.br/scielo.php?pid=S0034-71081999000200011&script=sci_arttext

SCHILLING, E.E.; PANERO, J.L. A revised classification os subtribe Helianthinae (Asteraceae: Heliantheae) II. Derived lineages. *Botanical Journal of the Linnean Society*, v.167,p.311-331, 2011. http://onlinelibrary. wiley.com/doi/10.1111/j.1095-8339.2011.01172.x/abstract

SILVA, L.M.M.; RODRIGUES, T.J.D.; AGUIAR, I.B. Efeito da luz e da temperatura na germinação de sementes de aroeira (*Myracrodruon urundeuva* Allemão). *Revista Árvore*, v.26, n.6, p.691-697, 2002. http://www.scielo.br/ pdf/rarv/v26n6/a06v26n6.pdf

SPRING, O.; ZIPPER, R.; CONRAD, J.; VOGLER, B.; KLAIBER, I.; COSTA, F.B. Sesquiterpene lactones from glandular trichomes of *Viguiera radula* (Heliantheae; Asteraceae). *Phytochemistry*, v.62, p.1185-1189, 2003. http://www.sciencedirect.com/science/article/pii/S0031942202007471

VELTEN, S.B.; GARCIA, Q.S. Efeitos da luz e da temperatura na germinação de sementes de *Eremanthus* (Asteraceae), ocorrentes na Serra do Cipó, MG, Brasil. *Acta Botanica Brasilica*, v.19, n.4, p.753-761, 2005. http://www. scielo.br/scielo.php?pid=S0102-33062005000400010&script=sci_arttext

VILHALVA, D.A.A.; APPEZZATO-DA-GLÓRIA, B. Morfoanatomia do sistema subterrâneo de *Callea verticillata* (Klatt) Pruski e *Isostigma megapotamicum* (Spreng.) Sherff – Asteraceae. *Revista Brasileira de Botânica*, v.29, n.1, p.29-47, 2006. http://www.scielo.br/scielo. php?script=sci_arttext&pid=S0100-84042006000100005&lng=en&nrm=iso &tlng=pt

ZAIDAN, L.B.P.; CARREIRA, R.C. Germination ecophysiology of Cerrado seeds. *Brazilian Journal of Plant Physiology*, v.20, n.3, p.167-181, 2008. http://www.scielo.br/scielo.php?script=sci_arttext&pid=S1677-04202008000300002&lng=en&nrm=iso&tlng=en

Permissions

The contributors of this book come from diverse backgrounds, making this book a truly international effort. This book will bring forth new frontiers with its revolutionizing research information and detailed analysis of the nascent developments around the world.

We would like to thank all the contributing authors for lending their expertise to make the book truly unique. They have played a crucial role in the development of this book. Without their invaluable contributions this book wouldn't have been possible. They have made vital efforts to compile up to date information on the varied aspects of this subject to make this book a valuable addition to the collection of many professionals and students.

This book was conceptualized with the vision of imparting up-to-date information and advanced data in this field. To ensure the same, a matchless editorial board was set up. Every individual on the board went through rigorous rounds of assessment to prove their worth. After which they invested a large part of their time researching and compiling the most relevant data for our readers.

The editorial board has been involved in producing this book since its inception. They have spent rigorous hours researching and exploring the diverse topics which have resulted in the successful publishing of this book. They have passed on their knowledge of decades through this book. To expedite this challenging task, the publisher supported the team at every step. A small team of assistant editors was also appointed to further simplify the editing procedure and attain best results for the readers.

Apart from the editorial board, the designing team has also invested a significant amount of their time in understanding the subject and creating the most relevant covers. They scrutinized every image to scout for the most suitable representation of the subject and create an appropriate cover for the book.

The publishing team has been an ardent support to the editorial, designing and production team. Their endless efforts to recruit the best for this project, has resulted in the accomplishment of this book. They are a veteran in the field of academics and their pool of knowledge is as vast as their experience in printing. Their expertise and guidance has proved useful at every step. Their uncompromising quality standards have made this book an exceptional effort. Their encouragement from time to time has been an inspiration for everyone.

The publisher and the editorial board hope that this book will prove to be a valuable piece of knowledge for researchers, students, practitioners and scholars across the globe.

List of Contributors

Carolina Maria Luzia Delgado, Cileide Maria Medeiros de Coelho and Gesieli Priscila Buba
Centro de Ciências Agroveterinárias, Universidade do Estado de Santa Catarina Luiz de Camões, 88520-000 – Lages, SC, Brasil

Breno Marques da Silva e Silva and Camila de Oliveira e Silva
Universidade do Estado do Amapá, 68906-970 – Macapá, AP, Brasil

Fabíola Vitti Môro
Departamento de Biologia Aplicada a Agropecuária, UNESP, 14884-900 – Jaboticabal, SP, Brasil

Roberval Daiton Vieira
Departamento de Produção Vegetal, UNESP, 14884-900 – Jaboticabal, SP, Brasil

Ronan Carlos Colombo, Vanessa Favetta, Lilian Yukari Yamamoto, Guilherme Augusto Cito Alves, Julia Abati, Lúcia Sadayo Assari Takahashi and Ricardo Tadeu de Faria
Universidade Estadual de Londrina, Caixa Postal 1011, 86057-970 - Londrina, PR, Brasil

Márcia Adriana Carvalho dos Santos, Manoel Abílio de Queiróz and Jaciara de Souza Bispo
Departamento de Tecnologia e Ciências Sociais, Universidade do Estado da
Bahia, Caixa Postal 171, 48905-680 - Juazeiro, BA, Brasil

Aparecida Leonir da Silva and Dimas Mendes Ribeiro
Departamento de Biologia Vegetal, Universidade Federal de Viçosa, 36570-000 – Viçosa, MG, Brasil

Eduardo Euclydes de Lima e Borges
Departamento de Engenharia Florestal, Universidade Federal de Viçosa, 36570-000 – Viçosa, MG, Brasil

Nestor Martini Neto and Claudio José Barbedo
Instituto de Botânica, Núcleo de Pesquisa em Sementes, Caixa Postal, 68041, 04301012 – São Paulo, SP, Brasil

Priscilla Brites Xavier, Henrique Duarte Vieira and Cynthia Pires Guimarães
Universidade Estadual do Norte Fluminense Darcy Ribeiro, Setor de Tecnologia de Sementes, 28013-602 – Campos dos Goytacazes, RJ, Brasil

Cesar Pedro Hartmann Filho, André Luís Duarte Goneli, Tathiana Elisa Masetto, Elton Aparecido Siqueira Martins and Guilherme Cardoso Oba
Universidade Federal da Grande Dourados, Faculdade de Ciências Agrárias, Caixa Postal 533, 79804-970 - Dourados, MS, Brasil

Karine Sousa Carsten Borges
Centro Universitário Leonardo da Vinci, Grupo UNIASSELVI, 89130-000 – Indaial, SC, Brasil

Raquel Custódio D'Avila
Fundação Municipal do Meio Ambiente, 88160-126 – Biguaçu, SC, Brasil

Mari Lúcia Campos
Departamento de Solos e Recursos Naturais, Universidade Estadual de Santa Catarina, 88520-000 – Lages, SC, Brasil

Cileide Maria Medeiros Coelho
Departamento de Agronomia, Universidade Estadual de Santa Catarina, 88520-000 – Lages, SC, Brasil

David José Miquelluti
Departamento de Solos e Recursos Naturais, Universidade Estadual de Santa Catarina, 88520-000 – Lages, SC, Brasil

Natiele da Silva Galvan
Universidade Estadual de Santa Catarina, 88520-000 – Lages, SC, Brasil

Raquel Maria de Oliveira Pires
Departamento de Agricultura, UFLA, Caixa Postal 3037, 37200-000 – Lavras, MG, Brasil

Janete Rodrigues Matias, Renata Conduru Ribeiro, Gherman Garcia Leal Araújo and Bárbara França Dantas
Embrapa Semiárido, Caixa Postal 23, 56302-970 - Petrolina, PE, Brasil

Carlos Alberto Aragão
Departamento de Tecnologia e Ciências Sociais Universidade do Estado da Bahia, 48900-000 - Juazeiro, BA, Brasil

Genaina Aparecida de Souza, Leonardo Araujo Oliveira and Amanda Ávila Cardoso
Departamento de Biologia Vegetal, UFV, 36570-000 – Viçosa, MG, Brasil

Fernanda Bernardo Cripa, Laura Cristiane Nascimento de Freitas, Andrieli Cristine Grings and Michele Fernanda Bortolini
Pontifícia Universidade Católica do Paraná, Escola de Saúde e Biociências, 85802-532 - Toledo, PR, Brasil

Luciana Aparecida de Souza Abreu and Édila Vilela de Resende Von Pinho
Departamento de Agricultura, UFLA, Caixa Postal, 3037, 37200 – Lavras, MG, Brasil

Adriano Delly Veiga
Embrapa Café, UFLA, Caixa Postal 3037, 37200-000,Lavras- MG, Brasil

Fiorita Faria Monteiro
Universidade Federal de Lavras, Caixa Postal, 3037, 37200-000, Lavras- MG, Brasil

Sttela Dellyzette Veiga Franco da Rosa
Embrapa Café, Departamento de Agricultura, UFLA, Caixa Postal 3037, 37200-000 – Lavras, MG, Brasil

Edmir Vicente Lamarca
Instituto de Botânica, Núcleo de Pesquisa em Sementes, Caixa Postal 68041, 04301012 – São Paulo, SP, Brasil

Narjara Walessa Nogueira, Rômulo Magno Oliveira de Freitas, Salvador Barros Torres and Caio César Pereira Leal
Laboratório de Análise de Sementes, Departamento de Fitotecnia, UFERSA, 59625-900 - Mossoró, RN, Brasil

Liana Baptista de Lima
Universidade Federal de Mato Grosso do Sul, Centro de Ciências Biológicas
e da Saúde, Laboratório de Botânica, 79070-900-Campo Grande, MS, Brasil

Glaucia Almeida de Morais
Universidade Estadual de Mato Grosso do Sul, 79740-000- Ivinhema, MS, Brasil

Tatiana Carvalho de Castro, Claudia Simões-Gurgel, Ivan Gonçalves Ribeiro and Norma Albarello
Departamento de Biologia Vegetal, Universidade do Estado do Rio de
Janeiro, Instituto de Biologia Roberto Alcântara Gomes, 20550-013 - Rio de
Janeiro, RJ, Brasil

Marsen Garcia Pinto Coelho
Departamento de Bioquímica , Universidade do Estado do Rio de Janeiro

Instituto de Biologia Roberto Alcântara Gomes, 20550-170 - Rio de Janeiro, RJ, Brasil

Glauter Lima Oliveira, Denise Cunha Fernandes dos Santos Dias, Paulo Cesar Hilst, Laércio Junio da Silva and Luiz Antônio dos Santos Dias
Departamento de Fitotecnia, Universidade Federal de Viçosa, 36570-000 - Viçosa, MG, Brasil

Patrícia Ribeiro Cursi
Coordenadoria de Assistência Técnica Integral - CATI, Caixa Postal 962, 13070-172 - Campinas, SP, Brasil

Silvio Moure Cicero
Departamento de Produção Vegetal, USP/ESALQ, Caixa Postal, 9, 13418-900 - Piracicaba, SP, Brasil

Patricia Pereira da Silva
Universidade Federal de Pelotas, 70351970 – Pelotas, RS, Brasil

Antônio Carlos Souza Albuquerque Barros
Universidade Federal de Pelotas, Caixa Postal 354, 96010-900 – Pelotas, RS, Brasil

Edila Vilela de Resende Von Pinho
Departamento de Agricultura, UFLA, Caixa Postal 3037, 37200-000 – Lavras, MG, Brasil

Warley Marcos Nascimento
Embrapa Hortaliças, Caixa Postal 218, 70359-970 – Brasília, DF, Brasil

Vitor Oliveira Rodrigues, Fabiano Ramos Costa, Marcela Carlota Nery, Sara Michelly Cruz and Soryana Gonçalves Ferreira de Melo
Departamento de Agronomia, Universidade Federal dos Vales do Jequitinhonha e Mucuri, 39100-000 – Diamantina, MG, Brasil

Maria Laene Moreira de Carvalho
Departamento de Agricultura, Universidade Federal de Lavras, Caixa Postal 3037, 37200-000 – Lavras, MG, Brasil

Aline Bertolosi Bombo, Tuane Santos de Oliveira and Beatriz Appezzato-da- Glória
Departamento de Ciências Biológicas, USP/ESALQ, Caixa Postal 9,13418-900 – Piracicaba, SP, Brasil

Ana Dionísia da Luz Coelho Novembre
Departamento de Produção Vegetal, USP/ESALQ, Caixa Postal 9,13418-900 – Piracicaba, SP, Brasil

Index

A
Adenium Obesum, 16, 18-23
Aldama Arenaria, 182-186
Antioxidative System, 75
Apocynaceae, 16

C
Cadmium Chloride, 93
Cerrado, 25, 27, 128, 182-183, 185, 187-188
Chemical Scarification, 129-132, 134-135
Cherry, 30, 100-105
Cleome Dendroides, 137, 139, 142-143
Cleome Rosea, 137, 139, 141, 143, 146
Cleome Spinosa, 137, 143, 145
Coating Machine, 49-51
Coffea Arabica, 107-108, 111-113
Compositae, 182
Consumption, 115, 156, 173
Contaminated Soil, 66
Crop Production, 57, 84
Cucurbita Maxima, 166-167
Cucurbita Moschata, 166-167
Curcubitaceae, 83
Cytoprotective, 76, 93

D
Desiccation Tolerance, 25, 30, 40, 45, 107-109, 111, 120, 163, 167, 172
Drying, 10, 24-25, 27-28, 30, 40-41, 44-48, 52, 57-65, 68, 84, 91, 106-115, 117-118, 120-121, 125, 129, 134-135, 138, 155-157, 159-162, 164, 167-168, 171-172

E
Embryo Growth, 26, 75
Exploitation, 155-156

F
Fabaceae, 9, 13, 15, 32, 36, 49, 54, 106, 129, 132, 136, 187
Fedegoso, 32, 39
Fertilization, 58, 123-124
Forest Seeds, 93, 115
Forest Species, 32-34, 36, 40, 100-101, 105, 116, 156, 158

G
Genetic Resources, 24
Germination, 1-3, 5-10, 12-20, 23-39, 42, 46, 48-51, 53, 55, 57-61, 64, 66-70, 78, 80-96, 98-105, 111, 116, 118, 121, 123-127, 129-154, 156-159, 163, 166-177, 188

Glycine Max, 7, 57, 91

H
Heavy Metals, 39, 66, 73-74, 76, 82, 92-93, 99
Helianthus Annuus, 7, 116, 173-174, 176-179
Herbicide, 1-3, 6-7
Homogenized, 3, 108, 183
Hygroscopic Equilibrium, 40-43
Hypocotyl, 12-13, 19, 45, 57-59, 61-63, 75-76, 78-79, 86-90, 92-93, 96, 123, 125-127, 139-140, 142-143, 151, 156, 183

I
Inhibition, 66-70, 72, 94, 96-97

L
Lipid Peroxidation, 76, 93, 99, 111
Longevity, 46-47, 107, 112, 162

M
Mechanical Scarification, 129-132, 163
Methodology, 2, 10, 42, 51, 68, 76, 93, 98, 100-101, 103, 115-117, 137, 145, 147-148, 175
Morphology, 9-10, 36, 139, 185
Myrtaceae, 24, 27-28, 30-31, 105-106, 115, 121

N
Nacl, 32-39, 42-44, 81, 108
Native Species, 32-33, 36, 137-138
Nitric Oxide, 32-34, 36, 38-39, 73, 75-76, 81-82, 92, 98-99
Nitrogen Fixation, 9

O
Ormosia Arborea, 13, 129-136
Ornamental Value, 16
Oxidative Stress, 39, 69, 76, 80, 82, 88-89, 91, 93, 95, 98-99, 107, 112-113, 172
Ozone, 7, 173-181

P
Physic Nut, 147-150, 152-154
Physiological Quality, 1-3, 8, 10, 17, 24, 26-30, 41, 55, 57-58, 61, 63, 101, 109, 115-116, 118, 120-121, 123-124, 154-156, 164-167, 169-171, 173-175, 187
Phytotoxicity, 66, 73-74, 176
Plant Extraction, 155
Post-harvest, 25, 27, 57, 107, 172
Protein, 1-2, 4-8, 30, 49, 69-70, 77, 79-81, 84, 86, 90, 94, 97-98, 111-112, 156, 159, 162, 166-167, 171-172
Pulped Fruit, 155, 158

R

Recalcitrant Seeds, 47, 101, 104-105, 115, 162

Reforestation, 9, 32, 130, 156

Reproduction, 23, 50, 182-183

S

Seed Coating, 49-51, 53-54

Seed Propagation, 25

Seed Quality, 1, 8, 25, 30, 33, 41, 49, 58, 61, 64, 107, 111, 116, 124-125, 127, 148, 166, 168-169, 172

Seed Vigor, 2-3, 7, 27-29, 36, 57, 61, 81, 109-110, 116, 169

Seedling Development, 46, 62, 66, 83, 137-140, 150

Seeds, 1-38, 40-47, 49-64, 66-70, 73, 75-118, 120-121, 123-127, 129-188

Seeds Storage, 16, 100-101, 104, 110

Seminal Development, 9-10, 13, 137, 145, 148

Sesamum Indicum, 75-76, 78-81, 92-93, 96

Sexual Propagation, 182

Soaking Curve, 2-3, 129, 131

Sodium Nitroprusside, 32-34, 75-76, 81, 92-93

Soluble Sugar, 1, 3-6, 87

Storage, 4, 7, 16-20, 23, 28, 30, 40-42, 44-48, 55, 57-65, 87-88, 91, 100-102, 104-114, 116, 120-121, 134, 146, 156, 163, 166-172, 175, 181, 183, 186

Stylosanthes Capitata, 49-50

Synthesis, 2, 5, 28, 69-70, 87, 97, 111, 123-124, 144-145, 167, 171

T

Toxicity, 33, 67, 72-73, 76, 91, 93, 95, 99

U

Uvaia, 100-106

V

Viability, 15-16, 23, 30, 34, 40-41, 47, 58, 64-65, 100-105, 107, 111, 115-118, 120-121, 134, 138-140, 144, 147, 163, 173, 175, 177, 180, 182, 184, 186

Viability Test, 100, 115

Vigna Unguiculata, 123

Vigor, 1-3, 5-8, 14, 24, 27-30, 36, 47, 49, 55, 58, 61, 64, 67, 70, 73, 75, 77, 82, 85, 92, 94, 99, 101, 105, 111, 116, 125, 127, 134, 154, 157, 163, 170, 173, 175, 177, 180, 188

Viguiera, 182-183, 187-188

Vulnerability, 24

W

Water Reuse, 83

X

Xylopodium, 183, 186